Network-on-Chip

The Next Generation of System-on-Chip Integration

Network-on-Chip

The Next Generation of System-on-Chip Integration

Santanu Kundu
Santanu Chattopadhyay

CRC Press
Taylor & Francis Group
Boca Raton   London   New York

CRC Press is an imprint of the
Taylor & Francis Group, an **informa** business

CRC Press
Taylor & Francis Group
6000 Broken Sound Parkway NW, Suite 300
Boca Raton, FL 33487-2742

© 2015 by Taylor & Francis Group, LLC
CRC Press is an imprint of Taylor & Francis Group, an Informa business

No claim to original U.S. Government works

Printed on acid-free paper
Version Date: 20141014

International Standard Book Number-13: 978-1-4665-6526-5 (Hardback)

Visit the Taylor & Francis Web site at
http://www.taylorandfrancis.com

and the CRC Press Web site at
http://www.crcpress.com

Contents

Preface...xiii
Authors..xvii

1. Introduction...1
   1.1   System-on-Chip Integration and Its Challenges 1
   1.2   SoC to Network-on-Chip: A Paradigm Shift.......................3
   1.3   Research Issues in NoC Development.................................5
   1.4   Existing NoC Examples 8
   1.5   Summary..10
   References 10

2. Interconnection Networks in Network-on-Chip 13
   2.1   Introduction 13
   2.2   Network Topologies...14
         2.2.1   Number of Edges 25
         2.2.2   Average Distance 25
   2.3   Switching Techniques...29
   2.4   Routing Strategies 30
         2.4.1   Routing-Dependent Deadlock 31
              2.4.1.1   Deterministic Routing in $M \times N$ MoT
                      Network 33
         2.4.2   Avoidance of Message-Dependent Deadlock 41
   2.5   Flow Control Protocol...43
   2.6   Quality-of-Service Support..45
   2.7   NI Module 46
   2.8   Summary..48
   References 48

3. Architecture Design of Network-on-Chip 53
   3.1   Introduction 53
   3.2   Switching Techniques and Packet Format 53
   3.3   Asynchronous FIFO Design...54
   3.4   GALS Style of Communication 57
   3.5   Wormhole Router Architecture Design...............................57
         3.5.1   Input Channel Module...58
         3.5.2   Output Channel Module 58
   3.6   VC Router Architecture Design 63
         3.6.1   Input Channel Module...65
         3.6.2   Output Links 66

                    3.6.2.1   VC Allocator 66
                    3.6.2.2   Switch Allocator 69
        3.7   Adaptive Router Architecture Design 70
        3.8   Summary 73
        References 73

4. Evaluation of Network-on-Chip Architectures 75
        4.1   Evaluation Methodologies of NoC 75
                    4.1.1   Performance Metrics 78
                    4.1.2   Cost Metrics 80
        4.2   Traffic Modeling 81
        4.3   Selection of Channel Width and Flit Size 84
        4.4   Simulation Results and Analysis of MoT Network
                with WH Router 84
                    4.4.1   Accepted Traffic versus Offered Load 85
                    4.4.2   Throughput versus Locality Factor 85
                    4.4.3   Average Overall Latency at Different Locality Factors 86
                    4.4.4   Energy Consumption at Different Locality Factors 88
        4.5   Impact of FIFO Size and Placement in Energy and
                Performance of a Network 90
        4.6   Performance and Cost Comparison of MoT with Other NoC
                Structures Having WH Router under Self-Similar Traffic 93
                    4.6.1   Network Area Estimation 94
                    4.6.2   Network Aspect Ratio 96
                    4.6.3   Performance Comparison 97
                                4.6.3.1   Accepted Traffic versus Offered Load 97
                                4.6.3.2   Throughput versus Locality Factor 98
                                4.6.3.3   Average Overall Latency under
                                            Localized Traffic 99
                    4.6.4   Comparison of Energy Consumption 102
        4.7   Simulation Results and Analysis of MoT
                Network with Virtual Channel Router 103
                    4.7.1   Throughput versus Offered Load 104
                    4.7.2   Latency versus Offered Load 104
                    4.7.3   Energy Consumption 105
                    4.7.4   Area Required 108
        4.8   Performance and Cost Comparison of MoT with Other
                NoC Structures Having VC Router 109
                    4.8.1   Accepted Traffic versus Offered Load 109
                    4.8.2   Throughput versus Locality Factor 109
                    4.8.3   Average Overall Latency under Localized Traffic 110
                    4.8.4   Energy Consumption 111
                    4.8.5   Area Overhead 113
        4.9   Limitations of Tree-Based Topologies 114

4.10 Summary 115
References 116

5. **Application Mapping on Network-on-Chip** 119
   5.1 Introduction 119
   5.2 Mapping Problem 120
   5.3 ILP Formulation 123
        5.3.1 Other ILP Formulations 127
   5.4 Constructive Heuristics for Application Mapping 128
        5.4.1 Binomial Merging Iteration 130
        5.4.2 Topology Mapping and Traffic Surface Creation 131
        5.4.3 Hardware Cost Optimization 132
   5.5 Constructive Heuristics with Iterative Improvement 134
        5.5.1 Initialization Phase 134
        5.5.2 Shortest Path Computation 135
        5.5.3 Iterative Improvement Phase 136
        5.5.4 Other Constructive Strategies 137
   5.6 Mapping Using Discrete PSO 141
        5.6.1 Particle Structure 141
        5.6.2 Evolution of Generations 142
        5.6.3 Convergence of DPSO 143
        5.6.4 Overall PSO Algorithm 144
        5.6.5 Augmentations to the DPSO 144
               5.6.5.1 Multiple PSO 144
               5.6.5.2 Initial Population Generation 145
        5.6.6 Other Evolutionary Approaches 148
   5.7 Summary 150
   References 150

6. **Low-Power Techniques for Network-on-Chip** 155
   6.1 Introduction 155
   6.2 Standard Low-Power Methods for NoC Routers 158
        6.2.1 Clock Gating 158
        6.2.2 Gate Level Power Optimization 159
        6.2.3 Multivoltage Design 160
               6.2.3.1 Challenges in Multivoltage Design 161
        6.2.4 Multi-V_T Design 164
        6.2.5 Power Gating 165
   6.3 Standard Low-Power Methods for NoC Links 166
        6.3.1 Bus Energy Model 167
        6.3.2 Low-Power Coding 168
        6.3.3 On-Chip Serialization 170
        6.3.4 Low-Swing Signaling 171

6.4   System-Level Power Reduction.. 172
       6.4.1   Dynamic Voltage Scaling... 172
               6.4.1.1   History-Based DVS... 174
               6.4.1.2   Hardware Implementation 178
               6.4.1.3   Results and Discussions............................. 179
       6.4.2   Dynamic Frequency Scaling 179
               6.4.2.1   History-Based DFS....................................... 181
               6.4.2.2   DFS Algorithm... 183
               6.4.2.3   Link Controller 183
               6.4.2.4   Results and Discussions............................. 184
       6.4.3   VFI Partitioning 185
       6.4.4   Runtime Power Gating... 186
6.5   Summary... 188
References 188

7. **Signal Integrity and Reliability of Network-on-Chip**...................... 191
   7.1   Introduction 191
   7.2   Sources of Faults in NoC Fabric 193
          7.2.1   Permanent Faults 194
          7.2.2   Faults due to Aging Effects... 194
                  7.2.2.1   Negative-Bias Temperature Instability 194
                  7.2.2.2   Hot Carrier Injection.................................. 195
          7.2.3   Transient Faults 195
                  7.2.3.1   Capacitive Crosstalk 195
                  7.2.3.2   Soft Errors.. 199
                  7.2.3.3   Some Other Sources of Transient Faults 203
   7.3   Permanent Fault Controlling Techniques................................ 204
   7.4   Transient Fault Controlling Techniques 205
          7.4.1   Intra-Router Error Control... 205
                  7.4.1.1   Soft Error Correction 206
          7.4.2   Inter-Router Link Error Control 210
                  7.4.2.1   Capacitive Crosstalk Avoidance Techniques...... 210
                  7.4.2.2   Error Detection and Retransmission.............. 216
                  7.4.2.3   Error Correction 220
   7.5   Unified Coding Framework... 221
          7.5.1   Joint CAC and LPC Scheme (CAC + LPC) 222
          7.5.2   Joint LPC and ECC Scheme (LPC + ECC) 223
          7.5.3   Joint CAC and ECC Scheme (CAC + ECC)................... 224
          7.5.4   Joint CAC, LPC, and ECC Scheme
                  (CAC + LPC + ECC) 227
   7.6   Energy and Reliability Trade-Off in Coding Technique............ 227
   7.7   Summary... 230
   References 231

8. Testing of Network-on-Chip Architectures 235
8.1   Introduction 235
8.2   Testing Communication Fabric 236
       8.2.1   Testing NoC Links 237
       8.2.2   Testing NoC Switches 238
       8.2.3   Test Data Transport 239
       8.2.4   Test Transport Time Minimization—A Graph
              Theoretic Formulation 241
              8.2.4.1   Unicast Test Scheduling 242
              8.2.4.2   Multicast Test Scheduling 244
8.3   Testing Cores 245
       8.3.1   Core Wrapper Design 246
       8.3.2   ILP Formulation 250
       8.3.3   Heuristic Algorithms 253
       8.3.4   PSO-Based Strategy 258
              8.3.4.1   Particle Structure and Fitness 258
              8.3.4.2   Evolution of Generations 259
8.4   Summary 260
References 260

9. Application-Specific Network-on-Chip Synthesis 263
9.1   Introduction 263
9.2   ASNoC Synthesis Problem 264
9.3   Literature Survey 265
9.4   System-Level Floorplanning 268
       9.4.1   Variables 268
              9.4.1.1   Independent Variables 268
              9.4.1.2   Dependent Variables 268
       9.4.2   Objective Function 269
       9.4.3   Constraints 269
       9.4.4   Constraints for Mesh Topology 270
9.5   Custom Interconnection Topology and Route Generation 271
       9.5.1   Variables 272
              9.5.1.1   Independent Variables 272
              9.5.1.2   Derived Variables 273
       9.5.2   Objective Function 273
       9.5.3   Constraints 274
9.6   ASNoC Synthesis with Flexible Router Placement 277
       9.6.1   ILP for Flexible Router Placement 278
              9.6.1.1   Variables 278
              9.6.1.2   Objective Function 279
              9.6.1.3   Constraints 279

            9.6.2   PSO for Flexible Router Placement.............................. 281
                     9.6.2.1   Particle Structure and Fitness Function 282
                     9.6.2.2   Local and Global Bests.............................. 282
                     9.6.2.3   Evolution of Generation 283
                     9.6.2.4   Swap Operator.. 283
                     9.6.2.5   Swap Sequence 283
      9.7   Summary... 284
      References 284

10. **Reconfigurable Network-on-Chip Design** 289
      10.1   Introduction 289
      10.2   Literature Review... 290
      10.3   Local Reconfiguration Approach................................... 291
            10.3.1   Routers... 292
            10.3.2   Multiplexers... 293
            10.3.3   Selection Logic.. 294
            10.3.4   Area Overhead 294
            10.3.5   Design Flow 296
                     10.3.5.1   Construction of CCG 298
                     10.3.5.2   Mapping of CCG 299
                     10.3.5.3   Configuration Generation..................... 299
            10.3.6   ILP-Based Approach...................................... 299
                     10.3.6.1   Parameters and Variables 300
                     10.3.6.2   Objective Function.................................... 300
                     10.3.6.3   Constraints... 300
            10.3.7   PSO Formulation... 301
                     10.3.7.1   Particle Formulation and Fitness Function ... 302
            10.3.8   Iterative Reconfiguration 303
      10.4   Topology Reconfiguration 304
            10.4.1   Modification around Routers 305
            10.4.2   Reconfiguration Architecture 306
                     10.4.2.1   Application Mapping 307
                     10.4.2.2   Core-to-Network Mapping..................... 309
                     10.4.2.3   Topology and Route Generation............. 310
      10.5   Link Reconfiguration... 311
            10.5.1   Estimating Channel Bandwidth Utilization 311
      10.6   Summary... 312
      References 314

11. **Three-Dimensional Integration of Network-on-Chip** 317
      11.1   Introduction 317
      11.2   3D Integration: Pros and Cons 318
            11.2.1   Opportunities of 3D Integration................. 319
            11.2.2   Challenges of 3D Integration 321

11.3 Design and Evaluation of 3D NoC Architecture 323
  11.3.1 3D Mesh-of-Tree Topology 326
    11.3.1.1 Number of Directed Edges 326
    11.3.1.2 Average Distance 327
  11.3.2 Performance and Cost Evaluation 331
    11.3.2.1 Network Area Estimation 336
    11.3.2.2 Network Aspect Ratio 339
  11.3.3 Simulation Results with Self-Similar Traffic 340
    11.3.3.1 Accepted Traffic versus Offered Load 340
    11.3.3.2 Throughput versus Locality Factor 341
    11.3.3.3 Average Overall Latency under
             Localized Traffic 342
    11.3.3.4 Energy Consumption 345
  11.3.4 Simulation Results with Application-Specific Traffic 349
11.4 Summary 350
References 351

12. Conclusions and Future Trends 353
12.1 Conclusions 353
12.2 Future Trends 354
  12.2.1 Photonic NoC 354
  12.2.2 Wireless NoC 354
12.3 Comparison between Alternatives 355
References 357

Index 359

Preface

System-on-chip (SoC) is a paradigm for designing today's integrated circuit (IC) chips that put an entire system onto a single silicon floor (instead of printed circuit boards containing a number of chips accomplishing the system task). With the increasing number of cores integrated on such a chip, on-chip communication efficiency has become one of the key factors in determining the overall system performance and cost. The communication medium used in most of the modern SoCs is a shared global bus. In spite of its fairly simple structure, extensibility, and low area cost, at the system level, it can be used for only up to tens of cores on a single chip. This restriction is mainly due to the following reasons: nonscalable wire delay with technology shrinking, nonscalable system performance with number of cores attached, decrease in operating frequency with each additional core attached, high power consumption in long wires, and so on. In many-core-based SoCs, the major challenge that designers face today is to come up with a scalable, reusable, and high-performance communication backbone.

Network-on-chip (NoC) is an emerging alternative that overcomes the above-mentioned bottlenecks for integrating a large number of cores on a single SoC. NoC is a specific flavor of interconnection networks where the cores communicate with each other using a router-based packet-switched network. Interconnection networks have been studied for more than the past two decades and a solid foundation of design techniques has been reported in the literature. NoC is today becoming an emerging research and development topic including hardware communication infrastructure design, software and operating system services, computer aided design (CAD) tools for NoC synthesis, NoC testing, and so on.

However, two-dimensional (2D) IC design has limited floorplanning choices with increasing number of cores attached. An attractive solution to this problem is the three-dimensional (3D) IC technology that stacks multiple layers of active silicon using special vertical interconnects, known as *through-silicon via* (TSV). The introduction of 2D NoC in a 3D IC platform is a gradual process and is known as 3D NoC. Although a number of 2D NoC implementations have already been fabricated in industries (e.g., Intel, IBM, Arteris, Tilera, etc.), research in 3D NoC is still in its infancy and demands more concentration from academia and industries.

Aim and scope: This book aims to cover the important aspects of NoC design: communication infrastructure design, communication methodology, evaluation framework, mapping of applications onto NoC, and so on. Apart from these, it also proposes to focus on other upcoming NoC issues, such as low-power NoC design, signal integrity issues, NoC testing, synthesis, reconfiguration, and 3D NoC design.

Organization: The book consists of 12 chapters. The contents of various chapters are as follows:

- Chapter 1 presents the evolution of NoC from SoC—its research and developmental challenges.
- Chapter 2 discusses NoC protocols, elaborating flow control, available network topologies, routing mechanisms, fault tolerance, quality-of-service support, and the design of network interfaces.
- Chapter 3 presents the router design strategies followed in NoCs. It elaborates on clocking strategies, first-in first-out (FIFO) design, globally asynchronous and locally synchronous style of communication, router architecture design for both single- and virtual channel wormhole routers, adaptive router design, and so on.
- Chapter 4 describes the evaluation mechanism of NoC architectures. After introducing the performance and cost metrics, it presents a detailed discussion on traffic modeling, simulator design, and performance evaluation and comparison between different NoC structures.
- Chapter 5 presents the application mapping strategies followed in NoCs. Given an application task graph, several mapping strategies have been developed to associate the intellectual properties (IPs) carrying out these tasks with the routers. The chapter enumerates various strategies such as integer linear programming, constructive and iterative heuristics, and meta-search techniques for the mapping problem.
- Chapter 6 reports on low-power design techniques specifically followed in NoCs. These include various low-power approaches adopted for NoC design, for example, low-power encoding, on-chip serialization, low-swing signaling, static voltage scaling, dynamic voltage scaling, dynamic frequency scaling, voltage–frequency island partitioning, clock gating, and so on. This chapter also includes energy–performance trade-offs.
- Chapter 7 discusses on the signal integrity and reliability issues of NoC. As technology shrinks toward ultra-deep submicron level, crosstalk, electromagnetic interference, synchronization failures, and soft errors are the most important factors affecting the system reliability. This chapter surveys different protection techniques that have been adopted for NoC design until now. It also focuses on energy–reliability trade-offs.
- Chapter 8 presents the details of NoC testing strategies reported so far. NoC testing can be broadly classified into three subproblems: testing the IP cores, testing the routers, and testing the links. It has a detailed discussion on each of the three issues.

- Chapter 9 discusses the problem of synthesizing application-specific NoCs. The NoC synthesis problem addresses the issue of evolving the best possible NoC topology for a given application task graph. It includes the issues such as topology generation, router placement, and scheduling algorithm development on the designed topology.

- Chapter 10 deals with reconfigurable NoC design issues. The subtopics include using field programmable gate array (FPGA) for NoC reconfiguration, designing a router architecture that aids in dynamic change of interconnection pattern between the routers, reconfigurable link design, and revisiting the application mapping problem from the reconfiguration viewpoint.

- Chapter 11 highlights the limited floorplanning choices of 2D NoC and also focuses on 3D NoC design, which is the amalgamation of 2D NoC and 3D IC. In 3D IC, multiple layers of active silicon are stacked using special vertical interconnects, known as *through-silicon via*. The actual benefit of 3D IC relies on the fact that the relatively long wires (approximately in millimeters) of 2D IC can be replaced by these TSVs whose lengths are about tens of microns. This chapter explores the design space of integrating multiple cores onto different silicon layers focusing on the performance and cost metrics.

- Finally, Chapter 12 presents the conclusions and enumerates the directions for future research and development in the field of NoC.

Santanu Kundu
LSI India Research and Development Pvt. Ltd.
(An Avago Technologies Company)

Santanu Chattopadhyay
Indian Institute of Technology, Kharagpur

Authors

Santanu Kundu received his BTech degree in instrumentation engineering from Vidyasagar University, Medinipur, West Bengal, India, in 2002. Thereafter, he served in industry for a couple of years as an electronics engineer and returned to academia for pursuing higher studies in 2004. He received his MTech in instrumentation and electronics engineering from Jadavpur University, Kolkata, West Bengal, India, in 2006. Immediately after that he joined the electronics and electrical communication engineering department at the Indian Institute of Technology, Kharagpur, West Bengal, India, for pursuing a PhD with specialization in microelectronics and very large scale integration (VLSI) design. He received his PhD degree in 2011. Currently he is working as a system-on-chip (SoC) senior design engineer at LSI India R&D Pvt. Ltd., Bangalore, Karnataka, India. His research interests include network-on-chip architecture design in 2D and 3D environments, performance and cost evaluation, signal integrity in nanometer regime, fault-tolerant schemes, and power–performance–reliability trade-off.

Santanu Chattopadhyay received his BE in computer science and technology from Calcutta University (BE College), Kolkata, West Bengal, in 1990. In 1992 and 1996, he received his MTech in computer and information technology and PhD in computer science and engineering, respectively, both from the Indian Institute of Technology (IIT), Kharagpur, West Bengal, India. Before joining the IIT, Kharagpur, he was a faculty member at BE College, Howrah, West Bengal, India, and the IIT, Guwahati, Assam, India. He is currently a professor in the electronics and electrical communication engineering department at the IIT, Kharagpur. His research interests include CAD tools for low-power circuit design and test, system-on-chip testing, and network-on-chip design and test. He has more than hundred publications in refereed international journals and conferences. He is the coauthor of the book *Additive Cellular Automata—Theory and Applications* published by the IEEE Computer Society Press in 1997. He has also written textbooks such as *Compiler Design, System Software,* and *Embedded System Design,* all published by PHI Learning, New Delhi, India. He is a member of the editorial board of the journal *IET Circuits, Devices and Systems.*

1

Introduction

1.1 System-on-Chip Integration and Its Challenges

Continuous reduction in time to market, required by the multimedia and consumer electronics commodities, makes full-custom design inappropriate. It has led to the design based on reuse of intellectual property (IP) cores. With the growing complexity in consumer-embedded products, a single-chip implementation integrating numerous IP cores performing various functions and possibly operating at different clock frequencies is now a well-established one. Such an implementation is conveniently known as *system-on-chip* (SoC). Depending on application domains and versatility, SoC can be classified into two categories: (1) general-purpose *multiprocessor SoC* (MPSoC) and (2) application-specific SoC.

Improving the performance and efficiency of a traditional large uniprocessor architecture is no longer achievable, thus enhancing the demand for parallel processing. This, in turn, has resulted in a revolution in microprocessor architecture—*chip multiprocessing* (CMP) system. For boosting up the performance of CMP-based systems, researchers have adopted SoC platform to build a general-purpose MPSoC for supporting a wide range of applications. This type of SoC is categorized by having a homogeneous set of processing elements and storage arrays. Application-specific SoC, as the name suggests, is dedicated to a specific application. This type of SoC, in many cases, contains heterogeneous processing elements (e.g., processors, controllers, and digital signal processors) and a number of domain-specific hardware accelerators. This heterogeneity may lead to a specific traffic pattern requirement. Hence, a prior knowledge of traffic pattern is required when the system is designed.

Shared medium arbitrated bus is the commonly used communication backbone in modern SoCs. Although this architecture has the advantages of simple topology, extensibility, and low area cost, a shared bus allows only one communication at a time that may block all other buses in the hierarchy. Thus, bus-based SoC does not scale the system performance with the number of cores attached. Its bandwidth is also shared by all the cores (Grecu et al. 2004).

Usage of segmented bus architecture where a shared bus is segmented to multiple buses using bridges also suffers from the same problem of bandwidth sharing. There is also a problem of distributing a synchronous clock signal over the whole chip. In deep submicron (DSM) technologies, according to the International Technology Roadmap for Semiconductors (ITRS) report (ITRS 2001), the delay of local wires and logic gates reduces with every process generation, whereas global wire delay increases exponentially, or at best linearly, by inserting repeaters as shown in Figure 1.1. For a relatively long bus, this delay is significant due to its high intrinsic parasitic resistance and capacitance. As the IP blocks are connected to the bus, they will add more capacitance to it, which may enhance the delay. In ultra-DSM processes, it has been observed that long wires mostly fall in the critical path of the design (Sylvester and Keutzer 2000; Kapur et al. 2002). The long wires in DSM regime also introduce many signal integrity problems, such as crosstalk noise, crosstalk delay, IR drop, and electromagnetic interference (EMI). Moreover, the power consumption of the global wires along with their drivers and repeaters can be a significant portion of the overall SoC power budget. Therefore, in DSM technologies, on-chip communication efficiency has become one of the key factors determining the overall system performance and cost. The major challenge, SoC researchers face today, is to come up with structured, scalable, reusable, and high-performance interconnection architectures.

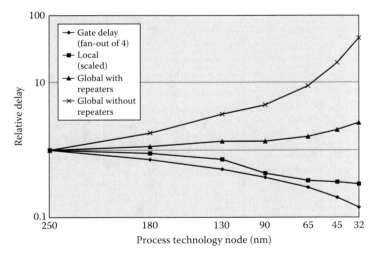

FIGURE 1.1
Projected relative delay for local and global wires and for logic gates at different technologies. (Data from ITRS, International technology roadmap for semiconductors, Technical report, International Technology Roadmap for Semiconductors, 2001.)

1.2 SoC to Network-on-Chip: A Paradigm Shift

Several research groups from academia and industry have started to find out the communication backbone of next-generation many-core-based SoCs for supporting the new inter-core communication demands. Point-to-point dedicated links can be a good alternative to global bus for a limited number of cores in a SoC in terms of bandwidth, latency, and power consumption. However, the number of links needed increases exponentially as the number of cores increases. Thus, for a large system, it may create a routing problem (Bjerregaard and Mahadevan 2006). A centralized crossbar switch overcomes some of the limitations of the buses. Again, connecting large number of cores to a single switch is not very effective as it is not ultimately scalable and, thus, is an intermediate solution only (Bjerregaard and Mahadevan 2006). At the system level, up to a certain number of cores on a single chip, the performance of traditional bus-based SoCs are expected to be satisfactory. But in a many-core regime, as the number of cores residing on a SoC increases significantly, it has a profound effect in shifting the focus from computation to communication.

To overcome the above-mentioned problems, several research groups have started to investigate systematic platform-based approaches to design the communication backbone of MPSoC. On-chip interconnection network is one solution to integrate IPs in complex SoCs. Network-on-chip (NoC) has emerged as the viable alternative for the design of modular and scalable communication architectures. The IP cores communicate with each other via the router-based network. A core is attached to a router through a network interface (NI) module (Benini and Micheli 2002). The network is used for packet-switched on-chip communication among routers, whereas the NIs enable seamless communication between various cores and the network. The need for global synchronization can thus disappear. NoC supports the *globally asynchronous locally synchronous* (GALS) style for multicore communication in SoCs.

The concept of on-chip network has been borrowed from off-chip interconnection networks where a single router is implemented per chip (Gratz et al. 2006). The bandwidth of off-chip networks is typically lower than that of on-chip networks. Off-chip networks are constrained by bit width, as each extra bit incurs one more pin. Also, the off-chip routers need to be connected by explicit board traces. This affects the overall system latency and aggravates the synchronization problem (Jerger and Peh 2009).

The introduction of on-chip networks in SoC design is an evolution of bus interconnect technology. Figure 1.2 shows a NoC structure where heterogeneous IP cores (CPU, DSP, etc.) communicate with each other via a network and NI modules. The function of NI is to isolate the computation from communication. The network consists of switches (routers) and point-to-point

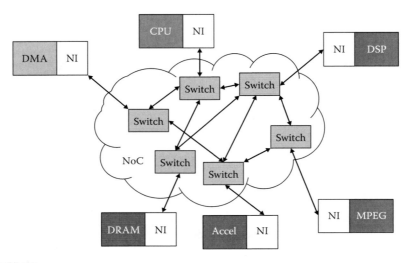

FIGURE 1.2
The NoC paradigm. (Data from Angiolini, F., NoC Architectures, n.d., http://www-micrel.deis
.unibo.it/MPHS/slidecorso0607/nocsynth.pdf.)

communication links between them. Routers route the packets from the
source node to the destination node depending on the underlying network
topology and routing strategy. The length of the point-to-point links should
be small to reduce wire delay.

To mitigate the ever increasing design productivity gap and to meet the
time-to-market requirement, reuse of IP cores is widely used in SoC development. Besides IP cores, the bus interface protocol can also be reused to integrate the IPs. While *reuse* is one of the key challenges that IC design houses
try to address, reuse of IPs, NI, and communication infrastructure such as
routers, underlying network, and flow control protocols can be adopted
in the NoC paradigm. Although selection of network topology and router
architecture is purely application specific, reusing these in different applications will not give the optimal solution. Hence, the reusability is limited
to a particular type of applications. For example, the network topology and
router architecture used for mobile application cannot be same as those of
video processing application. For similar applications, the design and verifi-
cation effort due to reuse will be drastically reduced.

NoC is a specific flavor of interconnection networks and involves several
abstraction layers such as physical, data link, network, and transport layers
(Jantsch and Tenhunen 2003), which are described as follows:

- The *physical layer* determines the number and length of wires connecting resources and switches.
- The *data link layer* defines the protocol of communication between
  a resource and a switch, and between the two switches. Both the

physical and data link layers are dependent on the technology. Thus, for each new technology, these layers are defined.

- The *network layer* defines how a *packet* is transmitted over the network from an arbitrary sender to an arbitrary receiver directed by the receiver's network address. This layer is also technology dependent.
- The *transport layer* is technology independent. In this layer, message size can be variable. This layer breaks the message into network layer packets.

Interconnection networks have been studied for more than the past two decades and a solid foundation of design techniques has been described in several text books (Duato et al. 2003; Dally and Towles 2004). With increasing communication demand, the introduction of interconnection network in SoC design has paved the route to NoC research almost a decade ago. Mullins (2009) has listed more than 400 related articles addressing all these aspects. NoC is today becoming an emerging research topic including hardware communication infrastructure, software and operating system services, CAD tools for NoC synthesis, and so on.

1.3 Research Issues in NoC Development

The major research problems in NoC design can be broadly classified into four different dimensions—communication infrastructure, communication paradigm, evaluation framework, and application mapping—as addressed in the works of Ogras et al. (2005) and Marculescu et al. (2009). This section first highlights these issues briefly followed by other associated issues.

The first dimension of research is focused on choice of *communication infrastructure*. The communication infrastructure design essentially points to the design of underlying hardware acting as the backbone for the on-chip communication network. Selection of network topology, design of router architecture with proper buffer organization, determining inter-router link width, clocking strategies, floorplanning, and layout design are the key design aspects of this dimension. The routers are often connected in certain topologies whose performance behaviors are well known to the distributed system design community and suit well for on-chip realizations. Individual routers are designed using some specific switching techniques, such as wormhole and virtual cut-through. Flow control is performed via handshaking signals between adjacent routers. The router's buffer space minimization and simplified buffer control mechanisms are two important features of the NoC design, as they directly affect the overall area–power overheads and network latency. To solve the problem of clock skew, the individual cores and routers are allowed to operate at their own clocks, giving rise to a GALS scheme. Another important hardware aspect in designing a complete NoC is the integration of cores with the routers. This needs the design of NI modules between the two.

The second dimension of research deals with the *communication paradigm* on a given NoC platform. Once the infrastructure has been finalized, the next important task is to design the communication methodology between the cores via the established network. Routing policies, switching techniques, congestion control, power and thermal management, and fault tolerance and reliability issues are the main focus of this set. It, first of all, necessitates the fixing of routing strategy. This is one of the very rich areas of research in NoC design. It has profound effect on the performance of the NoC as this chiefly determines the number of hops to be traversed in each communication, congestion, traffic load distribution in different routers, and so on. The domain is often complicated by the requirement to support the quality-of-service (QoS). Arbitration of network resources in terms of FIFOs and channels between the contending simultaneous communications is essential to ensure freedom from problems such as livelock and deadlock. Like off-chip communications, on-chip communications also suffer from capacitive crosstalk and electromagnetic radiations, corrupting the data being transmitted. This makes it essential to adopt some fault-tolerant schemes in the communication. As all designs are now invariably power aware, the same is the requirement for NoC as well. It is required to judge very critically the voltages and frequencies at which individual cores and routers are made to operate to satisfy the overall performance requirement with a minimum power budget.

The third dimension of research is paying attention to the design of an *evaluation framework* for NoC by applying stochastic and application-specific traffic. As the MPSoCs contain a large number of cores connected in some topology via routers and interconnection links, it is mandatory to have a clear idea about their performance before any investment is made in manufacturing the systems. The potential faults and drawbacks, if any, must be identified at the design phase to avoid huge loss after getting the silicon chips. Though many theoretical studies exist that can predict the behavior of such a system, they are mostly for congestion-free environment and under the assumption that all cores are equally active in producing traffic load to the network. Both of these assumptions are highly optimistic for any practical design of moderate size. This necessitates the design of high-quality NoC simulators to produce a behavior similar to that of the actual NoC. The simulator should model the network at the granularity of individual hardware blocks and wires in terms of functionality, delay, power, and so on. In the absence of the actual traffic pattern for applications, often synthetic traffic is used. This synthetic traffic should mimic the behavior of the actual core that it corresponds to. With confidence gained after determining the throughput, latency, and bandwidth of the network through simulation, the designer can quickly proceed to accurate estimation of area and power consumption of the network, as it can be a significant portion of the overall SoC cost budget.

The fourth dimension of research is related to *application mapping*. Mapping of cores with regular and irregular sizes onto an underlying NoC platform to achieve the required performance for a specific application is the major issue of this dimension. Performance and energy-aware task scheduling for heterogeneous NoC is another important problem of this class of research. Figure 1.3 summarizes the major dimensions of NoC research as discussed above.

Another important aspect is NoC testing. In any system development process, testing occupies a major part of its turnaround time. The problem is further complicated by the fact that the test volume becomes huge for a NoC. It is necessary to apply test patterns to all the cores and get their responses. The test patterns are to be transported from the system inputs to the core inputs and the responses are to be carried through the network from the core outputs to the system outputs. This gives rise to test schedule optimization problems. The NoC infrastructure itself needs to be tested. The power consumption during test is also a major concern.

While attempting to realize an application, or a set of applications, in NoC, it is imperative to use a NoC infrastructure most suitable for the application(s). This gives rise to the issue of application-specific NoC synthesis. Unlike general standard topologies (such as mesh), NoC synthesis approaches an attempt to derive the topology, routing policy, and so on to obtain the best possible performance of the NoC implementation. While the architecture may be synthesized for a single application, for a set of applications it is quite common to evolve a reconfigurable architecture. Depending upon the communication needs of various applications running at different points in time, a reconfigurable architecture can adapt itself to make it suitable for the currently running application. The reconfiguration may be in the form of link reconfiguration, router port reconfiguration, buffer reconfiguration, and so on.

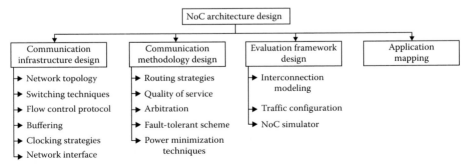

FIGURE 1.3
Major NoC research dimensions.

In the many-core era, integrating large number of cores on a two-dimensional integrated circuit (2D IC) has limited the floor planning choice. Although the size of an individual core is reduced up to a certain level due to technology shrinking, chip sizes may become larger for incorporating huge number of cores on a single silicon die. After the advent of three-dimensional (3D) IC (Davis et al. 2005) that stacks multiple layers of active silicon using special vertical interconnects, known as through-silicon vias (TSVs), the above-mentioned problem of long interconnects can be solved. The actual benefit of 3D IC relies on the fact that the relatively long wires (approximately in millimeters) of 2D IC can be replaced by these TSVs whose lengths are about tens of microns. These shorter TSVs minimize the link delay and link energy consumption significantly and at the same time more immunity to noise (Topol et al. 2006; Flic and Bertozzi 2010). Due to increased connectivity, 3D ICs have the potential for enhancing system performance, achieving better functionality, and producing higher packaging density compared to their traditional 2D counterpart (Davis et al. 2005). Combining these two emerging paradigms, NoC and 3D IC, a new area of research, 3D NoC, has evolved (Pavlidis and Friedman 2007). In a 3D NoC, an entire 2D NoC is divided into a number of blocks, and each block is placed on a separate silicon layer. The 3D NoC research is still in its infancy and needs attention of more researchers to exploit its full potential for using as communication backbone for future many-core-based SoCs.

1.4 Existing NoC Examples

Several research groups from academia and industry have implemented NoC to support MPSoC platform. Intel has introduced 80-core-based *Teraflops* research chip (Vangal et al. 2008) where each core is placed inside a tile of dimension 2 mm × 1.5 mm. The cores are connected in a 2D mesh topology and support wormhole switching of 32-bit flit size with two virtual channels. The routers have been implemented in 65-nm technology with five-stage pipelining. The operating frequency of the router has been found to be 4.27 GHz when implemented on a chip. IBM launched *Cyclops-64* (C64), a peta-flop supercomputer, built on a multicore system-on-a-chip technology. Each C64 chip has 80 custom-designed 64-bit processor cores, which are connected in a 3D mesh fashion (Zhang et al. 2006). The routers have been implemented using two virtual channels to support two service classes. It uses both input and output queuing with seven-stage pipelining and operates at 533 MHz. It can transfer bidirectional data in parallel. Tilera Inc. has introduced a 64-core-based *TILE64* processor (Wentzlaff et al. 2007). The routers are connected in an 8 × 8 2D mesh fashion and follow *XY* routing having a 32-bit link width with no virtual channel. The routers are working at 1 GHz when implemented on silicon in 90-nm technology having both input and output buffering. For supporting highly local traffic inside a node, Intel has introduced *single-chip cloud computer* (SCC) having 48 cores (SCC 2010). Two cores are connected with each router of a 6 × 4 2D mesh. The operating frequency of each core is 1 GHz, whereas the routers are targeted to work with 2 GHz in 45-nm technology. The routers have been implemented with eight virtual channels and four-cycle latency. The link width has been taken as 128 bits. ST Microelectronics have implemented *STNoC* (Coppola et al. 2004), a spidergon topology-based NoC that follows a credit-based flow control. Philips have developed a topology-independent NoC, *Æthereal* (Rijpkema et al. 2003), for supporting guaranteed throughput (GT) and best effort (BE) services. The router has been implemented by an input-buffering scheme with first-in first-out (FIFO) depth of 8 bits and width of 32 bits. It uses a standard credit-based end-to-end flow control. Both the routers and the NI operate at 500 MHz in 130-nm technology at the layout level. *Arteris* is another custom NoC that operates at 750 MHz in 90-nm technology (Arteris 2005). It has a set of configuration and modeling tools—NoC compiler, NoC verifier, and NoC explorer—for getting optimized performance and power result for any application.

Kumar et al. (2007) implemented a 36-core shared memory chip multi-processing (CMP) system in 65-nm technology targeting 3.6 GHz router with single-cycle latency. The cores are connected in a 6 × 6 2D mesh having a flit size of 128 bits. The router has 12 unreserved virtual channels and 1 reserved virtual channel for each of three message classes. It has been implemented with single-stage pipelining. Lee et al. (2004) implemented a hierarchical star-connected on-chip network by using a 16:1 serialized link. The routers and cores operate at 1.6 GHz and 100 MHz, respectively, in 180-nm technology. The authors have also implemented a custom NoC, *Slim-spider* (Lee et al. 2006), ensuring low-power consumption where each router operates at 1.6 GHz in 180-nm technology taking a flit size of 8 bits. Adriahantenaina et al. (2003) implemented a fat tree-based NoC, scalable, programmable, integrated network (*SPIN*), in 130-nm technology taking a flit size of 32 bits. The operating frequency of routers is found to be 200 MHz at the layout level. Another fat tree-based NoC, extended generalized fat-tree (*XGFT*) (Kariniemi et al. 2006), uses a flit size of 32 bits and operates at 400 MHz. Xpipes (Bertozzi et al. 2005), a custom NoC, consists of soft macros of switches, NIs, and links. It takes a flit width of 32 bits and supports error detection and retransmission. Kavaldjiev et al. (2006) modified the traditional virtual channel router and the new router is working at 500 MHz in 180-nm technology supporting the 2D mesh topology with 16-bit flit size. Pande et al. (2005) reported that the area overhead of the routers is reasonably low compared to that of full SoC. Feero and Pande (2009) designed a 3D NoC architecture based on 3D mesh, 3D butterfly fat-tree (BFT), and 3D fat tree topologies having 64 IP cores of size 2.5 mm × 2.5 mm each. They used a flit size of 32 bits and four virtual channels each of two flits deep. The frequency of each router is found to be 1.66 GHz in 90-nm technology after synthesis.

Some asynchronous NoCs have also been reported in the literature. *MANGO* (Bjerregaard and Sparsoe 2005), a clock-less NoC, uses the 2D mesh topology with a flit size of 32 bits. The NIs synchronize the clocked open core protocol (OCP) interfaces to the clock-less network in a GALS fashion and the overall network is running at 795 MHz in 130-nm technology at the register transfer level (RTL) level. Silistix Inc. has introduced its industry leading asynchronous NoC, CHAINworks (Rostislav et al. 2005), for the design and synthesis of complex devices. *FAUST* (Lattard et al. 2008), another asynchronous NoC implemented in 130-nm technology for telecom requirements, uses the 2D mesh technology with a flit size of 32 bits. In the work of Salminen et al. (2008), a list of NoC proposals has been presented in a tabular form that effectively characterizes many of the NoCs that are not covered here.

1.5 Summary

NoC is a very active research field with many practical applications in industry as it is expected to be an efficient communication backbone of next-generation many-core-based SoCs. This chapter focuses on the upcoming technology trends and the needs of NoC in designing many-core-based SoCs. It also briefly covers different horizons of research in the field of NoC design. Finally, a set of NoCs that has been designed till date from the industry and academia has also been covered.

The research dimensions of NoC noted in this chapter have been taken up in subsequent chapters and discussed in detail.

References

Adriahantenaina, A., Charlery, H., Greiner, A., Mortiez, L., and Zeferino, C. A. 2003. SPIN: A scalable, packet switched, on-chip micro-network. *Proceedings of the IEEE Conference on Design, Automation and Test in Europe*, pp. 70–73, Munich, Germany.

Angiolini, F. n.d. NoC architectures. http://www-micrel.deis.unibo.it/MPHS/slide-corso0607/nocsynth.pdf.

Arteris, 2005. A comparison of network-on-chip and buses. *White Paper*. http://www.arteris.com/noc-whitepaper.pdf.

Benini, L. and Micheli, G. D. 2002. Network on chips: A new SoC paradigm. *IEEE Computer*, vol. 35, no. 1, pp. 70–78.

Bertozzi, D., Jalabert, A., Murali, S., Tamhankar, R., Stergiou, S., Benini, L., and Micheli, G. D. 2005. NoC synthesis flow for customized domain specific multiprocessor systems-on-chip. *IEEE Transactions on Parallel and Distributed Systems*, vol. 16, no. 2, pp. 113–129.

Bjerregaard, T. and Mahadevan, S. 2006. A survey of research and practices of network-on-chip. *ACM Computing Surveys*, vol. 38, no. 1, pp. 1–51.

Bjerregaard, T. and Sparsoe, J. 2005. A router architecture for connection-oriented service guarantees in the MANGO clockless network-on-chip. *Proceedings of the Design, Automation and Test in Europe Conference,* pp. 1226–1231, Munich, Germany.

Coppola, M., Locatelli, R., Maruccia, G., Pieralisi, L., and Scandurra, A. 2004. Spidergon: A novel on-chip communication network. *Proceedings of the International Symposium on System on Chip,* p. 15, Tampere, Finland.

Dally, W. J. and Towles, B. 2004. *Principles and Practices of Interconnection Networks.* Morgan Kaufmann Publishers, San Francisco, CA.

Davis, W. R., Wilson, J., Mick, S., Xu, J., Hua, H., Mineo, C., Sule, A. M., Steer, M., and Franzon, P. D. 2005. Demystifying 3D ICs: The pros and cons of going vertical. *IEEE Design and Test of Computers,* vol. 22, no. 6, pp. 498–510.

Duato, J., Yalamanchili, S., and Ni, L. 2003. *Interconnection Networks: An Engineering Approach.* Morgan Kaufmann Publishers, San Francisco, CA.

Feero, B. S. and Pande, P. P. 2009. Networks-on-chip in a three dimensional environment: A performance evaluation. *IEEE Transactions on Computers,* vol. 58, no. 1, pp. 32–45.

Flic, J. and Bertozzi, D. 2010. *Designing Network On-Chip Architectures in the Nanoscale Era.* Chapman & Hall/CRC Computational Science, Boca Raton, FL.

Gratz, P., Changkyu, K., McDonald, R., Keckler, S. W., and Burger, D. 2006. Implementation and evaluation of on-chip network architectures. *Proceedings of the IEEE International Conference on Computer Design,* pp. 477–484, San Jose, CA.

Grecu, C., Pande, P. P., Ivanov, A., and Saleh, R. 2004. Structured interconnect architecture: A solution for the non-scalability of bus-based SoCs. *Proceedings of the ACM Great Lakes Symposium on VLSI,* pp. 192–195, Boston, MA.

ITRS. 2001. International technology roadmap for semiconductors. Technical report, International Technology Roadmap for Semiconductors.

Jantsch, A. and Tenhunen, H. 2003. *Networks on Chip.* Kluwer Academic Publishers, Boston, MA.

Jerger, N. E. and Peh, L. S. 2009. *On-Chip Networks (Synthesis Lectures on Computer Architectures).* Morgan & Claypool Publishers, San Rafael, CA.

Kapur, P., Chandra, G., McVittie, J. P., and Saraswat, K. C. 2002. Technology and reliability constrained future copper interconnects—Part II: Performance implications. *IEEE Transactions on Electron Devices,* vol. 49, no. 4, pp. 598–604.

Kariniemi, H. 2006. On-line reconfigurable extended generalized fat tree network-on-chip for multiprocessor system-on-chip circuits. PhD dissertation, Tampere University of Technology, Finland.

Kavaldjiev, N., Smit, G. J. M., Jansen, P. G., and Wolkotte, P. T. 2006. A virtual channel network-on-chip for GT and BE traffic. *Proceedings of the IEEE Computer Society Annual Symposium on Emerging VLSI Technologies and Architectures,* Karlsruhe, Germany.

Kumar, A., Kundu, P., Singh, A. P., Peh, L. S., and Jha, N. K. 2007. A 4.6Tbits/s 3.6GHz single-cycle NoC router with a novel switch allocator in 65nm CMOS. *Proceedings of the IEEE International Conference on Computer Design,* pp. 63–70, Lake Tahoe, CA.

Lattard, D., Beigne, E., Clermidy, F., Durand, Y., Lemaire, R., Vivet, P., and Berens, F. 2008. A reconfigurable baseband platform based on an asynchronous network-on-chip. *IEEE Journal of Solid-State Circuits,* vol. 43, no. 1, pp. 223–235.

Lee, K., Lee, S. J., Kim, S. E., Chol, H. M., Kim, D., Kim, S., Lee, M. W., and Yoo, H. J. 2004. A 51mW 1.6GHz on-chip network for low-power heterogeneous SoC platform. *Proceedings of the IEEE International Solid-State Circuits Conference*, San Francisco, CA.

Lee, K., Lee, S. J., and Yoo, H. J. 2006. Low-power network-on-chip for high-performance SoC design. *IEEE Transactions on Very Large Scale Integration (VLSI) Systems*, vol. 14, no. 2, pp. 148–160.

Marculescu, R., Ogras, U. Y., Peh, L. S., Jerger, N. E., and Hoskote, Y. 2009. Outstanding research problems in NoC design: Systems, microarchitecture, and circuit perspectives. *IEEE Transactions on Computer-Aided Design of Integrated Circuits and Systems*, vol. 28, no. 1, pp. 3–21.

Mullins, R. D. 2009. On-chip network bibliography. http://www.cl.cam.ac.uk/~rdm34 /onChipNetBib/onChipNetwork.pdf.

Ogras, U. Y., Hu, J., and Marculescu, R. 2005. Key research problems in NoC design: A holistic perspective. *Proceedings of the IEEE/ACM/IFIP International Conference on Hardware/Software Codesign and System Synthesis*, pp. 69–74, Jersey City, NJ.

Pande, P. P., Grecu, C., Jones, M., Ivanov, A., and Saleh, R. 2005. Performance evaluation and design trade-offs for MP-SOC interconnect architectures. *IEEE Transactions on Computers*, vol. 54, no. 8, pp. 1025–1040.

Pavlidis, V. F. and Friedman, E. G. 2007. 3-D Topologies for networks-on-chip. *IEEE Transactions on VLSI Systems*, vol. 15, no. 10, pp. 1081–1090.

Rijpkema, E., Goossens, K. G. W., and Radulescu, A. 2003. Trade offs in the design of a router with both guaranteed and best-effort services for network on chip (extended version). *IEE Proceedings of the Computers and Digital Techniques*, vol. 150, no. 5, pp. 294–302, Munich, Germany.

Rostislav, D., Vishnyakov, V., Friedman, E., and Ginosar, R. 2005. An asynchronous router for multiple service levels networks on chip. *Proceedings of the IEEE International Symposium on Asynchronous Circuits and Systems*, pp. 44–53, New York.

Salminen, E., Kulmala, A., and Hamalainen, T. D. 2008. Survey of network-on-chip proposals. *White Paper*, © OCP-IP ns2.ocpip-server.com/uploads/documents/ OCP-IP_Survey_of_NoC_Proposals_White_Paper_April_2008.pdf.

SCC. 2010. Single-chip cloud computer. http://techresearch.intel.com/UserFiles/en-us /File/SCC_Sympossium_Mar162010_GML_final.pdf.

Sylvester, D. and Keutzer, K. 2000. A global wiring paradigm for deep submicron design. *IEEE Transactions on Computer Aided Design of Integrated Circuits and Systems*, vol. 19, no. 2, pp. 242–252.

Topol, A. W., Tulipe, D. C. L., Shi, L., Frank, D. J., Bernstein, K., Steen, S. E., Kumar, A., Singco, G. U., Young, A. M., Guarini, K. W., and Ieong, M. 2006. Three-dimensional integrated circuits. *IBM Journal of Research and Development*, vol. 50, nos. 4/5, p. 491.

Vangal, S. R., Howard, J., Ruhl, G., Dighe, S., Wilson, H., Tschanz, J., Finan, D., Singh, A., Jacob, T., Jain, S., Erraguntla, V., Roberts, C., Hoskote, Y., Borkar, N., and Borkar, S. 2008. An 80-tile sub-100-W TeraFLOPS processor in 65-nm CMOS. *IEEE Journal of Solid-State Circuits*, vol. 43, no. 1, pp. 29–41.

Wentzlaff, D., Griffin, P., Hoffmann, H., Bao, L., Edwards, B., Ramey, C., Mattina, M., Miao, C. C., Brown, J. F., and Agarwal, A. 2007. On-chip interconnection architecture of the TILE processor. *IEEE Micro*, vol. 27, no. 5, pp. 15–31.

Zhang, Y. P., Jeong, T., Chen, F., and Wu, H. 2006. A study of the on-chip interconnection network for the IBM Cyclops64 multi-core architecture. *Proceedings of the IEEE International Parallel and Distributed Symposium*, Rhode Island.

2

Interconnection Networks in Network-on-Chip

2.1 Introduction

In most modern multiprocessor system-on-chip (MPSoC) architectures, processors and memories are combined in an integrated node. With this arrangement, each processor can access its local memory without using the network. Interconnection networks are also used to connect I/O devices such as disk drives and displays as shown in Figure 2.1. To meet the performance requirement of a specific application, network designer must work within technology constraints to implement the topology, routing, and flow control mechanisms of the network.

In a network topology, the nodes are connected in a different fashion such as mesh and tree. Once a topology has been chosen, routing determines the path through which packets will traverse to the destination. If there are multiple

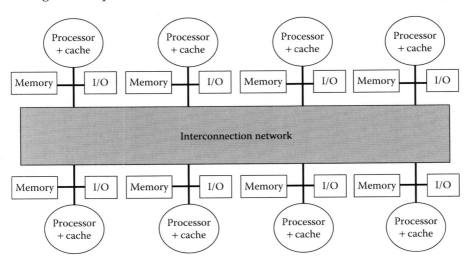

FIGURE 2.1
Interconnection network.

paths exist from source to destination, a good routing mechanism selects a path through which the number of hops will be minimized. Another important aspect in routing is the load balancing. If a particular path is overutilized while another sits idle, known as load imbalance, the total bandwidth of messages being delivered by the network is reduced. Flow control, however, manages the allocation of resources to packets as they progress along their route. A good flow control mechanism forwards packets with minimum delay and is also capable of handling faults in communication. Each of these aspects has been described in detail in the subsequent sections as follows.

Section 2.2 focuses on the basics of network topology, the parameters to consider while selecting a topology, and also the merits and demerits of selecting a topology in network-on-chip (NoC) paradigm. Section 2.3 depicts different switching techniques applicable to NoC. Section 2.4 describes the routing strategies of NoC. It shows how a deadlock can occur in a network and also the deadlock avoidance techniques. Section 2.5 and Section 2.6 discusses the flow control technique and the quality of service, respectively. Section 2.7 describes the design of network interface module, whereas Section 2.8 summarizes the chapter.

2.2 Network Topologies

Selecting a network topology is the most important step of NoC design as it deals with the wire length, the node degree, the routing strategies, and so on. The interconnection architectures having smaller diameter, lower average distance, smaller node degree, more number of links, and larger bisection width are preferable (Dally and Towles 2004). A network diameter is defined as the maximum shortest distance (in terms of the number of hops) between any pair of nodes in a network graph, whereas an average distance is the average of the distances (hop count) between all pairs of nodes in a network graph. A large diameter signifies that packets have to cross more number of hops to reach their farthest destinations, whereas a large average distance denotes the higher average overall latency. A bisection width is defined as the minimum number of wires to be removed to bisect the network. A larger bisection width enables faster information exchange. A node degree can be defined as the number of channels connecting the node to its neighbors. Lower the number of node degree is easier to build the network. The number of links is another important parameter for choosing any topology. A topology with large number of links can support high bandwidth.

In the NoC paradigm, researchers have come up with a number of interconnection architectures with their pros and cons. The *mesh* architecture having a single core connected with each router is the most common interconnection topology. A mesh-based interconnection architecture called *Chip-Level*

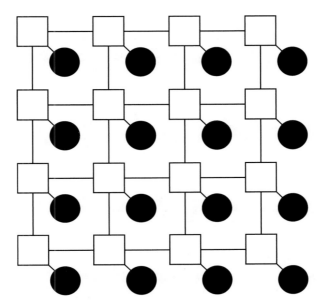

FIGURE 2.2
A 4 × 4 2D mesh with single core connected to each router.

Integration of Communicating Heterogeneous Elements (CLICHÉ) was proposed by Kumar et al. (2002). Mesh structures have large bisection width, but with a drawback of large diameter. Every switch, except those at the corners and boundaries, is connected to four neighboring switches and one intellectual property (IP) block as shown in Figure 2.2. A mesh network having M rows and N columns has the following parameters:

Diameter: $(M + N - 2)$
Average distance: $(M + N)/3$
Bisection width: $min(M,N)$
Number of links: $2 \times [M \times (N - 1) + N \times (M - 1)]$
Number of routers required: $(M \times N)$
Node degree: 3 (corner), 4 (boundary), 5 (center)

The *torus* interconnection architecture has been proposed to solve the large diameter problem of mesh by connecting the routers at the edges via wrap-around links (Dally and Towles 2001). In the torus architecture, the difference with mesh is that the switches at the edges are connected to the switches at the opposite edges through wraparound channels as shown in Figure 2.3. A torus network having M rows and N columns has the following parameters:

Diameter: $\lfloor M/2 \rfloor + \lfloor N/2 \rfloor$
Bisection width: $2 \times min(M,N)$

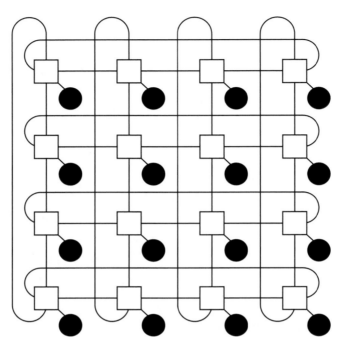

FIGURE 2.3
A 4 × 4 2D torus with single core connected to each router.

Number of routers required: $(M \times N)$

Node degree: 5

For a larger network, this wraparound link will be long enough and will cause excessive delay.

A *folded torus* solves the problem of excessive delay in the long wraparound connections of torus by folding it (Dally and Seitz 1986). Figure 2.4 shows a 4 × 4 folded torus network. A folded torus network having M rows and N columns has the following parameters:

Diameter: $\lfloor M/2 \rfloor + \lfloor N/2 \rfloor$

Bisection width: $2 \times min(M,N)$

Number of routers required: $(M \times N)$

Node degree: 5

To reduce the average hop count of a mesh structure, a *concentrated mesh* (*CMESH*) topology (Balfour and Dally 2006) has been proposed where four cores are connected to each router. Thus, the number of routers required

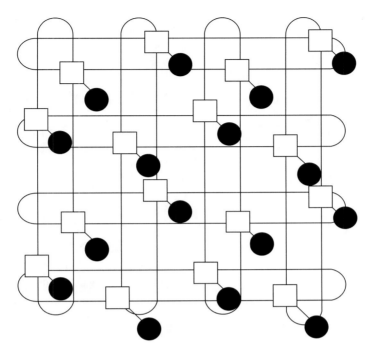

FIGURE 2.4
A 4 × 4 2D folded torus with single core connected to each router.

for implementing the CMESH network is one-fourth of that of a traditional mesh structure as shown in Figure 2.5. To make the bisection width same as that of the mesh structure, additional long interconnection links are attached along the perimeter of the network. The node degree of each router in the CMESH network is 8, much higher than in the mesh, torus, and folded structures. A CMESH network having $M \times N$ IP blocks has the following parameters:

Diameter: $(M/2 + N/2 - 4)$
Bisection width: min $[\{(M/2 + (2 \times \lfloor (\log_2 N) - 1 \rfloor)\}, \{N/2 + (2 \times \lfloor (\log_2 M) - 1 \rfloor)\}]$
Number of routers required: $(M \times N)/4$

Another interesting network is the *octagon* structure, in which connection between any two nodes (within an octagon subnetwork of eight nodes) requires at most two hops (Karim et al. 2002). Each node in this network is associated with an IP and a switch as shown in Figure 2.6. For embedding more than eight processors, more octagons can be combined together by using bridge nodes. For a system consisting of more than eight nodes,

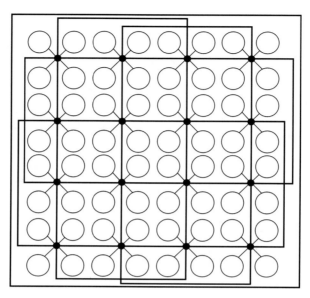

FIGURE 2.5
A 2D concentrated mesh with four cores to each router.

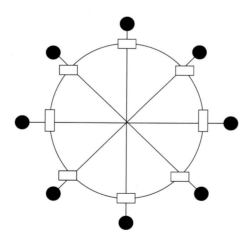

FIGURE 2.6
A 2D octagon network with single core to each router.

the network is extended to a multidimensional space. A network having N IP blocks has the following parameters:

Diameter: $2 \times \lceil (N/8) \rceil$

Bisection width: 6 for $N \leq 8$ or $6 \times (1 + \lfloor N/8 \rfloor)$ for $N > 8$

Number of routers required: 8 for $N \leq 8$ or $(8 + 7 \lfloor N/8 \rfloor)$ for $N > 8$

Node degree: 4 (member node), 7 (bridge node)

The concept of octagon network can be extended to any arbitrary even number of nodes using a *spidergon* topology (Coppola et al. 2004). However, both octagon and spidergon may lead to a significant increase in the wiring complexity for large-sized networks. In the spidergon topology, all nodes are connected to three neighbors and an IP as shown in Figure 2.7. A spidergon network having N IP blocks has the following parameters:

Diameter: $\lceil N/4 \rceil$

Bisection width: $N/2 + 2$

Number of routers required: N

Node degree: 4

A *binary tree* architecture has also been proposed for NoC (Jeang et al. 2004). It has the advantages of having nice recursive structure and desired low diameter but with a drawback of having small bisection width. In the binary tree architecture, four IPs are connected at the *leaf*-level node, but none at the others as shown in Figure 2.8. In particular, tree-based topologies require long interconnection links between the routers toward the root of the tree, which increase the delay and power consumption of links. A binary tree-based network with N IP blocks ($N = 2^i$, where $i = 2, 3, 4, \ldots$) has the following parameters:

Diameter: $2 \times (\log_2 N - 2)$

Bisection width: 1

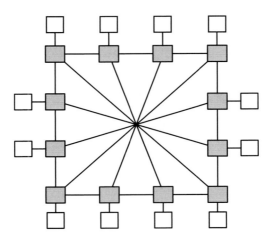

FIGURE 2.7
A 2D spidergon network with single core to each router.

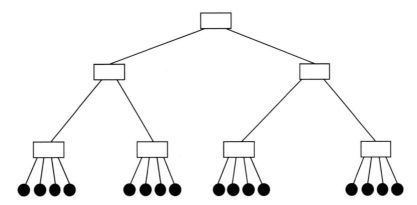

FIGURE 2.8
A 2D binary tree network with four cores to each leaf level router.

Number of routers required: $(N/2 - 1)$

Node degree: 5 (leaf), 3 (stem), 2 (root)

A *fat tree*-based generic interconnect template called *Scalable, Programmable Integrated Network (SPIN)* has been proposed for on-chip interconnection (Guerrier and Greiner 2000). Every node has four children and the parent is replicated four times at any level of the tree as shown in Figure 2.9. The functional IP blocks reside at the leaves and the switches reside at the vertices. The disadvantages of a fat tree architecture are its large switch size and high node degree. A fat tree-based network with N IP blocks ($N = 2^i$, where $i = 4$, 5, 6, ...) has the following parameters:

Diameter: $2 \times \left(\lceil (\log_2 N)/2 \rceil \right) - 2$

Bisection width: $N/2$ when i is even, $N/4$ when i is odd

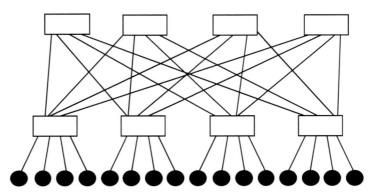

FIGURE 2.9
A 2D SPIN network with four cores to each level router.

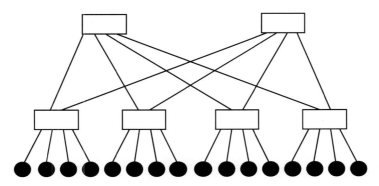

FIGURE 2.10
A 2D BFT network with four cores to each router.

Number of routers required: $(N/4) \times \left(\lceil (\log_2 N)/2 \rceil \right)$
Node degree: 8 (non-root node), 4 (root node)

Pande et al. (2003b) proposed a *butterfly fat tree* (BFT) interconnection architecture in which four IP cores are placed at each leaf as shown in Figure 2.10. BFT has the advantages of having large bisection width and low diameter. It uses lesser number of switches to build large networks. However, the number of links in BFT based network is lesser than other available topologies, which leads to more congestion and lesser throughput in a real traffic scenario. A BFT-based network with N IP blocks ($N = 2^i$, where $i = 4, 5, 6, \ldots$) has the following parameters:

Diameter: $2 \times \left(\lceil (\log_2 N)/2 \rceil \right) - 2$

Bisection width: $N \times (0.5)^{\lceil \log_2 N/2 \rceil}$ for i is even, $(N/2) \times (0.5)^{\lceil \log_2 N/2 \rceil}$ for i is odd

Number of routers needed: $(N/2) \times \left[1 - (0.5)^{\lceil \log_2 N/2 \rceil} \right]$
Node degree: 6 (non-root), 4 (root)

A derivative of BFT, *extended-BFT interconnection* (EFTI) (Hossain et al. 2005), has been proposed for improving the packet latency and throughput over BFT. The node degree of EFTI is higher than that of BFT and it has long wraparound interconnection wires as shown in Figure 2.11. An EFTI-based network with N IP blocks ($N = 4^i$, where $i = 2, 3, 4, \ldots$) has the following parameters:

Diameter: $\log_2 N - 2$
Bisection width: $2 + N \times (0.5)^{\log_2 N/2}$
Number of routers needed: $(N/2) \times \left[1 - (0.5)^{\log_2 N/2} \right]$
Node degree: 8 (non-root), 4 (root)

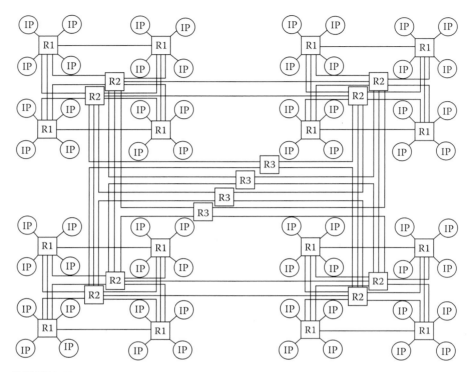

FIGURE 2.11
EFTI network.

Kundu and Chattopadhyay (2008a, 2008b) proposed a *mesh-of-tree* (MoT) interconnection network for NoC. MoT enjoys the advantages of having smaller diameter and node degree compared to mesh. Compared to BFT, it has more number of edges and hence reduced congestion. A QUOTE($M \times N$) MoT-based network (M and N denoting the number of row trees and column trees, respectively) has the following parameters:

Number of nodes = $3 \times (M \times N) - (M + N)$
Diameter = $2 \lfloor \log_2 M \rfloor + 2 \lfloor \log_2 N \rfloor$
Bisection width = *min* (M, N);
Node degree = 4 (leaf), 3 (stem), 2 (root)
Symmetric and recursive structure

Figure 2.12 shows a 4 × 4 MoT structure, having four row and four column trees. The row and column trees are formed by the white and black nodes, respectively, as shown in Figure 2.12a. The leaf level nodes are common to both the trees. Two cores (shown as white circles in Figure 2.12b) have been attached to each leaf level node, whereas the *stem* and *root* nodes are not

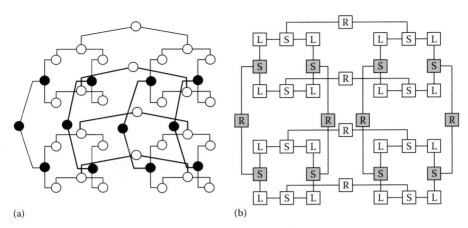

FIGURE 2.12
MoT topology (a) and its simplified graph (b).

having any core attached to them. The simplified 4 × 4 model of MoT graph is shown in Figure 2.12b in which L, S, and R denote the leaf, stem, and root level nodes, respectively.

Kim et al. (2007) proposed a *flattened butterfly* topology for NoC implementation in which four cores are connected to each router as shown in Figure 2.13. The routers are oriented in a two-dimensional (2D) grid fashion such that each of them is connected to all other routers in the same row and also in the same column by exploiting the nature of *express cubes* (Dally 1991). Express cube requires long wires and high connectivity routers. The channels can be increased to the point that wire delays dominate node delay. Moreover, the number of links increases quadratically with the number of interconnected nodes. A flattened butterfly-based network with N IP blocks has the following parameters:

Diameter: 2
Bisection width: $N \times (0.5)^{\log_2 N/2}$
Number of routers needed: $N/4$
Node degree: 10

Another express cube-based topology, *multidrop express channels* (MECS) (Grot et al. 2009), eliminates the problem of increasing the number of links of flattened butterfly quadratically by introducing point-to-multipoint communication links. But in point-to-multipoint links, every additional node adds to parasitic capacitance of the links and causes system performance and frequency degradation.

The performance of an on-chip communication network is characterized by its throughput. It is directly proportional to the number of directed edges (E) and inversely proportional to the average distance (D) of the network

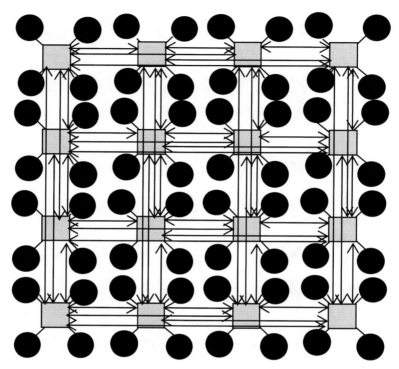

FIGURE 2.13
Flattened BFT interconnection network.

(Decina et al. 1991). Thus, theoretically, the *E/D* ratio is a good indicator for the throughput of a network, without considering contention among packets. Another important performance metric is latency of the network. Theoretically, zero-load latency is a widely used parameter to illustrate the impact of topology in which no contention among packets has been considered. In wormhole switching, taking equal clock cycle delay of the routers, zero-load latency is defined as (Dally and Towles 2004)

$$T_{\text{zero load}} = Dt_{\text{r}} + t_{\text{c}} + \frac{L_{\text{p}}}{b} \tag{2.1}$$

where:
   D is the average distance
   t_{r} is the delay (in clock cycle) of each router
   t_{c} is the link delay (in clock cycle)

The third term of the above equation signifies the serialization delay of the packet, where L_{p} is the length of the packet in bits and b is the communication channel bandwidth. Thus, taking t_{c} and (L_{p}/b) as fixed quantities, the zero-load latency of a network is proportional to the average distance of the network. Thus, for any topology, knowledge of the number of directed edges (E) and the average distance (D) are equally important as its diameter and bisection width. Kundu et al. (2012) presented a general formulation to find both of these parameters for an $M \times N$ MoT structure having two cores connected to each leaf level node, where each row tree and column tree is a complete binary tree. This formulation has been presented in Sections 2.2.1 and 2.2.2.

2.2.1 Number of Edges

The number of edges in each complete binary tree having k number of leaf nodes is ($2k - 2$) (West 2002). In an $M \times N$ MoT topology, each row tree and column tree is a complete binary tree having N and M leaf nodes, respectively. Thus, the number of edges $É$ of an undirected MoT graph can be formulated as follows:

$$É = \left[M(2N-2) + N(2M-2) \right] = 4MN - 2(M+N) \qquad (2.2)$$

As in the MoT structure, adjacent routers are connected by two unidirectional opposite edges, the number of directed edges will be

$$E_{\text{MoT}}(M \times N) = 2É = 8MN - 4(M+N) \qquad (2.3)$$

2.2.2 Average Distance

The average distance of a network is the average of the minimum distances (in hop count) between all pairs of IP cores. In a complete binary tree having N number of leaf nodes with two cores connected to each router, the distribution of destination cores from a specific source core is shown in Table 2.1. Thus, the summation of minimum distances to all the destination cores from

TABLE 2.1

Distribution of Destination Cores from a Specific Core in a Complete Binary Tree

Minimum Distance	Number of Cores
0	2^0
2	2^1
4	2^2
6	2^3
...	...
...	...
...	...
$2\log_2 N$	$2^{(\log_2 N)}$

a specific source core in a complete binary tree having two cores in each leaf level router can be written as

$$S_b = \left[0.2^0 + 2.2^1 + 4.2^2 + 6.2^3 + \cdots + (2\log_2 N) \times 2^{(\log_2 N)} \right]$$

$$= \left[4N\log_2 N - 4(N-1) \right]$$
(2.4)

A $(2^1 \times N)$ MoT consists of two row-wise binary trees of depth $\log_2 N$ and N column-wise binary trees of depth 1 (which can be written as $\log_2 2^1$. Each row tree consists of $(2^1 \times N)$ cores. Thus, the summation of distances of $(2^1 \times N)$ cores lying in the second row tree from any specific core of the first row tree is $[2^0 \times S_b + (2^1 \times N) \times (2\log_2 2^1)]$. Hence, the summation of minimum distances to all the destination cores from a specific source core in a $(2^1 \times N)$ MoT can be written as

$$S_{MoT}\left(2^1 \times N\right) = S_b + \left[2^0 \times S_b + \left(2^1 \times N\right) \times \left(2\log_2 2^1\right) \right]$$
(2.5)

In the same way, a $(2^2 \times N)$ MoT can be split into two $(2^1 \times N)$ MoTs (as shown in Figure 2.14) where each $(2^1 \times N)$ MoT consists of 2^1 row-wise binary trees. The depth of each column tree of $(2^2 \times N)$ MoT is 2 (which can be written as $\log_2 2^2$. Thus, the summation of distances of $(2^2 \times N)$ cores lying in the second $(2^1 \times N)$ MoT (shown as dotted box in the figure) from any specific core of the first $(2^1 \times N)$ MoT (shown as firm box in the figure) is

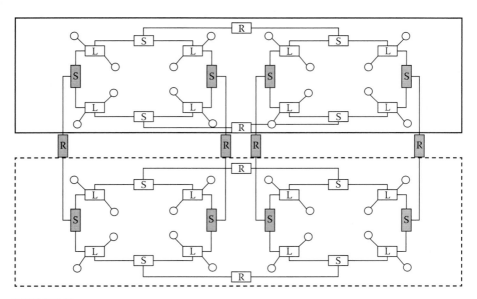

FIGURE 2.14
Splitting of a $(2^2 \times 2^2)$ MoT into two $(2^1 \times 2^2)$ MoTs.

$[2^1 \times S_b + (2^2 \times N) \times (2 \log_2 2^2)]$. The following equation shows the summation of minimum distances to all the destination cores from a specific source core in a $(2^2 \times N)$ MoT:

$$S_{\text{MoT}}\left(2^2 \times N\right) = S_{\text{MoT}}\left(2^1 \times N\right) + \left[2^1 \times S_b + 2^2 \times \left(2\log_2 2^2\right) \times N\right] \quad (2.6)$$

Similarly, Equations 2.7 and 2.8 show the summation of minimum distances to all the destination cores from a specific source core in $(2^3 \times N)$ and $(2^4 \times N)$ MoTs, respectively:

$$S_{\text{MoT}}\left(2^3 \times N\right) = S_{\text{MoT}}\left(2^2 \times N\right) + \left[2^2 \times S_b + 2^3 \times \left(2\log_2 2^3\right)N\right] \quad (2.7)$$

$$S_{\text{MoT}}\left(2^4 \times N\right) = S_{\text{MoT}}\left(2^3 \times N\right) + \left[2^3 \times S_b + 2^4 \times \left(2\log_2 2^4\right)N\right] \quad (2.8)$$

In general, a $(2^{\log_2 M} \times N)$ MoT can be split into two $[2^{(\log_2 M)-1} \times N]$ MoTs where each $[2^{(\log_2 M)-1} \times N]$ MoT consists of $2^{(\log_2 M)-1}$ row-wise binary trees. The depth of each column tree of the $(2^{\log_2 M} \times N)$ MoT is $\log_2(2^{\log_2 M})$. Thus, the summation of minimum distances of $(2^{\log_2 M} \times N)$ cores lying in the second $[2^{(\log_2 M)-1} \times N]$ MoT from any specific core of the first $[2^{(\log_2 M)-1} \times N]$ MoT is $[2^{(\log_2 M)-1} \times S_b + (2^{\log_2 M} \times N)] \times [2\log_2(2^{\log_2 M})]$. For a $(2^{\log_2 M} \times N)$ MoT, the summation of minimum distances to all the destination cores from a specific source core can be written as

$$S_{\text{MoT}}\left(2^{\log_2 M} \times N\right) = S_{\text{MoT}}\left[2^{(\log_2 M)-1} \times N\right] +$$
$$\left\{2^{(\log_2 M)-1} \times S_b + \left(2^{\log_2 M} \times N\right) \times \left[2\log_2\left(2^{\log_2 M}\right)\right]\right\} \quad (2.9)$$

After simplification, the above equation becomes

$$S_{\text{MoT}}\left(2^{\log_2 M} \times N\right) = S_{\text{MoT}}(M \times N) = \left[4MN \times \log_2(MN) - 8MN + 4(M+N)\right] \quad (2.10)$$

It is noticeable that due to the symmetrical structure of MoT topology, the summation of minimum distances to all the destination cores from any source core is always the same. Hence, the average distance of $M \times N$ MoT network connecting C number of cores can be written as

$$D_{\text{MoT}}\left(M \times N\right) = \frac{\left[4MN \times \log_2(MN) - 8MN + 4(M+N)\right]}{C-1} \quad (2.11)$$

In general, for an $M \times (C/2M)$ MoT network (where M is the number of row trees and C is the total number of cores attached) having two cores connected to each leaf node, the number of directed edges (Equation 2.4) and the average distance (Equation 2.11), respectively, can be written as

$$E_{\text{MoT}}\left(M \times \left\lceil \frac{C}{2M} \right\rceil \right) = 8M \times \left\lceil \frac{C}{2M} \right\rceil - 4\left(M + \left\lceil \frac{C}{2M} \right\rceil \right)$$

$$D_{\text{MoT}}\left(M \times \left\lceil \frac{C}{2M} \right\rceil \right) = \frac{\left[\begin{array}{l} 4M \times \lceil C/2M \rceil \times \log_2 \left(M \times \lceil C/2M \rceil \right) - \\ 8M \times \lceil C/2M \rceil + 4\left(M + \lceil C/2M \rceil \right) \end{array} \right]}{(C - 1)}$$

Theoretically, *E/D* is a good indicator for the throughput of a network, without considering contention among packets (Decina et al. 1991). It can be shown that the value of $(E_{\text{MoT}}/D_{\text{MoT}})$ reaches its maximum and D_{MoT} reaches its minimum when the condition $M = \lceil C/2M \rceil$ holds. This implies that the MoT network will show maximum throughput and minimum latency in a congestion-free environment when the number of row trees and that of column trees are same.

Similarly, for an $M = \lceil C/M \rceil$ mesh network (where *M* is the number of nodes in each row and *C* is the total number of cores attached in the network), the average distance (*D*) and the number of directed edges (*E*), respectively, can be written as follows (Pavlidis and Friedman 2007):

$$D = \frac{\left(M + \lceil C/M \rceil \right)}{3} \tag{2.12}$$

$$E = 2 \times \left\{ M \times \left(\left\lceil \frac{C}{M} \right\rceil - 1 \right) + \left\lceil \frac{C}{M} \right\rceil \times (M - 1) \right\} \tag{2.13}$$

$$\left(\frac{E}{D} \right) = 2 \times \left(\frac{6C}{M + \lceil C/M \rceil} - 3 \right) \tag{2.14}$$

It can be shown that for a mesh network having a single core connected to each router, the value of (*E/D*) reaches its maximum and the value of *D* reaches its minimum when the condition $M = \lceil C/M \rceil$ holds. This signifies that a square mesh network with an equal number of row and column trees is expected to show the best performance. The performance will degrade as the network becomes more and more rectangular in nature. Thus, to connect 2^n cores, where *n* is odd, a mesh network that connects a single core to each router may not be the ideal choice to the NoC designers due to its rectangular shape. This statement is also true for a mesh network connecting two cores to each router for 2^n cores, where *n* is even.

Pande et al. (2005) compared a set of network topologies with 256 cores in terms of throughput, latency, energy consumption, and area overhead. They have reported that SPIN and octagon networks have very high throughput, but their energy consumption and silicon area overheads are much higher than both mesh and BFT networks. Folded torus shows almost similar results as mesh does. In deep submicron era where low-power design is a major goal, for a NoC designer, it is always preferable to choose a topology with lower average energy profile. Thus, although the performance of mesh network is comparatively inferior to that of SPIN- and Octagon-based networks, it is widely used in the industry (Vangal et al. 2008; Wentzlaff et al. 2007).

2.3 Switching Techniques

Switching techniques determine when and how internal switches of the network are set to connect router inputs to outputs for transferring the messages. They can be classified as circuit switching and packet switching. In circuit switching, a physical path from source to destination is reserved prior to the transmission of the data. The base latency of a circuit-switched message is determined by the time to set up a path and the time to transmit the data. Banerjee et al. (2007) designed a circuit switching-based NoC router. However, this switching technique is inefficient as it produces excessive blocking, which in turn affects the network bandwidth and also leads to excessive communication latency.

In the packet switching technique, each message is partitioned into fixed length packets and the packets are transmitted without reserving the entire path. Packet-switched networks can further be classified as *store-and-forward* (SAF), *virtual cut-through* (VCT), and *wormhole*. In SAF switching, a packet is completely buffered at each immediate node before it is forwarded to the next node. Therefore, it needs huge silicon area. The latency in communication depends on the size of the packet. CLICHÉ is an example of SAF switching (Kumar et al. 2002). In VCT switching, a packet is forwarded to the next router as soon as there is enough space to store the packet. VCT switching overcomes the latency penalty of SAF switching but also requires huge silicon area to store the entire packet. In both SAF and VCT switching, message flow control is performed at packet level. In *Proteo* network, VCT switching technique has been adopted (Tortosa et al. 2004).

In wormhole switching, packets are divided into *flow control units* (flits) such as header flit, payload flit, and tailer flit. The header flit contains information about source and destination addresses. The payload flit consists of data, whereas the tailer flit contains the end of packet information. The buffers are expected to store only a few flits. As a result, the buffer space requirement in the switches can be small, compared to SAF and VCT switching. Header flit decoding enables the switches to establish the path, whereas payload and tailer flits simply follow this path in a pipelined fashion. If a certain flit faces a busy channel, subsequent flits also have to wait at their current locations.

In the absence of contention, VCT and wormhole switching have the same latency. Otherwise, VCT has lower latency and higher acceptance rate compared to wormhole switching (Banerjee et al. 2004). However, VCT requires larger silicon area and consumes higher energy due to larger buffer size. Therefore, a trade-off is necessary between energy consumption, area, and performance. Wormhole switching is preferable for large packet size, whereas VCT switching is a better choice for short packets. When wormhole switching is employed, the header flit gets blocked if the output channel is already assigned to another packet. This problem is known as *head-of-line* (HoL) blocking. This affects the overall system performance. To mitigate this effect, Dally (1992) proposed the usage of several virtual channels (VCs) within each physical channel. When a particular packet is blocked, VC allows other packets to use the link that would otherwise be left idle. Usage of VC improves the overall performance at the cost of increased energy consumption and silicon area overhead. However, VC cannot eliminate the HoL blocking problem completely. Duato et al. (2003) compared the performance of VCT- and VC-based wormhole switching, both having an equal buffer capacity. It has been observed that the VC-based wormhole switching achieves much higher throughput, whereas the average latency in both the cases is almost identical before saturation. In NoC design, wormhole routers with limited number of VCs are preferable. The optimum number of VCs per physical channel is determined by power–performance trade-off of the overall system. Pande et al. (2005) reported that the optimum value of VCs per physical channel is 4, beyond which the throughput increment is marginal, whereas the energy consumption and area overhead increase. To reduce zero-load latency and router energy in a VC-based network, Kumar et al. (2007) introduced the express cube structure that allows packets to virtually bypass the intermediate routers along their path in a completely nonspeculative fashion.

2.4 Routing Strategies

Routing strategies determine the traversal path of a packet from the source to the destination. Depending on the number of destinations of a single packet, routing algorithm can be classified as *unicast* and *multicast*. In unicast routing, each packet has a single destination, whereas in multicast routing, a single packet has multiple destinations. For on-chip communication, unicast routing seems to be a practical approach due to the presence of point-to-point communication links between various components inside a chip (Agarwal et al. 2009). Routing techniques can be further classified as *source* and *distributed* based on the position at which routing decision takes place. In source routing, precomputed routing table is stored in the network interface (NI). *Æthereal* uses source routing (Goossens et al. 2005). In distributed routing, each packet carries the source and destination addresses. The routing decision is implemented in each router either by a routing table or by executing a finite-state machine. *SoCIN* is an example of distributed routing (Zeferino and Susin 2003). Depending on the adaptability, both source and distributed routing can further be classified as *deterministic, oblivious,* and *adaptive* (Duato et al. 2003). In deterministic (or static) routing, packets always follow a specific path from source to destination. This assures in-order delivery of packets. Oblivious routing, however, selects the path randomly or cyclically. Both deterministic and oblivious routing do not consider the current state of the network. In adaptive (or dynamic) routing, the routing decisions are made according to the current state of the network (congestion, available links, etc.) and alternative paths are chosen dynamically to avoid congested or faulty regions of the network. Therefore, in-order delivery is not guaranteed and reordering of packets at destination NI is a necessity. Adaptive routing can be classified as *progressive* and *backtracking.* Progressive routing moves the header forward, reserving a new channel at each routing operation. Backtracking routing allows the header to backtrack as well, releasing previously reserved channels. Backtracking algorithms are mainly used for fault-tolerant routing. Both deterministic and adaptive routing can be *minimal* and *non-minimal,* based on the number of hops traversed from source to destination. Delay and power consumption in communication are higher in non-minimal routing than in minimal routing as it traverses more number of hops. Adaptive routing that follows a minimal path from source to destination is further classified as *minimal fully adaptive* and *partially adaptive.* The challenges of any routing scheme are that the routing should be *livelock* and *deadlock* free.

Livelock arises when packets travel around their destination node, but unable to reach it because the channels to do so are occupied by other packets. It can only occur in adaptive routing when packets are allowed to follow non-minimal paths. *Deadlock* occurs when a set of messages is blocked forever because each message in the set holds one or more resources needed by another message in the set. There are two ways in which a deadlock can occur in a network—routing-dependent deadlock and message-dependent deadlock.

2.4.1 Routing-Dependent Deadlock

Depending on the routing information, a deadlock situation will arise if there exists any cycle in its *channel dependency graph* (Dally and Seitz 1987). A channel dependency graph is a directed graph whose vertices are the channels of the interconnection network, whereas edges show the dependency between any pair of channels. Figure 2.15 shows a scenario of routing-dependent deadlock in a mesh network. Figure 2.15a depicts that s1, s2, s3,

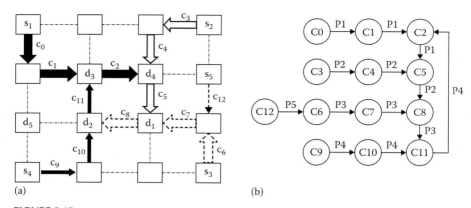

FIGURE 2.15
Routing-dependent deadlock in mesh network: (a) packet traversal path; (b) channel dependency graph.

s4, and s5 are the sources of packets, whereas d1, d2, d3, d4, and d5 are their respective destinations. Packet P1 originated from s1 traversed through the channels c0–c1–c2 and blocked by the packet P2 originated from s2. Packet P2 originated from s2 traversed through the channels c3–c4–c5 and blocked by the packet P3 originated from s3. Packet P3 originated from s3 traversed through the channels c6–c7–c8 and blocked by the packet P4 originated from s4. Similarly, the packet P4 originated from s4 traversed through the channels c9–c10–c11 and blocked by the packet P1 originated from s1. Figure 2.15b shows the formation of a cycle (c2–c5–c8–c11–c2) in the channel dependency graph of the above communication. Due to this routing-dependent deadlock, none of the packet can reach to its destination.

For handling routing-dependent deadlock, any of the two techniques—deadlock avoidance and deadlock recovery—can be adopted. In order to avoid routing-dependent deadlock in a network, the sufficient condition is to show the nonexistence of a cycle in the channel dependency graph.

Dimension-ordered routing is the most simplistic approach to avoid deadlock in deterministic routing. In a 2D mesh, this routing is known as *XY* routing where a packet is forwarded first in *X*-dimension until it reaches the *X*-coordinate of the destination and then it is forwarded in *Y*-dimension until it reaches the destination. Figure 2.16a depicts the scenario and Figure 2.16b shows its channel dependency graph where no cycle exists, and hence this routing can avoid deadlock in the network. Like *XY* routing, *YX* routing is also deadlock free.

For SAF and VCT switching, as the buffer has the capacity to store the entire packet, it is much simpler to avoid deadlock. Deflection routing or hot potato routing (Greenberg and Hajek 1992) is used for these switching techniques.

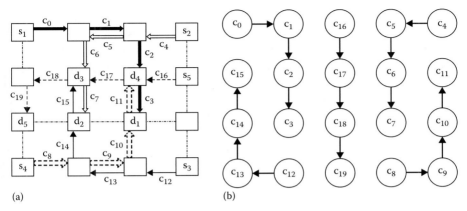

FIGURE 2.16
XY routing in 2D mesh topology (a) and its channel dependency graph (b).

For ring, torus, and folded torus, Dally and Seitz (1987) proposed a deadlock avoidance technique by splitting each physical channel into a group of VCs. For a ring network, Figure 2.17 shows that the packets at a node numbered less than their destination node are routed on the firm channels (C_{00}, C_{01}, and C_{02}), whereas the packets at a node numbered greater than their destination node are routed on the dotted channels (C_{11}, C_{12}, and C_{13}). For example, if a packet originated from n3 and destined for n2, it traversed through the following channels: C_{13}–C_{00}–C_{01}.

For routing in BFT, a *least common ancestor* (LCA) algorithm was proposed by Pande et al. (2003a). For deadlock free routing in any irregular topology or faulty regular topology, instead of using VC, *up*/down** routing is used (Schroeder et al. 1991). A deterministic routing in an $M \times N$ MoT network was proposed by Kundu and Chattopadhyay (2008a, 2008b) and presented here in detail.

2.4.1.1 Deterministic Routing in M × N MoT Network

To propose a routing algorithm for any network, the following steps are required: (1) addressing of each node, (2) proof of livelock free, and (3) proof of deadlock free routing.

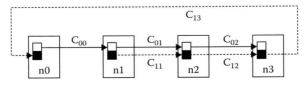

FIGURE 2.17
Routing-dependent deadlock avoidance in ring network using VC.

2.4.1.1.1 Addressing Scheme

The address of each node in an $M \times N$ MoT network consists of four fields: (1) *row number* (RN), (2) *column level* (CL), (3) *column number* (CN), and (4) *row level* (RL). For each row tree, RN is fixed; thus, for a 4×4 MoT, RNs are 00, 01, 10, and 11. RLs are gradually increased by 1 from leaf level to root level of the row tree. In a row tree, CL is 00 for all the nodes. CN is assigned taking parent–child relationship as shown in Figure 2.18. For example, 00–00–10–00 and 00–00–11–00 are the children and 00–00–1X–01 is the row parent where X denotes "don't care." Similarly, for each column tree, CN is fixed. For a 4×4 MoT, CNs are 00, 01, 10, and 11. CLs are gradually increased by 1 from leaf level to root level of each column tree. In a column tree, RL is 00 for all the nodes and RNs are assigned taking parent–child relationship as shown in Figure 2.18. For example, 00–00–00–00 and 01–00–00–00 are the children and 0X–01–00–00 is the column parent where X denotes "don't care." A core has the same RN, CL, CN, and RL similar to its associated router node. A core has only one additional *Core-ID* bit. For example, the addresses of *Core1* and *Core2* attached to the leaf-node 11–00–11–00 are 0–11–00–11–00 and 1–11–00–11–00, respectively.

In the above addressing scheme, Xs (don't cares) are used in the addresses of stem and root nodes. As all the nodes have different addresses, the RN and CN are represented using 4-bit numbers ($0 = 01$; $1 = 10$; $X = 11$). Therefore, each node has 12-bit address; for example, node address 10–00–XX–10 becomes 1001–00–1111–10. Therefore, in 4×4 MoT, each core has a 13-bit address. The bit size required for addressing a core in $M \times N$ MoT is given below:

Core-ID = 1 bit

Row number = $2\log_2 M$ bit

Column level = $\log_2(\log_2 2M)$ bit

Column number = $2\log_2 N$ bit

Row level = $\log_2(\log_2 2N)$ bit

2.4.1.1.2 Routing Algorithm

The routing algorithm follows the deterministic approach. The algorithm ensures that the packet will reach its destination always through a specified shortest path. Thus, the proposed network is always livelock free. The following abbreviations have been used to describe the algorithm:

- *addr (curr)* denotes the address of the current node.
- *addr (dest)* denotes the address of the destination node.

Each leaf and stem router executes the same algorithm as proposed below. In root routers, no routing is performed and routers are replaced by first-in first-out (FIFOs).

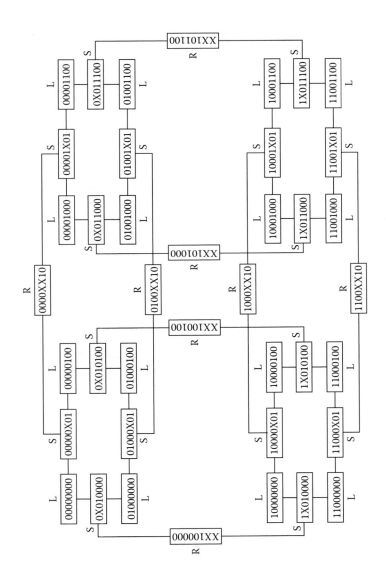

FIGURE 2.18
Addressing of a 4 × 4 MoT network.

Algorithm

If (RN of *addr (curr)* ≠ RN of *addr (dest)*) // **Step1**
   Route to Column Parent;
   Else If (CL of *addr (curr)* ≠ CL of *addr (dest)*) // **Step2**
      Route to Column Child having equal RN as *addr (dest)*;
      Else If (CN of *addr (curr)* ≠ CN of *addr (dest)*) // **Step3**
         Route to Row Parent;
         Else If (RL of *addr (curr)* ≠ RL of *addr (dest)*) // **Step4**
            Route to Row Child having equal CN as *addr (dest)*;
            Else If (Destination Core-ID field = 0) // **Step5**
               Route to Core1;
               Else Route to Core2;

In step 1, a difference in RN of current and destination addresses signifies that current and destination routers are at different row trees. Therefore, the packet needs to be routed toward the root of the column tree until RN of the current router becomes equal to that of the destination router. For example, if router addresses associated with source and destination cores are 00–00–00–00 and 11–00–11–00, respectively, the path traversed by the packet is 00–00–00–00 → 0X–01–00–00 → XX–10–00–00. Therefore, after step 1, the packet will reach a node whose RN of the current router is the same as that of the destination router.

In step 2, a difference in CL of current and destination addresses signifies that the current router is not at the leaf level as all destination routers are. Therefore, the packet needs to traverse toward the leaf level of the column tree for which RN is equal to the destination. For the above example, the path traversed by the packet is XX–10–00–00 → 1X–01–00–00 → 11–00–00–00. Therefore, after step 2, the packet reaches a node whose RN and CL are same as those of destination, that is, the packet has arrived at a node which is on the same row tree as the destination.

In step 3, the difference in CN of current and destination addresses signifies that current and destination routers are at different column trees. Therefore, the packet should be routed toward the root of the row tree until CN of the current router becomes equal to that of the destination router. For the above example, the path traversed by the packet is 11–00–00–00 → 11–00–0X–01 → 11–00–XX–10. Therefore, after step 3, the packet will reach a node whose first three fields are same as destination.

In step 4, a difference in RL of current and destination addresses signifies that the current router is not at leaf level. Therefore, the packet should traverse toward the leaf level of the row tree, where CN is same as destination. For the above example, the path traversed by the packet is 11–00–XX–10 → 11–00–1X–01 → 11–00–11–00. Therefore, after step 4, the packet reaches a router whose all four fields are same as destination; in other words, the packet reaches a router to which the destination core is attached.

In step 5, based on the Core-ID bit, the packet gets forwarded to the destination core. Therefore, the proposed routing algorithm always governs the packet to reach the destination in a specified path.

2.4.1.1.3 Proof for Shortest Path

Here, a general proof has been given to show that the above algorithm will always govern the packet to traverse in a shortest path from source to destination. From the addressing scheme of $M \times N$ MoT, the RN and CN fields are of $\log_2 M$ and $\log_2 N$ bits. According to step 1 of the algorithm, the packet will first follow that path leading to a node where RN of the current address is the same as that of the destination address. That is,

```
if (RN of addr (curr) ≠ RN of addr (dest))
{ for (i = log₂M; i > = 1; i-- ) {
      if (iᵗʰ bit position of the RN of addr (curr) ≠ iᵗʰ bit
position of the RN of addr (dest))
            {k = i; break ;} }
Route upwards by k hops in column tree. }
```

Therefore, the packet will traverse k hops in upward direction through a column tree and will reach to a node where RN becomes the same as the destination. However, CL of that node is equal to k. As the cores are attached only at the leaf level, the CL fields of the destination router are always zero. According to step 2 of the algorithm, the packet will follow that path where CL is gradually decreasing to zero, having RN same as destination. Therefore, the packet will traverse k hops in downward direction through a column tree. Thus, the packet will traverse a total of $2k$ hops and will reach a node having RN and CL fields same as those of destination. Now, according to step 3 of the algorithm, it will follow that path where CN of the current address is the same as that of the destination address. Arguing in the same way as above, the packet will traverse l hops in upward direction and then in downward direction through a row tree before reaching the destination, where l is the most significant bit position at which the column numbers of source and destination differ. Therefore, the packet will traverse a total of $(2k + 2l)$ hops and will reach a node to which the destination core is attached. We can consider this situation as if the source and the destination were the two extreme nodes of a $(2^k \times 2^l)$ MoT. In general, $(2^k \times 2^l)$ MoT has the diameter of $(2k + 2l)$. As the diameter signifies the minimum number of hops to be traversed between two nodes that are at maximum distance, the routing algorithm always governs the packet to traverse in a shortest path from source to destination.

2.4.1.1.4 Avoidance of Routing-Dependent Deadlock

In a multiprocessor on-chip network, communication channels and buffers constitute the set of permanent reusable resources. The processors that send or receive messages compete for these resources. Deadlock occurs when a set of messages is blocked forever because each message in the set holds one or more resources needed by another message in the set. A practical routing algorithm must be deadlock free. The sufficient condition to avoid deadlock in a network is that there should not be any cycle in the channel dependency graph.

Here, a general proof has been given to show that the proposed routing algorithm is deadlock free. Two opposite unidirectional links are used between two adjacent routers as shown in Figure 2.19. All the channels in the $M \times N$ MoT are labeled as shown in the figure (considering $M = 4, N = 4$) in different steps. Each channel has a unique label. The labeling scheme is mentioned as follows:

1. Label those channels in ascending order which are directed from the leaf levels to the root of the column trees, for example, labels 1–24 as shown in Figure 2.19.

2. Label those channels in ascending order which are directed from the root of the column trees to the leaf levels, for example, labels 25–48 as shown in Figure 2.19.

3. Label those channels in ascending order which are directed from the leaf levels to the root of the row trees, for example, labels 49–72 as shown in Figure 2.19.

4. Label those channels in ascending order which are directed from the root of the row trees to the leaf levels, for example, labels 73–96 as shown in Figure 2.19.

In step 1 of the routing algorithm, if RN of the source address is different from that of the destination address, the packet will be directed from the leaf level to the root of the column tree until the RN becomes the same as destination. Therefore, the packet will traverse from a lower labeled channel to a higher labeled channel as mentioned in step 1 of the labeling scheme, for example, labels 1–3 in Figure 2.19.

In step 2, if CL of the current node is different from that of the destination node, the packet will be directed from the root of the column tree to the leaf level where RN is same as destination. Therefore, the packet will traverse from a lower labeled channel to a higher labeled channel as mentioned in step 2 of the labeling scheme, for example, labels 28–30 in Figure 2.19.

In step 3, if CN of the current node is different from that of the destination node, the packet will be directed from the leaf level to the root of the row tree until CN becomes the same as destination. Therefore, the packet will again traverse from a lower labeled channel to a higher labeled channel as mentioned in step 3 of the labeling scheme, for example, labels 49–51 in Figure 2.19.

In step 4, if RL of the current node is different from that of the destination node, the packet will be directed from the root of the row tree to the leaf level where RN is same as destination. Therefore, the packet will traverse from a lower labeled channel to a higher labeled channel as mentioned in step 4

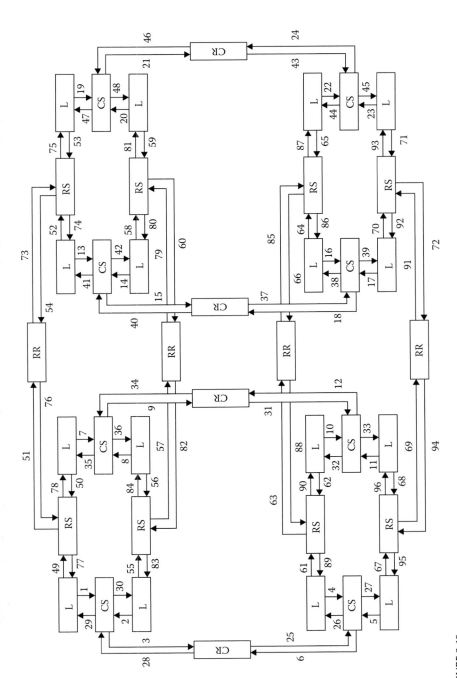

FIGURE 2.19
Labeling of channels in 4 × 4 MoT.

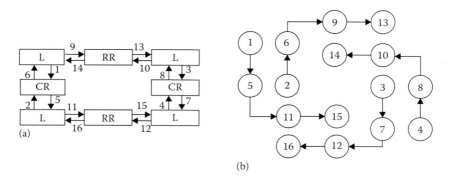

FIGURE 2.20
Labeling of 2 × 2 MoT (a) and its channel dependency graph (b).

of the labeling scheme, for example, labels 76–78 in Figure 2.19. Ultimately, the packet will reach to that node where the destination core is connected.

In step 5, based on the Core-ID bit, the packet will go to the destination core.

After tracing the labels of the channels from source to destination, it can be observed that the channel labels are always increasing, and thus the channel dependency graph is acyclic. Therefore, this network is always deadlock free. The labeling of the channels by applying the above-mentioned scheme in the 2 × 2 MoT and its channel dependency graph has been shown in Figure 2.20. In the figure, the channel label is always increasing in a particular direction from start to end and has not formed any cycle. Thus, the 2 × 2 MoT-based network is deadlock free. Similarly, the channel dependency graph of $M \times N$ MoT by applying the same labeling scheme will be acyclic, and hence deadlock free.

For wormhole switching, Glass and Ni (1992) proposed the turn model to solve the routing-dependent deadlock in adaptive routing in 2D mesh. For partially adaptive routing, west-first, north-last, and negative-first turn models were proposed by Glass and Ni (1994). In the west-first turn model, the packet first routed toward the west, if necessary, and then adaptively south, east, and north. The prohibited turns are the two to the west. Similarly, in the north-last turn model, the packet first routed adaptively toward the west, south, east, and then north. In the negative-first turn model, the packet is routed first toward adaptively the west and south, and then adaptively the east and north. Figure 2.21 depicts the scenarios.

Another deadlock-free turn model for partially adaptive routing, odd–even turn model, was proposed by Chiu (2000). In the odd-even turn model, two rules are being followed:

1. Any packet is not allowed to take an east-north or east-south turn at any nodes located in an even column.
2. Any packet is not allowed to take an north-west or south-west turn at any nodes located in an odd column.

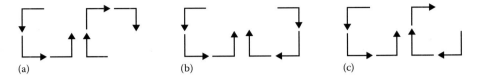

FIGURE 2.21
Turn model: (a) west-first; (b) north-last; (c) negative-first.

A VC-based deadlock-free fully and partially adaptive routing in 2D mesh was proposed by Duato (1993). When restricted to a minimal path, this routing algorithm is referred to *Duato's protocol*. Ascia et al. (2008) proposed a *neighbors-on-path* (NoP) congestion-aware selection in 2D mesh and used it in the odd–even turn model-based adaptive routing. The experimental results show that the latency of the system gets improved after adopting this selection policy with nonuniform traffic.

Deadlock recovery, however, is useful when the deadlock situation is rare. It allows a deadlock to occur, but once the deadlock situation is detected, it breaks at least one of the cyclic dependencies by using any of the recovery schemes, regressive (abort-and-retry) and progressive (preemptive), to gracefully recover. Regressive recovery scheme removes a packet from a dependency cycle by aborting and later reinjecting the packet into the network after some delay. Progressive recovery scheme removes a packet from a dependency cycle by rerouting it onto a deadlock-free lane.

2.4.2 Avoidance of Message-Dependent Deadlock

On-chip communication, depending on the behavior of the IP modules, can lead to four types of message dependencies—request–response, response–request, request–request, and response–response. Message-dependent deadlocks arise when dependency cycles on the resources (free from routing-dependent deadlock) exist due to message dependencies between the NoC and IP cores at the network end points (Song and Pinkston 2003). Figure 2.22 shows the coupling between the reception of request and the generation of responses, which introduces a dependency between the request and response buffers in the NI and thus causing deadlock in the network. In the figure, two master and slave pairs communicate via two shared input-buffered routers. The two connections between m1 and s1 are drawn with continuous lines and the connections of m2 and s2 with dashed lines. Responses from s1 enter the network, turn the east, and end up in b2. This buffer is shared by responses destined for m1 and requests going to s2. From b2, the dependencies continue through the slave s2, and the shared buffer b1, back to s1, closing the cycle. As a result, a deadlocked situation can occur.

In response–request dependency, the master sends a request, the slave responds to that request, and then the master reacts on the response from the slave by sending an additional request. Request–request dependencies

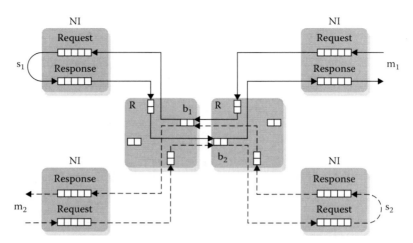

FIGURE 2.22
Request–response message-dependent deadlock.

are created when reception of a request on the slave side is coupled to the generation of a request on the master side. This occurs when IP modules process a certain input that is sent to them by the preceding module and then write their output to the succeeding module. In such protocols, an initial request passes through a number of intermediate IPs, generating new requests until the final destination is reached. Potentially, a response is travelling in the other direction, creating response–response dependencies on the way back. Two prominent examples of request–request and response–response protocols are cache coherency protocols and collective communication protocols.

Hansson et al. (2007) proposed four solutions to avoid message-dependent deadlock—increased buffer sizing, end-to-end flow control, strict ordering, and use of virtual circuit. Buffer sizing solves the deadlock problem by ensuring enough space by oversizing the buffers. This can be implemented by designing the NIs such that NI buffers are guaranteed to consume all messages sent to them. While extensively used in parallel computers, this method is prohibitively expensive in NoCs and is not used in any known architecture. Instead of adapting the buffer size to the maximum requirements, end-to-end flow control does the other way around: it assures that no more is ever injected than what can be consumed. This approach, end-to-end flow control, is used in the Æthereal (Radulescu et al. 2005) NoC. As illustrated in Figure 2.23, it removes a dependency edge from the network to the NI.

In strict ordering, deadlock avoidance is performed by introducing logically independent networks, physical or virtual, for each message type. A major drawback of the strict ordering is that buffers cannot be shared between the different message classes, increasing the amount of buffering required. The partitioning into logical networks leads to inefficient utilization of network resources and increased congestion due to unbalance.

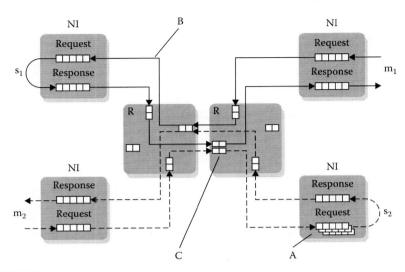

FIGURE 2.23
Solutions to message-dependent deadlock.

Having virtual instead of physical networks mitigates the aforementioned problem as shown in Figure 2.23. However, the router complexity increases as it must forward messages considering the message type. Virtual circuits represent the extreme case of strict ordering as every connection has its own logical network. This way of implementing unconditional delivery is found in the guaranteed service networks of Æthereal (Radulescu et al. 2005) and Nostrum (Millberg et al. 2004).

2.5 Flow Control Protocol

Flow control protocol determines how packets traverse through the network and reach from source to destination. It also supports error control scheme either at an end-to-end level or at a switch-to-switch level in the presence of transmission error. For the end-to-end case, a standard flow control protocol is credit based. In a credit-based flow control, an upstream node keeps count of data transfers. Available free slots are termed as credits. Once the transmitted data packet is either consumed in the receiver NI or further transmitted to the core, a credit is sent back. Æthereal (Radulescu et al. 2005), SPIN (Guerrier and Greiner 2000), and QoS architecture and design process for network-on-chip (*QNOC*) (Bolotin et al. 2004) use the end-to-end credit-based flow control technique.

Murali et al. (2005) reported that the average packet latency is higher in an end-to-end flow control compared to a switch-to-switch flow control. The later scheme can further be classified as *flit level* (*ssf*) and *packet level* (*ssp*).

In a switch-to-switch flow control, ssp shows more latency than ssf. In NoC, ssf flow control technique is widely used. Pullini et al. (2005) proposed three types of relevant switch-to-switch flit-level flow control protocols, namely, *STALL/GO, T-Error,* and *ACK/NACK.*

STALL/GO is a very simple realization of an *ON/OFF* flow control protocol. However, it cannot handle faults. Two wires are used for flow control between each pair of sender and receiver: one going forward and flagging data availability, and the other going backward and signaling either a condition of buffers filled (STALL) or buffers free (GO). When there is an empty buffer space, a GO signal is activated. Upon the unavailability of buffer space, a STALL signal is activated. Figure 2.24 depicts the STALL/GO protocol implementation for NoC.

The T-Error flow control, however, can be deployed to improve either link performance or system reliability by catching timing errors. The T-Error protocol (Figure 2.25) aggressively deals with communication over physical links, either stretching the distance among repeaters or increasing the operating frequency with respect to a conventional design. As a result, timing errors become likely on the link. Faults are handled by a repeater architecture leveraging upon a second delayed clock to resample input data, to detect any inconsistency, and to emit a VALID control signal. If the surrounding logic is to be kept unchanged, a resynchronization stage must be added between the end of the link and the receiving switch. This logic handles the offset among the original and delayed clocks, thus realigning the timing of DATA and VALID wires; this incurs a one-cycle latency penalty. The corresponding hardware was implemented by Tamhankar et al. (2005). However, T-Error lacks a really thorough fault handling in a real-time system operating in a noisy environment.

The ACK/NACK flow control protocol (Figure 2.26) is used for detection and retransmission purposes. While flits are sent on a link, a copy is kept locally at the sender. When flits are received, either an ACK or a NACK is sent back. Upon receipt of the ACK, the sender deletes the local copy of the flit, whereas upon receipt of the NACK, the sender stops transmitting from its queue and retransmits flits from its local copy starting from the corrupted one, with a *go-back-N* policy. The other flits which are on the fly in that time window will be discarded and resent. If NACKs were only due to sporadic errors, the impact on performance would be negligible.

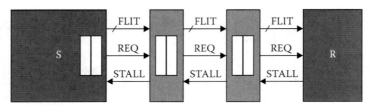

FIGURE 2.24
STALL/GO protocol implementation.

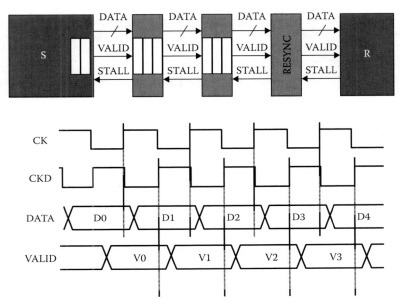

FIGURE 2.25
T-Error protocol implementation.

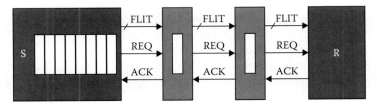

FIGURE 2.26
ACK/NACK protocol implementation.

This flow control technique is used in *XPIPES* (Bertozzi and Benini 2004; Bertozzi et al. 2005).

2.6 Quality-of-Service Support

Quality of service (QoS) is an important aspect of NoC architecture where instead of maximizing the average performance of a network, attention is given to a fair allocation policy such that each application gets sufficient resources to meet its throughput, latency, and bandwidth requirements. NoC traffic falls into two broad categories: guaranteed throughput (GT) and best effort (BE).

The GT service guarantees both latency and throughput over a finite time interval and also supports uncorrupted, lossless, and ordered data transfer. BE scheme, however, forwards packets as soon as possible, but no guarantees are given for latency and throughput in general. Vellanki et al. (2004) proposed a mesh-based router architecture for supporting QoS by modifying a traditional VC-based router design. In this scheme, two out of four VCs are reserved for supporting GT services. For high GT load, those traffics are allowed to transmit through the BE VCs, but not vice versa. Reservation-based schemes for GT traffic generally leads to degradation of average performance for BE traffic. Nostrum (Millberg et al. 2004) ensures bandwidth for guaranteed throughput traffic by reserving time slots for its transmission on inter-router links. If no guaranteed throughput traffic is injected into the network, the time slots are not utilized. Æthereal (Goossens et al. 2005), another mesh-based NoC, supports guaranteed throughput traffic by utilizing a centralized scheduler for allocation of link bandwidth. Andreasson and Kumar (2004) proposed a scheme where BE traffic may use the reserved path when there is no GT traffic present.

2.7 NI Module

An NI module is used to interface a core with the interconnection network. By definition, NI module decouples the computation from communication and performs a protocol conversion between the IP core and the router to which the core is connected. The *Open Core Protocol* (OCP) (OCP 2003) is a widely used interface standard for simplifying the integration task between the IP cores and the network fabric. Wrapping of IP cores with OCP interface exhibits a higher reusability and cost-effective plug-and-play-based system implementation. The NI is generic with respect to the network and performs different services. There are many works reported in the literature on NI design. This book has taken the work reported by Singh et al. (2007) as an example. They showed that the NI architecture can be divided into three parts: (1) generic core interface (GCI), (2) packet maker (PM), and (3) packet disassembler (PD), as shown in Figure 2.27. The function of each part has been described below.

1. GCI: It lies between the network and the core-specific wrapper like OCP. It abstracts the network communication protocol from the core-specific wrapper for heterogeneous system implementation. If a new core is added to the system, the core-specific wrapper views the NI as a black box.

2. PM: The core-specific wrapper transmits the message to PM memory. The PM performs the following tasks at the source core and maintains data integrity:

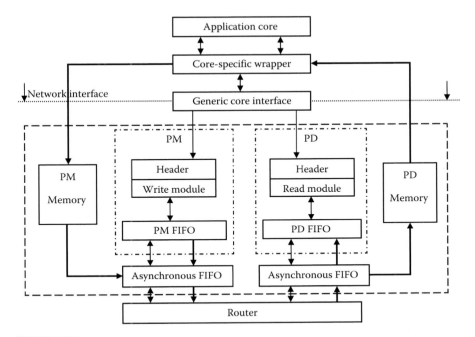

FIGURE 2.27
The NI architecture.

a. It packetizes the message stored in PM memory and breaks them into several flits (such as header, payload, tailer, and invalid flits) before queuing them into asynchronous FIFO having independent read and write clocks.

b. In case of source routing, NI maintains the routing information in a look-up table.

c. It inserts redundant bits (parity or cyclic redundancy [CRC]) in the packet tailer for supporting end-to-end flow control with proper retransmission mechanism.

3. PD: It performs the following tasks at the destination core:

a. It writes the incoming flits from the asynchronous FIFO to the PD memory.

b. It decodes the packet header from the PD memory, extracts the control information required by the core, and passes it to GCI.

c. The core wrapper reads the payload and tailer from the PD memory to obtain the total message. It also performs error detection for end-to-end flow control.

d. It ensures in-order delivery of packets, which is extremely important for adaptive routing.

As the routers and IP cores are operated by independent clocks, the asynchronous FIFO in NI performs the synchronization task in clock domain crossing. Apart from the above services, NI also performs cache coherency by implementing directory-based protocols (Acacio et al. 2004).

2.8 Summary

This chapter presents different aspects of NoC design. It has been observed that the 2D mesh topology with wormhole packet switching and deterministic routing is the most common and widely used in academia and industry so far. Besides mesh architecture, this chapter also considers MoT topology as an example and shows an addressing scheme and a deterministic routing strategy in it. It proves that the MoT-based network, with the proposed routing algorithm, is always livelock and deadlock free. It also proves that the packets will always follow a specific shortest path from source to destination. The number of edges of an $M \times N$ MoT graph and the average distance of the network have been formulated. It has been shown that the network can achieve the maximum throughput and the minimum latency when the number of row trees and that of column trees are equal.

Toward the design of a complete NoC-based SoC, Chapter 3 presents the design of wormhole and VC router architecture in detail.

References

Acacio, M. E., Gonzalez, J., Garcia, J. M., and Duato, J. 2004. An architecture for high-performance scalable shared-memory multiprocessors exploiting on-chip integration. *IEEE Transactions on Parallel and Distributed Systems*, vol. 15, no. 8, pp. 755–768.

Agarwal, A., Iskander, C., and Shankar, R. 2009. Survey of network on chip (NoC) architectures and contribution. *Journal of Engineering, Computing, and Architecture*, vol. 3, no. 1.

Andreasson, D. and Kumar, S. 2004. On improving best-effort throughput by better utilization of guaranteed-throughput channels in an on-chip communication system. *Proceedings of the IEEE Norchip Conference*, pp. 265–268, Oslo, Norway.

Ascia, G., Catania, V., Palesi, M., and Patti, D. 2008. Implementation and analysis of a new selection strategy for adaptive routing in networks-on-chip. *IEEE Transactions on Computers*, vol. 57, no. 6, pp. 809–820.

Balfour, J. and Dally, W. J. 2006. Design tradeoffs for tiled CMP on-chip networks. *Proceedings of the ACM International Conference on Supercomputing (ICS)*, pp. 187–198, Christchurch, New Zealand.

Banerjee, A., Mullins, R., and Moore, S. 2007. A power and energy exploration of network-on-chip architectures. *Proceedings of International Symposium on Networks-on-Chip*, pp. 163–172, Princeton, NJ.

Banerjee, N., Vellanki, P., and Chata, K. S. 2004. A power and performance model for network on chip architecture. *Proceedings of the IEEE Design, Automation and Test in Europe Conference and Exhibition*, pp. 1250–1255, Paris, France.

Bertozzi, D. and Benini, L. 2004. Xpipes: A network-on-chip architecture for gigascale systems-on-chip. *IEEE Circuits and Systems Magazine*, vol. 4, no. 2, pp. 18–31.

Bertozzi, D., Jalabert, A., Murali, S., Tamhankar, R., Stergiou, S., Benini, L., and Micheli, G. D. 2005. NoC synthesis flow for customized domain specific multiprocessor systems-on-chip. *IEEE Transactions on Parallel and Distributed Systems*, vol. 16, no. 2, pp. 113–129.

Bolotin, E., Cidon, I., Ginosar, R., and Kolodny, A. 2004. QNoC: QoS architecture and design process for network on chip. *Journal of Systems Architecture: the Euromicro Journal* (special issue), vol. 50, no. 2/3, pp. 105–128.

Chiu, G. M. 2000. The odd-even turn model for adaptive routing. *IEEE Transactions on Parallel and Distributed Systems*, vol. 11, no. 7, pp. 729–738.

Coppola, M., Locatelli, R., Maruccia, G., Pieralisi, L., and Scandurra, A. 2004. Spidergon: A novel on-chip communication network. *Proceedings of the International Symposium on System on Chip*, p. 15, Tampere, Finland.

Dally, W. J. 1991. Express cubes: Improving the performance of k-ary n-cube interconnection networks. *IEEE Transactions on Computers*, vol. 40, no. 9, pp. 1016–1023.

Dally, W. J. 1992. Virtual-channel flow control. *IEEE Transactions on Parallel and Distributed Systems*, vol. 3, no. 2, pp. 194–205.

Dally, W. J. and Seitz, C. L. 1986. The torus routing chip. *Journal of Distributed Computing*, vol. 1, no. 4, pp. 187–196.

Dally, W. J. and Seitz, C. L. 1987. Deadlock-free message routing in multiprocessor interconnection networks. *IEEE Transactions on Computers*, vol. C-36, no. 5, pp. 547–553.

Dally, W. J. and Towles, B. 2001. Route packets, not wires: On-chip interconnection networks. *Proceedings of the Design Automation Conference*, pp. 683–689, Las Vegas, NV.

Dally, W. J. and Towles, B. 2004. *Principles and Practices of Interconnection Networks*. Morgan Kaufmann Publishers, San Francisco, CA.

Decina, M., Trecordi, V., and Zanolini, G. 1991. Throughput and packet loss in deflection routing multichannel-metropolitan area networks. *IEEE GLOBECOM*, pp. 1200–1208, Phoenix, AZ.

Duato, J. 1993. A new theory of deadlock-free adaptive routing in wormhole networks. *IEEE Transactions On Parallel and Distributed Systems*, vol. 4, no. 12, pp. 1320–1331.

Duato, J., Yalamanchili, S., and Ni, L. 2003. *Interconnection Networks: An Engineering Approach*. Morgan Kaufmann Publishers, San Francisco, CA.

Glass, C. J. and Ni, L. M. 1992. The turn model for adaptive routing. *International Symposium on Computer Architecture*, pp. 278–287, Gold Coast, Australia.

Glass, C. J. and Ni, L. M. 1994. The turn model for adaptive routing. *Journal of the ACM*, vol. 41, no. 5, pp. 874–902.

Goossens, K., Dielissen, J., and Radulescu, A. 2005. The Æthereal network on chip: Concepts, architectures, and implementations. *IEEE Design and Test of Computers*, vol. 22, no. 5, pp. 21–31.

Greenberg, H. J. and Hajek, B. 1992. Deflection routing in hypercube networks. *IEEE Transactions on Communications*, vol. COM-40, no. 6, pp. 1070–1081.

Grot, B., Hestness, J., Keckler, S. W., and Mutlu, O. 2009. Express cube topologies for on-chip interconnects. *Proceedings of the International Symposium on High-Performance Computer Architecture*, pp. 163–174, Raleigh, NC.

Guerrier, P. and Greiner, A. 2000. A generic architecture for on-chip packet-switched interconnections. *Proceedings of the Design, Automation and Test in Europe*, pp. 250–256, Paris, France.

Hansson, A., Goossens, K., and Radulescu, A. 2007. Avoiding message-dependent deadlock in network-based systems on chip. *VLSI Design, Hindawi Publishing Corporation*, vol. 2007, article id. 95859, pp. 1–10.

Hossain, H., Akbar, M., and Islam, M. 2005. Extended-butterfly fat-tree interconnection (EFTI) architecture for network-on-chip. *IEEE Pacific Rim Conference on Communication, Computers, and Signal Processing*, pp. 613–616, Victoria, British Columbia, Canada.

Jeang, Y. L., Huang, W. H., and Fang, W. F. 2004. A binary tree architecture for application specific network on chip (ASNOC) design. *Proceedings of the IEEE Asia-Pacific Conference on Circuits and Systems*, pp. 877–880, Tainan, Taiwan.

Karim, F., Nguyen, A., and Dey, S. 2002. An interconnect architecture for networking systems on chips. *IEEE Micro*, vol. 22, no. 5, pp. 36–45.

Kim, J., Balfour, J., and Dally, W. J. 2007. Flattened butterfly topology for on-chip networks. *Computer Architecture Letters*, vol. 6, no. 2, pp. 3–40.

Kumar, A., Peh, L. S., Kundu, P., and Jha, N. K. 2007. Express virtual channels: Towards the ideal interconnection fabric. *Proceedings of the International Symposium on Computer Architecture*, pp. 433–436, San Diego, CA.

Kumar, S., Jantsch, A., Soininen, J. P., Forsell, M., Millberg, M., Oberg, J., Tiensyrja, K., and Hemani, A. 2002. A network on chip architecture and design methodology. *Proceedings of the IEEE Computer Society Annual Symposium on VLSI*, pp. 117–124, Pittsburgh, PA.

Kundu, S. and Chattopadhyay, S. 2008a. Mesh-of-tree deterministic routing for network-on-chip architecture. *ACM Great Lake Symposium on VLSI*, pp. 343–346, Orlando, FL.

Kundu, S. and Chattopadhyay, S. 2008b. Network-on-chip architecture design based on mesh-of-tree deterministic routing topology. *International Journal for High Performance Systems Architecture*, vol. 1, no. 3, pp. 163–182.

Kundu, S., Soumya, J., and Chattopadhyay, S. 2012. Design and evaluation of mesh-of-tree based network-on-chip using virtual channel router. *Microprocessors and Microsystems*, vol. 36, no. 6, pp. 471–488.

Millberg, M., Nilsson, E., Thid, R., and Jantsch, A. 2004. Guaranteed bandwidth using looped containers in temporally disjoint networks within the Nostrum network on chip. *Proceedings of the Design, Automation, and Test in Europe*, pp. 890–895, Paris, France.

Murali, S., Theocharides, T., Vijaykrishnan, N., Irwin, M. J., Benini, L., and Micheli, G. D. 2005. Analysis of error recovery schemes for networks on chips. *Proceedings of the IEEE Design and Test of Computers*, pp. 434–442.

OCP. 2003. Open Core Protocol. Available at: http://www.ocpip.org.

Pande, P. P., Grecu, C., Ivanov, A., and Saleh, R. 2003a. Design of a switch for network on chip applications. *Proceedings of the International Symposium on Circuits and Systems*, vol. 5, pp. 217–220, Bangkok, Thailand.

Pande, P. P., Grecu, C., Ivanov, A., and Saleh, R. 2003b. High-throughput switch-based interconnect for future SoCs. *Proceedings of the IEEE International Workshop on System-on-Chip for Real Time Applications*, pp. 304–310, Calgary, Alberta, Canada.

Pande, P. P., Grecu, C., Jones, M., Ivanov, A., and Saleh, R. 2005. Performance evaluation and design trade-offs for MP-SOC interconnect architectures. *IEEE Transactions on Computers*, vol. 54, no. 8, pp. 1025–1040.

Pavlidis, V. F. and Friedman, E. G. 2007. 3-D topologies for networks-on-chip. *IEEE Transactions on VLSI Systems*, vol. 15, no. 10, pp. 1081–1090.

Pullini, A., Angiolini, F., Bertozzi, D., and Benini, L. 2005. Fault tolerance overhead in network-on-chip flow control schemes. *Proceedings of the ACM Symposium on Integrated Circuits and Systems Design*, pp. 224–229, Florianópolis, Brazil.

Radulescu, A., Dielissen, J., Pestana, S. G., Gangwal, O. P., Rijpkema, E., Wielage, P., and Goossens, K. 2005. An efficient on-chip network interface offering guaranteed services, shared memory abstraction, flexible network programming. *IEEE Transactions on CAD of Integrated Circuits and Systems*, vol. 24, no. 1, pp. 4–17.

Schroeder, M. D., Birrell, A. D., Burrows, M., Murray, H., Needham, R. M., Rodeheffer, T. L. 1991. Autonet: A high speed, self configuring, local area network using point-to-point links. *IEEE Journal on Selected Areas in Communications*, vol. 9, pp. 1318–1335.

Singh, S. P., Bhoj, S., Balasubramanian, D., Nagda, T., Bhatia, D., and Balsara, P. 2007. Network interface for NoC based architectures. *International Journal of Electronics*, vol. 94, no. 5, pp. 531–547.

Song, Y. H. and Pinkston, T. M. 2003. A progressive approach to handling message-dependent deadlock in parallel computer systems. *IEEE Transactions on Parallel and Distributed Systems*, vol. 14, no. 3, pp. 259–275.

Tamhankar, R. R., Murali, S. and Micheli, G. D. 2005. Performance driven reliable link design for network on chips. *Proceedings of the Asia South Pacific Design Automation Conference*, pp. 749–756, Shanghai, People's Republic of China.

Tortosa, D. S., Ahonen, T., and Nurmi, J. 2004. Issues in the development of a practical NoC: The Proteo concept. *Integration, the VLSI Journal*, vol. 38, no. 1, pp. 95–105.

Vangal, S. R., Howard, J., Ruhl, G., Dighe, S., Wilson, H., Tschanz, J., Finan, D., et al. 2008. An 80-tile sub-100-W TeraFLOPS processor in 65-nm CMOS. *IEEE Journal of Solid-State Circuits*, vol. 43, no. 1, pp. 29–41.

Vellanki, P., Banerjee, N., and Chata, K. S. 2004. Quality-of-service and error control techniques for mesh-based network-on-chip architectures. *Integration, the VLSI Journal*, vol. 38, pp. 353–382.

Wentzlaff, D., Griffin, P., Hoffmann, H., Bao, L., Edwards, B., Ramey, C., Mattina, M., Miao, C. C., Brown, J. F., and Agarwal, A. 2007. On-chip interconnection architecture of the TILE processor. *IEEE Micro*, vol. 27, no. 5, pp. 15–31.

West, D. B. 2002. *Introduction to Graph Theory*. 2nd Edition. Pearson Education, Upper Saddle River, NJ.

Zeferino, C. A. and Susin, A. A. 2003. SoCIN: A parametric and scalable network-on-chip. *Proceedings of the IEEE Symposium on Integrated Circuits and Systems Design*, pp. 169–175, São Paulo, Brazil.

3

Architecture Design of Network-on-Chip

3.1 Introduction

This chapter focuses on the architectural design of wormhole and virtual channel (VC) routers for NoC. The salient contributions of this chapter are as follows:

1. To support globally asynchronous locally synchronous style of communication, a gray counter-based dual-clock first-in first-out (FIFO) has been designed and used in NoC router.
2. Design of wormhole router for mesh-of-tree (MoT) topology has been described in detail.
3. To mitigate the head-of-line (HoL) blocking problem of wormhole router-based network, VC router has also been implemented.

The rest of the chapter is organized as follows: Section 3.2 describes the switching technique and packet format of 4 × 4 MoT-based NoC. Sections 3.3 and 3.4 discuss the design of dual-clock FIFO and globally asynchronous locally synchronous (GALS) style of communication, respectively. Sections 3.5 and 3.6 present a detailed architecture design of wormhole and VC routers, respectively. Section 3.7 describes an adaptive router architecture. Finally, Section 3.8 summarizes this chapter.

3.2 Switching Techniques and Packet Format

Both wormhole and VC routers follow the packet switching technique. Messages are sent in terms of packets, which are further decomposed into flits (flow control units) in the network interface. A flit is a smaller unit over which flow control is performed. A packet is composed of header, payload, and tailer flits. Header flit carries information about the source and destination core addresses, whereas payload and tailer flits contain the actual information. For a 4 × 4 MoT, a flit size of 32 bits has been considered. From the addressing scheme described in Chapter 2, each core address in a 4 × 4 MoT

eop (1-bit)	*bop* (1-bit)	(2-bit)	(2-bit)	(13-bit)	(13-bit)
0	1			Header flit	
		vc_id	Unused	Source core address	Destination core address
... 0 ...	... 0 ...			... Payload flit ...	
1	0			Tailer flit	
1	1			Invalid flit	

FIGURE 3.1
Packet format for a 4 × 4 MoT-based network.

requires 13 bits. Hence the header flit consists of 13-bit destination core address, 13-bit source core address, 2 optional bits for supporting different traffic classes (unused here), 2 bits for VC identification (*vc_id*), and 2 bits for packet framing: end-of-packet (*eop*) and begin-of-packet (*bop*). The *vc_id* bits are used only for VC-based routers. In wormhole router architecture, these two bits are also left unused. The packet format is shown in Figure 3.1. The *eop* and *bop* bits identify the type of the flits: *eop* = 0, *bop* = 1 denotes the header flit; *eop* = 0, *bop* = 0 denotes the payload flit; *eop* = 1, *bop* = 0 denotes the tailer flit; *eop* = 1, *bop* = 1 denotes the invalid flit.

3.3 Asynchronous FIFO Design

In NoC, the cores and routers are operating at their own clock frequencies and there is, as such, no dependency between these clocks. Moreover, the inter-router communication should support mesochronous clocking strategy where the clock frequency for each router is the same but phases may vary. Therefore, to start with the router design, it is essential to design a low-latency FIFO with independent read and write clocks. The major challenge of dual-clock asynchronous FIFO is that the *FULL* and *EMPTY* signals of the FIFO are dependent on both the clocks. Synchronization of a binary count value from one clock domain to another is problematic because more than one bit of an n-bit counter may change at a time, which may cause glitch in *FULL* and *EMPTY* signals. Thus a binary counter-based FIFO is inefficient. In gray code, only one bit changes in each clock. Thus synchronization from one clock domain to another becomes simpler. Cummings (2002) and Cummings and Alfke (2002) implemented the gray-code counter-based FIFO capable of handling metastability. In a *mod-N* gray counter, N is an integral power of 2. But if N is a nonintegral power of 2 ($2^m - N$), buffer locations will be wasted, where $m = \log_2 N$. A scalable gray code concept has been proposed by Jiang (2004) and Cheng (2004) to solve this problem. In this technique, for a *mod-N* gray counter, first the difference of ($2^m - N$) is obtained. Now there

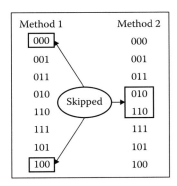

FIGURE 3.2
Scalable gray coding scheme.

are two ways to solve the problem if the difference is even. One method is to compute $(2^m - N)/2$ and skip these many locations from the top and the bottom. The other method is to skip these many locations on both sides from the center. These two techniques are shown in Figure 3.2 for a *mod-6* gray code. However, this scalable gray code concept fails if $(2^m - N)$ is odd (e.g., $N = 5$). In this chapter, FIFO depth has been taken as 6 where the patterns 010 and 110 are skipped to get a *mod-6* gray counter.

The block diagram of FIFO with a gray-code counter is shown in Figure 3.3. The FIFO memory has been designed as a stack of six write registers followed by a 6:1 multiplexer (MUX) and a single read register as shown in Figure 3.4. In the FIFO, read and write counters are two separate gray counters, synchronous with read and write clocks, respectively. The write pointer denotes the position to where the next incoming flit will be written and the read pointer denotes the position from where a flit will be read next. Initially write direction (*wr_dir*) and read direction (*rd_dir*) signals are reset. The individual direction bits are getting inverted if the respective counter wraps around after reaching the maximum count.

Asynchronous comparator module compares the write and read pointers and generates the asynchronous *full* and *empty* signals. This module contains only a combinational comparison logic. The functionality of the asynchronous comparator module is as follows:

$$full = ((wr\_addr == rd\_addr) \&\& (wr\_dir\ != rd\_dir))$$
$$empty = ((wr\_addr == rd\_addr) \&\& (wr\_dir == rd\_dir))$$

These asynchronous *full* and *empty* signals are synchronized by write clock and read clock, respectively, to generate *FULL* and *EMPTY* signals. Here, two D flip-flops (marked as *synchronizer* in Figure 3.3) are cascaded to reduce the probability of metastability. An attempt to read from an empty FIFO produces an invalid data at the output port.

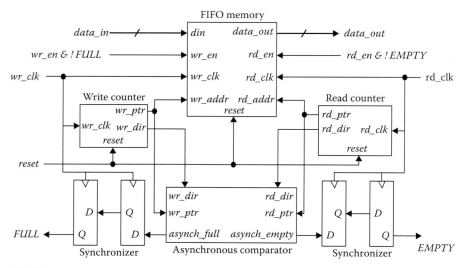

FIGURE 3.3
Dual-clock asynchronous FIFO.

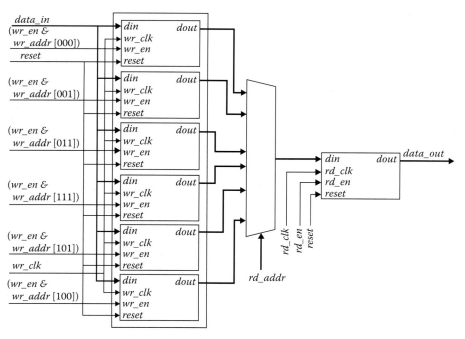

FIGURE 3.4
Design of FIFO memory.

3.4 GALS Style of Communication

The communication strategy in NoC follows the GALS style by using a dual-clock FIFO in each router. Each router in the network has a separate clock (e.g., *rd-clk1* for router-1, *rd-clk2* for router-2, etc.) as shown in Figure 3.5. As NoC supports mesochronous clocking, these clock frequencies have been assumed to be the same, whereas phases may differ. The input FIFO of router-2 sends a request signal (*in-req2* = 1) to router-1 until it is full. Router-1, after receiving the request signal (*in-req2*), sends 32-bit data (*flit2*) and a data valid signal (*in-val2*) to router-2 which is synchronous with router-1's own clock (*rd-clk1*). Router-1 also sends this clock signal to router-2, which uses it as the write clock of its input FIFO (*wr-clk2* in Figure 3.5). The *in-req2* signal is synchronous with *wr-clk2*, which is the same as *rd-clk1*. Therefore, all the signals (*in-req2*, *flit2*, and *in-val2*) between router-1 and router-2 in Figure 3.5 are synchronous with *rd-clk1*. Similarly, *in-req3*, *flit3*, and *in-val3* signals are synchronous with *rd-clk2*. As each router uses a separate local clock and these clocks are globally independent of each other throughout the network, this communication strategy leads to the GALS style.

3.5 Wormhole Router Architecture Design

The MoT-based router architecture is described in this section. Externally, each *leaf* level router has four links, whereas the *stem* and *root* level routers are having three and two links, respectively, as shown in Figure 3.6. Two cores are connected to each *leaf* level router via the local channels. No core is attached to the *stem* and *root* level routers. Each router is connected to its adjacent modules via two opposite dedicated unidirectional channels, each one

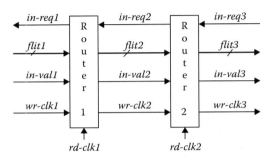

FIGURE 3.5
GALS style of communication in NoC.

with its data, framing, and flow control signals. Each channel includes n bits for data and two bits for packet framing (*eop* and *bop*) as shown in Figure 3.6. The flow control signals are used for the following (as shown in the figure).

- To request for incoming data (*in-req*) from the previous module
- To request for outgoing data (*out-req*) from the present module
- To validate the incoming (*in-val*) data
- To validate the outgoing (*out-val*) data

Next, we look into the circuitry inside the wormhole routers. The router consists of instances of two kinds of modules: input channel and output channel. A brief description of each module is given in the following section.

3.5.1 Input Channel Module

Each input channel module is composed of four architectural blocks as shown in Figure 3.7: (1) input flow controller (IFC), (2) input FIFO buffer (IB), (3) routing computation (RC) unit, and (4) input read switch (IRS).

The IFC block implements the logic that performs the translation between the handshake and the FIFO flow control protocols. It sends *in-req* signal to the previous router when its FIFO is not full. The *in-req* signal is connected with the *out-req* port of the output channel of the previous router. If the incoming flit is valid (*in-val* = 1) and FIFO is not full, flit will be written into the FIFO buffer synchronously with the *wr-clk*. The *in-val* signal is coming from the *out-val* port of the output channel of the previous router.

The IB block is designed as mentioned in Section 3.3. Data (*data-out*) is coming out from the FIFO synchronously with the router's own clock (*router-clk*).

The RC unit performs the routing function. It detects the header from the outgoing flits of FIFO and runs the routing algorithm to select an output channel. After detecting the header, it sends a *request* to one of the output channels. The selected output channel, in turn, sends a *grant* signal to that *request*, when it is free. This *grant* signal comes to the IRS block of the input channel module. The input channel module also sends a *read-ok* signal to the selected output channel. The *read-ok* signal is turned high if the outgoing flit is a valid one. The *request* signal remains high until the tailer comes out from FIFO.

The IRS block receives three *x-RD* signals and three *x-gnt* signals from other output channel modules (Figure 3.7) and generates a granted read enable signal (*rd-en*) for reading from FIFO, provided that the FIFO is not empty.

3.5.2 Output Channel Module

Each output channel consists of four blocks: (1) switch arbiter (SA), (2) output flow control (OFC), (3) output read switch (ORS), and (4) output FIFO buffer (OB).

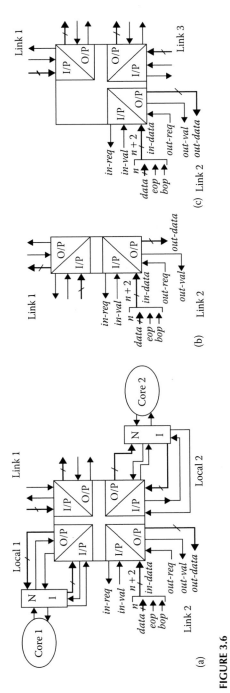

FIGURE 3.6
Connections for (a) *leaf*, (b) *root*, and (c) *stem* level routers.

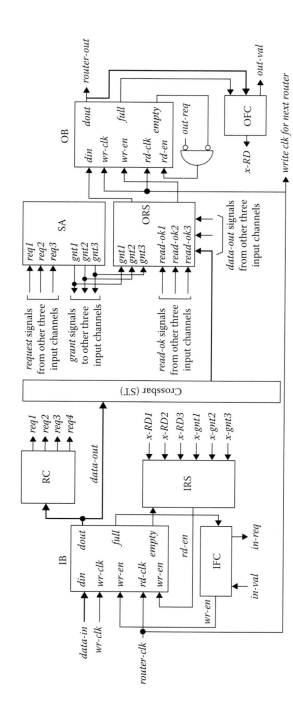

FIGURE 3.7
Wormhole router architecture for *leaf* level nodes having connectivity 4.

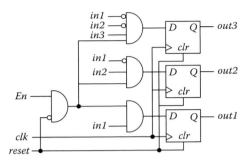

FIGURE 3.8
Implementation of priority logic.

When more than one input channel modules send their *request* signals to a particular output channel, the SA selects one *request* signal by following a round-robin strategy for avoiding the starvation problem. It sends a grant signal (*x-gnt*) to that *request* (Figure 3.7). Thus, the input channel receives the grant (*x-gnt*) signal and starts passing flits in a pipelined fashion till this *x-gnt* signal is high. This grant signal is also an input to the ORS block. The *read-ok* and *data-out* signals from the input channel are also input to the ORS block as shown in Figure 3.7. The ORS block just passes the selected data and the *read-ok* signal to the OB depending on the *x-gnt* signal. Depending on the *out-req* signal that represents the availability of the FIFO buffer of next router's input channel, *rd-en* is generated and data (*router-out*) comes out from OB synchronously with *router-clk*. The OFC block takes the *full* signal of OB and generates the *x-RD* signal by inverting it. It also takes the *router-out* signal and generates the *out-val* signal by checking the *eop* and *bop* bits.

The implementation of round-robin arbiter is presented in the following text. The round-robin arbiter is based on priority logic. Figure 3.8 shows a design of priority logic with three inputs and three outputs. The priority of inputs is in descending order from *in1* to *in3*. Thus, *in1* has the highest priority, *in2* has the next highest priority, and *in3* has the lowest priority. The D flip-flop is used for avoiding glitch due to ANDing. The round-robin arbiter consists of a single priority logic as shown in Figure 3.9. The input to the priority logic is selected by a MUX. Here, *in1* has the highest priority, *in2* has the next highest priority, and *in3* has the lowest priority. The inputs to the MUX are fed in a round-robin fashion. The output of the priority logic will generate the grant signals (*gnt1*, *gnt2*, *gnt3*) by using a look-up table (LUT), which is shown in Table 3.1. The select line of the MUX is generated by using a Moore finite state machine (FSM) and changes its value only when all the outputs of the priority logic block are at logic 0. The D flip-flops are used to generate glitch-free grant signals.

The data path of the above wormhole architecture is shown in Figure 3.10. One IB and one RC are connected to every incoming physical channel, whereas one OB and one SA are connected to every outgoing physical channel.

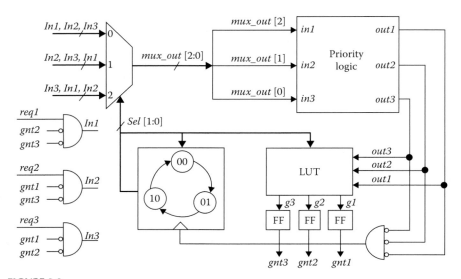

FIGURE 3.9
Round-robin arbiter.

TABLE 3.1

Grant Signal Generation Logic

sel [1]	sel [0]	out1	out2	out3	g1	g2	g3
0	0	0	0	0	0	0	0
0	0	1	0	0	1	0	0
0	0	0	1	0	0	1	0
0	0	0	0	1	0	0	1
0	1	0	0	0	0	0	0
0	1	1	0	0	0	1	0
0	1	0	1	0	0	0	1
0	1	0	0	1	1	0	0
1	0	0	0	0	0	0	0
1	0	1	0	0	0	0	1
1	0	0	1	0	1	0	0
1	0	0	0	1	0	1	0

The control units generate the read enable and write enable signals for the
FIFOs and also send a request signal to the previous router for incoming flits.
After coming out from the IB, the header flit of any packet goes to the RC for
generating a request signal by executing the routing algorithm to access an
outgoing physical channel. If more than one request signal tries to access
the same outgoing physical channel simultaneously, the SA selects one of
them in a round-robin fashion to avoid starvation and sends a grant signal

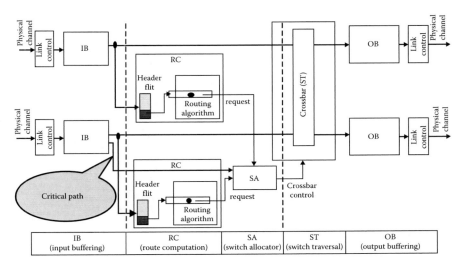

FIGURE 3.10
Wormhole router architecture data path.

to establish the path through the switch traversal (ST). The header flit gets forwarded through this established path and stored in the OB. Till that time, the payload and tailer flits wait in the IB. Depending on the availability of the buffer space in the OB, the payload and tailer flits pass through the established path sequentially and get stored in the OB.

The router has been designed in Verilog HDL and synthesized in Synopsys Design Vision supporting 90-nm complimentary metal oxide semiconductor (CMOS) technology with *Faraday* library to generate a gate-level netlist. For a leaf router in the MoT-based network having a node degree of 4, a synthesis of the design using single-stage pipelining inside the SA module shows that the critical path lies from the IB to the SA (Figure 3.10) and the delay of the critical path is found to be 600 ps. Hence, the router can be operated at 1.66 GHz. The overall router architecture has three-cycle latency, one cycle each in IB, SA, and OB.

3.6 VC Router Architecture Design

In the wormhole router architecture, a header flit gets blocked if the required output channel is already assigned to some other packet. Since only the header flit has the destination address, all the remaining flits must wait in their channels until the header flit can make its progress. The physical channels used by any of these blocked flits cannot be used to route other packets. This essentially slows down the whole network. A solution to this problem is to use VCs

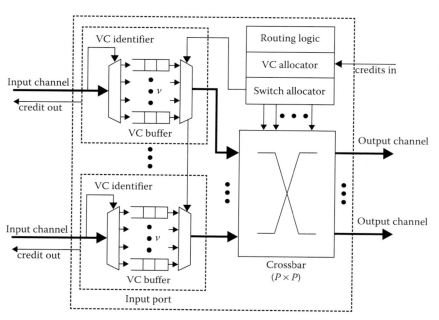

FIGURE 3.11
Generic nonspeculative VC router architecture.

(Dally 1992). Many VCs are multiplexed over a single physical channel. Thus, even if one VC gets blocked, other VCs can make use of this physical channel, and thus higher utilization of the physical resources can be made.

A generic nonspeculative VC router was implemented by Mullins et al. (2004) with P input/output ports and V number of VCs per input port as shown in Figure 3.11. Each incoming packet has separate VC identifier (*vc_id*) bits. These *vc_id* bits are the same for all the flits in a packet. According to the *vc_id* bits, the packets are demultiplexed and buffered in FIFOs. This stage is known as *VC allocation*. Next, each of these VCs is multiplexed again and passes through a $P \times P$ crossbar. Thus, only one VC can advance from an input port to the crossbar per cycle. This stage is known as *switch allocation*.

Kavaldjiev et al. (2006) reported that due to the conflicts at the input VCs, maximum throughput of a network cannot be achieved. To overcome this problem, they eliminated the multiplexer after the FIFOs in each input port and connected the channels directly to the crossbar. Although the description of above nonspeculative router implies that VC allocation and switch allocation are performed sequentially to reduce the number of pipeline stages in the router architecture. Peh and Dally (2001) described a speculative scheme where both VC allocation and switch allocation may be performed in parallel. Mullins et al. (2006) implemented a speculative single-cycle router clocked at 250 MHz. In NoC design, nonspeculative routers still attract a lot of attention from the research and development community (Kumar et al. 2007).

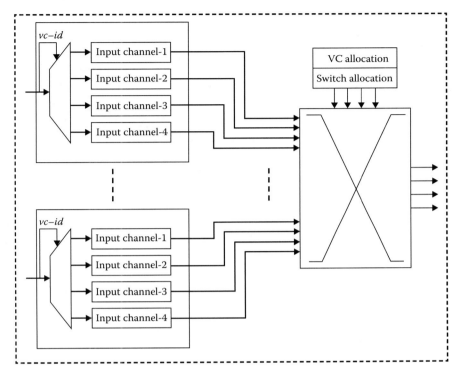

FIGURE 3.12
Modified VC router architecture.

In this book, we have implemented a nonspeculative router as discussed in the work of Kavaldjiev et al. (2006).

The router consists of two types of modules—input links and output links. Each input link module consists of four input channels. Incoming flits are written into the FIFO of the input channel depending on their *vc_id* bits by using a 1:4 demultiplexer (DEMUX) as shown in Figure 3.12. A brief description of the input channel is given in Section 3.6.1.

3.6.1 Input Channel Module

Each input channel has an single FIFO (IB), a routing computation (RC) unit, an IRS, and a control logic. The block diagram of the input channel module is shown in Figure 3.13.

The *full* signal generated from each IB is sent back to its previous router. The *rd-en* signals are coming from all the output link modules of the present router. The IRS module generates the read signal for the FIFO. The *header* module checks for the header flit. The RC module generates the *request* signals to access one of the output physical links. From each input channel, both *request* and *data* signals go to the output link module through crossbar.

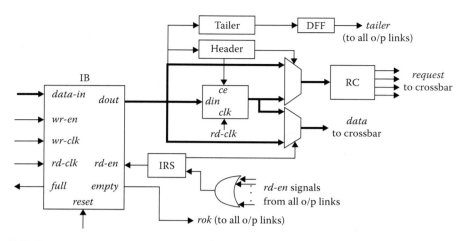

FIGURE 3.13
Input channel module for VC router.

3.6.2 Output Links

The output link module consists of two major architectural components—VC allocator and switch allocator. These two components are briefly described in Sections 3.6.2.1 and 3.6.2.2.

3.6.2.1 VC Allocator

In the generic VC router architecture (Mullins et al. 2004), VC allocator needs two stages of arbitration. The first stage has a V:1 arbiter at each input VC, followed by a second stage of $P*V$:1 arbiters for each output VC. In the modified router architecture (Figure 3.12), as all the request signals from each input channel are coming to the crossbar, the first-stage arbiters are eliminated. The number of request signals reaching each output link module is $(P-1)*V$. Thus, in each output link module, a $(P-1)*V$ input arbiter is required. For designing an arbiter with such a large number of inputs, a tree of smaller arbiters is used.

A single $(P-1)*V$ input arbiter is replaced by a single $(P-1)$ input arbiter and $(P-1)$ groups of V input arbiters. Each V input arbiter arbitrates in a round-robin fashion between the incoming requests from different input channels of a specific input link module. The $(P-1)$ input arbiters select the winner of them in a round-robin fashion, thus selecting a single request at a time from a set of $(P-1)*V$ inputs. The implementation of a $(P-1)*V$ input tree arbiter is shown in Figure 3.14. The *all_vc_busy* signal has the control to reset all the grant signals.

The VC allocator module uses the above tree arbiter to allocate the VCs. Figure 3.15 depicts the overall VC allocator architecture. The outputs of the control logic module ($g1$, $g2$, $g3$, etc.) are ANDed with the incoming request signals. Initially, $gnt1 = gnt2 = gnt3 = \ldots = 0$, and $g1 = g2 = g3 = \ldots = 1$. Now, if any of the request signals (say $req1$) is granted, the corresponding grant

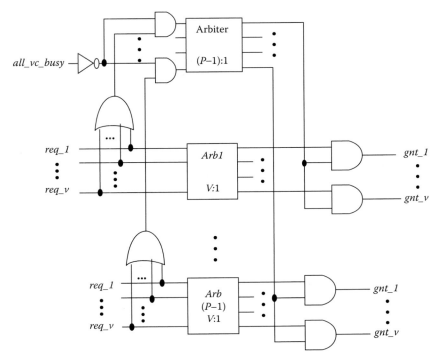

FIGURE 3.14
$(P - 1)*V$ input tree arbiter.

signal (*gnt1*) is at logic 1. The *g1* signal is latched at logic 0 one clock cycle after the *gnt1* becomes high. It remains at logic 0until *req1* is reset to 0. Thus, *gnt1* signal will remain high for one clock cycle. As the arbiter selects a single request signal at a time, no two grant signals will overlap with each other. The encoder module encodes all the grant signals. The width of the encoder output signal is $\log_2[(P-1)*V]$.

To store the status of each VC of the next router's input link module, a *V*-bit status register is used. Each bit (b_1, b_2, etc.) signifies whether the corresponding VC of the next router's input link is engaged or not. The status register has been implemented using D-type flip-flops. Initially, all the D flip-flops are reset. Hence, $b\_1 = b\_2 = \ldots = b\_v = 0$ and $ce\_1 = ce\_2 = \ldots = ce\_v = 0$. Now, if any grant signal is high, the ce_1 signal is set; as a result, b_1 gets latched at *logic-1*. When a particular status register is set by a grant signal, it will remain high until the *tailer* of the corresponding packet comes. The *tailer* signal from each input channel module is used as an input to a $(P - 1)*V{:}1$ MUX. The number of MUXes used in each VC allocator module is *V*. When ce_1 is set, the encoder output selects the corresponding *tailer* signal. For example, *tailer1* is selected if *gnt1* is set, *tailer2* is selected if *gnt2* is set, and so on. This selected *tailer* signal is used to reset the individual bits of the

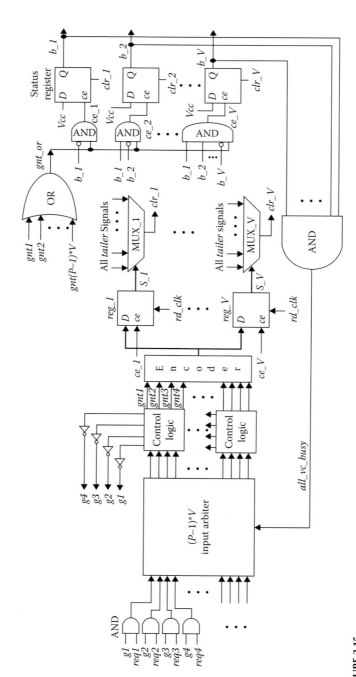

FIGURE 3.15
VC allocator module.

status register. If more than one request signal is high at a time, the output bits of the status register will be set on a cycle-by-cycle basis in the following order: $b\_1, b\_2, b\_3, \ldots b\_v$. When all the status register bits are set, the *all_vc_busy* signal will be at *logic-1*, which signifies that all the VCs of the next router's input link module are engaged. This *all_vc_busy* signal is used to disable the tree arbiter as shown in Figure 3.14.

3.6.2.2 Switch Allocator

Individual flits are arbitrated to access the outgoing physical channels via crossbar on each cycle by the switch allocator module. Switch allocator module needs two stages of arbitration (Mullins et al. 2004). The first stage has *V*:1 arbiter for each input VC.

It is followed by a second stage of *P*:1 arbiters for each output port. In the modified router architecture, as all the request signals from each input channel are coming to the crossbar, the first-stage arbiters are eliminated. As the status register stores the status of each VC of the next router's input link module (Figure 3.15), the switch allocator module needs a *V*:1 arbiter to access the outgoing physical channel. The architecture of the switch allocator module is shown in Figure 3.16. The inputs to the switch allocator module are all *rok* signals, incoming flits from each input channel, and the *full* signals (*full_vc_1, full_vc_2,* etc.) from each VC of the next router's input link module.

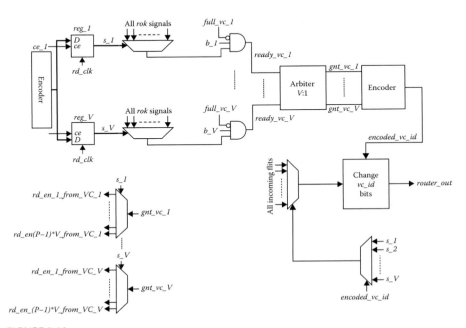

FIGURE 3.16
Switch allocator module.

The *rok* signals from each input channel module are used as input signals to a $(P - 1)*V$:1 MUX. The number of MUXes used in each switch allocator module is V. From the VC allocator module, when *ce_1* is set, the encoder output selects the corresponding *rok* signal. For example, *rok1* is selected if *gnt1* is set, *rok2* is selected if *gnt2* is set, and so on. Now if the *full* signal coming from the first VC of the next router's input link is at logic 0 and the status bit *b_1* is at logic 1, the *ready_vc_1* signal becomes set, which signifies that this VC is ready to accept the incoming flits. This signal is used as an input to the V:1 round-robin arbiter to access the outgoing physical link. In the same way, the other inputs of the arbiter (*ready_vc_2*, *ready_vc_3*, etc.) can send requests to access the outgoing physical link. Depending on the winner of the arbitration process, the encoder module generates the *encoded_vc_id* signal. For example, *encoded_vc_id* = 00 for *gnt_vc_1* = 1, *encoded_vc_id* = 01 for *gnt_vc_2* = 1, and so on. These *encoded_vc_id* bits are used to select the incoming flits to access the outgoing physical channel as shown in Figure 3.16. It also overwrites the existing *vc_id* bits such that the outgoing flits will be written into a newly allocated VC of the next router's input link module. The arbiter output signals are also used as *rd_en* signals to the IRS block of each input channel module.

The router is designed in Verilog HDL and synthesized in Synopsys Design Vision supporting 90-nm CMOS technology with *Faraday* library to generate a gate-level netlist. For a *leaf* router in the MoT-based network having a node degree of 4, a synthesis of the design shows that the critical path lies from the IB to $(P - 1)*V$ input arbiter inside the VC allocator module and the delay of the critical path is found to be 600 ps. Hence, the router can operate at 1.66 GHz. The overall router architecture has four-cycle latency, one cycle each in FIFO, $(P - 1)*V$ input arbiter of the VC allocator, status register of the VC allocator module, and V:1 arbiter of the switch allocator module.

3.7 Adaptive Router Architecture Design

A number of adaptive router architectures for NoC have been proposed in the literature. Among these architectures, this chapter adopts the work cited by Hu and Marculescu (2004a) for its simplicity in implementation. The authors proposed a dynamic adaptive–deterministic (*DyAD*) routing for a 2D mesh. *DyAD* is a new *paradigm* for NoC router design that exploits the advantages of deterministic and adaptive routing. Indeed, based on this idea, any suitable deterministic and adaptive routing scheme can be combined to form a *DyAD* router (although care must be taken for issues such as deadlock freedom). *DyAD* selects minimal odd–even routing as the adaptive routing and XY routing as the deterministic routing. The *odd–even* is able to achieve much higher saturation throughput compared to XY routing.

When the network is not congested, a *DyAD* router works in a deterministic mode and hence enjoying low routing latency. On the contrary, when the network becomes congested, the *DyAD* router switches back to the adaptive routing mode and thus avoids the congested links by exploiting other routing paths; this leads to higher network throughput.

This section presents the actual router design, *DyAD–OE*, which implements the concept of *DyAD* for odd–even routing. Combining odd–even and *XY* to form a *DyAD* router may lead to deadlock. Thus, a new routing scheme has been developed, called *oe-fixed*, as the deterministic routing mode in *DyAD–OE*. The *oe-fixed* is indeed a deterministic version of *odd–even* based on removing the *odd–even*'s adaptiveness. For instance, in the *odd–even* mode, if a packet with a given source and destination can be routed to two outputs, it will always be routed to a single output in *oe-fixed*. Figure 3.17 illustrates the architecture of the *DyAD–OE* implementation.

In Figure 3.17, each input controller has a separate FIFO (typically several flits implemented by registers for performance and power efficiency), which buffers the input packets before delivering them to the output ports. When a

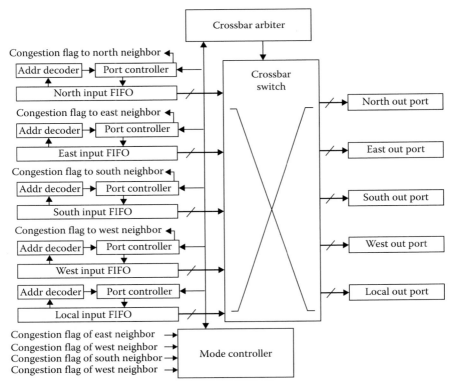

FIGURE 3.17
DyAD–OE router architecture.

new header flit is received, the *address decoder* first processes that flit and then sends the destination address to the *port controller*; this determines which output port the packet should be delivered to. In the *odd–even* mode, there can be more than one output direction to route packets. In this case, the port controller will choose the direction in which the corresponding downstream router has more empty slots in its input FIFO. Once the router has made its decision on which direction to route, the port controller sends the connection request to the crossbar arbiter in order to set up a path to the corresponding output port. Except for the local input controller, each input port controller also monitors its FIFO occupation ratio. If the ratio reaches the preset *congestion threshold* (~60%), a value of 1 will be asserted (indicating to the upstream router that the downstream router is congested) on the corresponding congestion flag wire. Otherwise, a value of 0 will be asserted, indicating to the upstream router that congestion is not an issue.

The *crossbar arbiter* maintains the status of the current crossbar connection and determines whether to grant connection permission to the port controller. When there are multiple input port controllers request for the same available output port, the crossbar arbiter uses the first-come-first-served policy to decide which input port to grant the access, such that the starvation at a particular port can be avoided.

The *mode controller* continuously monitors its neighboring congestion to determine if the deterministic or the adaptive routing mode needs to be used. Although more advanced techniques can be used to determine the optimal routing mode, we use the following simple policy: if any congestion flag from its neighboring routers is asserted, then the mode controller commands all the input port controllers to work in the adaptive (*odd–even*) mode; otherwise, it switches the port controllers to the deterministic (*oe-fixed*) mode.

It has been observed in simulation that *XY* routing performs better than both *odd–even* and *DyAD–OE* routing under *uniform* traffic load. The reason why *XY* performs best under uniform traffic is because it embodies global, long-term information about this traffic pattern. However, the adaptive algorithms select the routing paths based on *local*, short-term information. This type of decision benefits only the packets in the immediate future, which tend to interfere with other packets. Thus, the evenness of uniform traffic is not necessarily maintained in the long run. However, for most of the applications in the real world, each node will communicate with some nodes more frequently compared to others. *XY* routing has serious problems in dealing with such nonuniform traffic patterns because of its determinism. More precisely, *XY* routing blindly maintains the unevenness of the nonuniform traffic, just as it maintains the evenness for the uniform traffic. In this scenario, *XY* routing is clearly outperformed by *odd–even* and *DyAD–OE* under *transpose1* traffic. It has been observed in simulation that *odd–even* and *DyAD–OE* have 53.3% and 61.7% improvement over *XY*, respectively, in terms of sustainable throughput. In fact, for the same traffic pattern and injection rate, *DyAD–OE* achieves *shorter* average packet latency compared to *odd–even* throughout the experiments. Another interesting fact is that *DyAD–OE* does keep the advantage of deterministic routing when the network is not congested. However, the average latency a packet experiences in *odd–even* is 14% higher compared to that in *DyAD–OE*, when the network is lightly loaded. Other nonuniform traffic patterns (such as *transpose2* and *hot spot*) have been simulated as well and the results were similar to those under the *transpose1* traffic pattern. Simulation has been carried out for different network sizes (ranging from 4×4 to 8×8 tiles) and different FIFO sizes (ranging from three to eight flits) and also with realistic traffic such as multimedia traffic. All the results reflect the same characteristic (Hu and Marculescu 2004b).

3.8 Summary

In this chapter, the architecture design of wormhole and VC for NoC has been described. The implementation of GALS style of communication with dual-clock FIFO has also been discussed in detail. The evaluation of performance and cost of NoC by applying self-similar traffic with varying locality factors will be discussed in Chapter 4. The simulation results are compared with a well-known tree-based topology, butterfly fat-tree (BFT), and two variants of mesh topology connecting single/two cores to each router under the same bisection width constraint.

References

Cheng, Y. 2004. Gray code sequences. U.S. Patent 6703950, March 9, 2004.

Cummings, C. E. 2002. Simulation and synthesis techniques for asynchronous FIFO design. Synopsys User Group, San Jose, CA, http://www.sunburst-design.com/papers/CummingsSNUG2002J_FIFO1.pdf.

Cummings, C. E. and Alfke, P. 2002. Simulation and synthesis techniques for asynchronous FIFO design with asynchronous pointer comparisons. Synopsys User Group, San Jose, CA, http://www.sunburst-design.com/papers/CummingsSNUG2002J_FIFO2.pdf.

Dally, W. J. 1992. Virtual-channel flow control. *IEEE Transactions on Parallel and Distributed Systems*, vol. 3, no. 2, pp. 194–205.

Hu, J. and Marculescu, R. 2004a. DyAD-smart routing for networks-on-chip. *Proceedings of Design and Automation Conference*, pp. 260–263, July 7–11, San Diego, CA.

Hu, J. and Marculescu, R. 2004b. Smart routing for networks-on-chip. Technical report, ECE Department, Carnegie Mellon University, http://www.ece.cmu.edu/_sld/pubs. Accessed at September 2007.

Jiang, H. J. 2004. Scalable gray code counter and applications thereof. U.S. Patent 6762701, July 13, 2004.

Kavaldjiev, N., Smit, G. J. M., Jansen, P. G., and Wolkotte, P. T. 2006. A virtual channel network-on-chip for GT and BE traffic. *Proceedings of IEEE Computer Society Annual Symposium on Emerging VLSI Technologies and Architectures*, March 2–3, Karlsruhe, Germany.

Kumar, A., Kundu, P., Singh, A. P., Peh, L. S., and Jha, N. K. 2007. A 4.6Tbits/s 3.6GHz single-cycle NoC router with a novel switch allocator in 65nm CMOS. *Proceedings of IEEE International Conference on Computer Design*, pp. 63–70, October 7–10, Lake Tahoe, CA.

Kundu, S. and Chattopadhyay, S. 2008. Network-on-chip architecture design based on mesh-of-tree deterministic routing topology. *International Journal for High Performance Systems Architecture, Inderscience Publishers*, vol. 1, no. 3, pp. 163–182.

Kundu, S., Soumya, J., and Chattopadhyay, S. 2012. Design and evaluation of mesh-of-tree based network-on-chip using virtual channel router. *Microprocessors and Microsystems Journal*, vol. 36, pp. 471–488.

Mullins, R., West, A., and Moore, S. 2004. Low-latency virtual-channel routers for on-chip networks. *Proceedings of 31st Annual International Symposium on Computer Architecture*, pp. 188–197, June 19–23, Munich, Germany.

Mullins, R., West, A., and Moore, S. 2006. The design and implementation of a low-latency on-chip network. *Proceedings of Asia and South Pacific Design Automation Conference*, pp. 164–169, January 24–27, Yokohama, Japan.

Peh, L.-S. and Dally, W. J. 2001. A delay model and speculative architecture for pipelined routers. *Proceedings of International Symposium on High-Performance Computer Architecture*, pp. 255–266, Monterrey, Mexico.

4

Evaluation of Network-on-Chip Architectures

4.1 Evaluation Methodologies of NoC

This section presents the strategy to evaluate the performance and cost of networks-on-chip (NoCs). In the NoC paradigm, while evaluating the performance of an interconnect infrastructure, its energy consumption profile and silicon area overhead must also be considered, as it can be a significant portion of the overall system-on-chip (SoC) cost budget. It has been reported in the work of Pande et al. (2005) that Scalable, Programmable Integrated Network (SPIN) and octagon network have very high throughput, but their energy consumption and silicon area overhead are much higher than both mesh and butterfly fat tree (BFT). Folded torus shows almost similar results like mesh. In the deep submicron (DSM) era where high-performance and low-power design is a major goal, for a NoC designer it is always preferable to choose a topology with lower average energy per packet profile. Taking this fact into consideration, mesh topology is widely used in academia and industry.

In this chapter, a thorough comparative study of performance evaluation, estimation of energy consumptions and area overhead of different mesh- and tree-based NoC topologies have been shown with the same number of intellectual property (IP) cores under the same bisection width constraint as reported in Kundu et al. (2012). A bisection width is defined as the minimum number of wires to be removed in order to bisect the network into two equal halves. A network with higher bisection width is expected to show better performance. This chapter considers two variants of mesh structure—one core and two cores connected to each router, BFT, and mesh-of-tree (MoT) network structures—and compares their performance and cost for a 32-core-based system. The bisection width of all the above networks is taken as 4. Each IP core has been inserted into a tile of dimension 2.5 mm × 2.5 mm, similar to that discussed in the work of Feero and Pande (2009).

Although the total chip area can be obtained only after layout, Figures 4.1 through 4.4 show the possible distributions of cores, routers, and links for

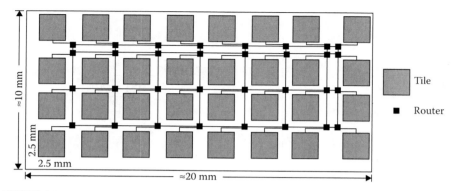

FIGURE 4.1
Possible distribution of cores, routers, and links in a 4 × 8 mesh structure with a single core to each router (Mesh-1).

chip area estimations of all the topologies with 32 tiles. These diagrams will enable us to compare the area overheads of NoC topologies under consideration. In all the figures, the routers are denoted by small black nodes. Here, it has been considered that a single link traversal of length 2.5 mm can be completed within a single clock cycle. The length of tile-to-router link (also termed as local link) is taken as 1.25 mm. For designing a mesh structure with 32 tiles, the probable distribution of cores, routers, and links of a 4 × 8 network is shown in Figure 4.1, where a single core has been attached to each router. Depending on the connectivity, the routers of this network can be classified into three types: (1) *center* having node degree 5, (2) *edge* having node degree 4, and (3) *corner* having node degree 3. The lengths of most of the inter-router links in Figure 4.1 are 2.5 mm, whereas links between the first two rows and between the last two columns are taken as few micrometers.

The second variation of mesh network has two cores connected to each router. A similar distribution of cores, routers, and links for such a 4 × 4 mesh structure has been shown in Figure 4.2. This structure also has three types of routers: (1) *center* having node degree 6, (2) *edge* having node degree 5, and (3) *corner* having node degree 4. As wire delay increases exponentially with its length, links more than 2.5 mm are pipelined. The registers used for pipelining are shown as small white nodes in Figure 4.2. The links between the first two rows are few micrometers long. In this book, mesh topology having a single core attached to each router is termed as Mesh-1 network (Figure 4.1), whereas the mesh with two cores connected to each router is called Mesh-2 network (Figure 4.2).

A 32-core BFT-based NoC (Pande et al. 2003) with four cores attached to each *leaf* level router is shown in Figure 4.3. This network also has three types of routers: (1) *leaf* having node degree 6, (2) *stem* having node degree 6, and (3) *root* having node degree 2. The MoT-based NoC with two cores

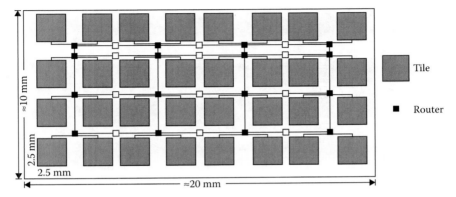

FIGURE 4.2
Possible distribution of cores, routers, and links in a 4 × 4 mesh structure with two cores to each router (Mesh-2).

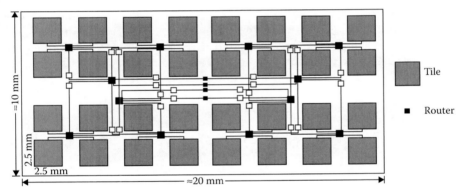

FIGURE 4.3
Possible distribution of cores, routers, and links in BFT networks with four cores to each *Leaf* router.

attached to each *leaf* level router is shown in Figure 4.4. The three types of routers in this network are (1) *leaf* with node degree 4, (2) *stem* with node degree 3, and (3) *root* with node degree 2. In MoT, as shown in Figure 4.4, the inter-router links between the *leaf* and the *stem* of the column tree are few micrometers long. In any tree-based topology, link length increases with growing network size. Hence, the links of BFT and MoT are pipelined after every 2.5 mm. The registers used for pipelining are shown as white nodes in Figures 4.3 and 4.4. From Figures 4.1 through 4.4, it can be observed that the wire density of mesh networks is more uniform across any cross section compared to that of BFT and MoT.

Routers for all the above-mentioned networks have been designed in Verilog HDL and synthesized using Synopsys Design Vision supporting 90-nm complementary metal oxide semiconductor (CMOS) technology. For a specific

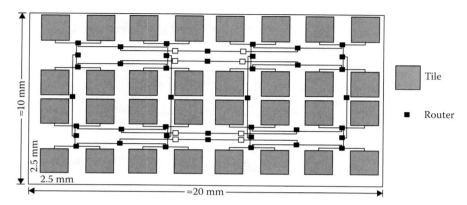

FIGURE 4.4
Possible distribution of cores, routers, and links in a 4 × 4 MoT structure with two cores to each *Leaf* router.

network structure, critical path delay of a router increases with increasing node degree. This happens as the routing logic and arbitration complexity increase with increasing node degree. Hence, in a network, the router with a highest node degree has the minimum frequency. To support mesochronous clocking, the clock having minimum frequency is applied to all the routers of a network. Table 4.1 shows the clock frequencies of different types of wormhole (WH) routers used in implementing the networks under consideration. Though MoT has the highest minimum frequency, in this work, to provide a consistent comparison with other networks, all the routers are driven at 1.5-GHz clock.

For evaluating the performance of these networks, a SystemC-based cycle-accurate NoC simulator has been developed. The simulator operates at the granularity of individual architectural components of the router. It supports mesochronous clocking strategy where the routers are driven by the same clock frequency with varying phase.

4.1.1 Performance Metrics

The performance of an on-chip communication network is characterized by its throughput and latency. Throughput is the maximum accepted traffic from the network and it is related to the peak data rate sustainable by the network. Although the ratio of the number of edges (E) and the average distance (D) of a particular network is a good indicator of throughput in contention-free traffic, in an actual traffic scenario, the average time spent by the flits in the network will increase due to traffic congestion, and hence throughput and latency values will differ from their theoretical counterparts. In this book, throughput is defined as (Pande et al. 2005)

$$\text{Throughput} = \frac{\text{Total packets completed} \times \text{Packet length}}{\text{Number of IP blocks} \times \text{Total time}} \quad (4.1)$$

TABLE 4.1

Connectivity, Number of WH Routers, and Frequency of Different Types of WH Routers to Implement the Networks under Consideration for Connecting 32 Cores

Networks	Type-1 Router				Type-2 Router				Type-3 Router			
	Position	Node Degree	Number of Routers	Frequency (GHz)	Position	Node Degree	Number of Routers	Frequency (GHz)	Position	Node Degree	Number of Routers	Frequency (GHz)
Mesh-1	Center	5	12	1.60	Edge	4	16	1.68	Corner	3	4	1.72
Mesh-2	Center	6	4	1.55	Edge	5	8	1.60	Corner	4	4	1.68
BFT	Leaf	6	8	1.52	Stem	6	4	1.52	Root	4	4	1.90
MoT	Leaf	4	16	1.66	Stem	3	16	1.70	Root	2	8	1.90

Note: In BFT network with 2^n cores, the node degree of root routers is 4 when n is even and 2 when n is odd.

where:
> *Total packets completed* refers to the number of packets that successfully arrive at their destination IP cores
>
> *Packet length* is measured in terms of flits
>
> *Number of IP blocks* refers to the number of IP blocks involved in the communication
>
> *Total time* denotes the simulation time (in clock cycles)

Hence, throughput is represented as flits/cycle/IP. Network *bandwidth* (BW) refers to the maximum number of bits that can be sent successfully to the destination through the network per unit time. It is represented as bits/sec (bps).

$$BW = \frac{\text{Throughput} \times \text{Number of IP cores} \times \text{Flit size}}{\text{Clock period}} \quad (4.2)$$

Depending on the source–destination pair and the routing algorithm, each packet may have a different latency. There is also some overhead in the source and destination that contributes to the overall latency. Therefore, for any packet i, the overall latency (L_i) is defined as

$$L_i = \text{Sender overhead} + \text{Transport latency} + \text{Receiver overhead} \quad (4.3)$$

Let P be the total number of packets reaching their destination IPs. The average overall latency, L_{avg}, is then calculated as (Pande et al. 2005)

$$L_{avg} = \frac{\sum_{i=1}^{P} i}{P} \quad (4.4)$$

4.1.2 Cost Metrics

Energy consumption by the network is one of the most important cost metrics in NoC design. Energy consumption of each router is determined by using Synopsys Prime Power in 90-nm CMOS technology with *Faraday* library by running their gate-level netlists. The clock frequency to each router is been fixed at 1.5 GHz. The number of toggles of every individual I/O pin of the router and their probability of remaining in logic 1 state for the entire simulation window is calculated from the NoC simulator. This information is then fed to Synopsys Prime Power tool to estimate the power of each router with the following parameters: process = normal, supply voltage = 1 V, and temperature = 75°C.

Energy consumption of the links is determined separately from that of the routers. Links can be modeled as semiglobal interconnects. Copper wire (resistivity = 17 nΩ·m) is chosen as interconnection link. The width and thickness of the wires are taken to be 0.25 and 0.5 µm, respectively. The spacing between two adjacent wires is kept at 0.25 µm. The spacing between two adjacent metal layers is fixed at 0.75 µm and is filled by a dielectric material

TABLE 4.2

Parasitic Capacitance, Inductance, and Resistance of a Three-Wire Model

Self-Capacitance (pF/m)		Coupling Capacitance (pF/m)		Self-Inductance (μH/m)		Mutual Inductance (μH/m)		Resistance (kΩ/m)	
Line 1	134.54	Lines 1–2	70.33	Line 1	0.33	Lines 1–2	0.17	Line 1	137.93
Line 2	175.92	Lines 2–3	70.33	Line 2	0.32	Lines 2–3	0.17	Line 2	137.93
Line 3	134.54	Lines 1–3	1.99	Line 3	0.33	Lines 1–3	0.09	Line 3	137.93

having a relative permittivity of 2.9. Table 4.2 shows the parasitic resistance, capacitance, and inductance per meter length of the wires, extracted by the *Field Solver* tool from *HSPICE* supporting 90-nm CMOS technology with a three-wire model.

In a three-wire model, the middle wire is considered as the victim line, whereas the other two wires are known as aggressor lines. In an n-bit channel, coupling effect on a wire by the nonadjacent lines is negligible. A nonideal input signal is supplied to the link driver and a load capacitance of 5 fF is connected to the other end of the link. Here, repeater is placed exactly at the middle of a 2.5-mm long wire. It has been observed that the worst-case link delay is much lesser than the router clock period of 666 ps (frequency = 1.5 GHz). The delay of the links having a length of tens of microns is also very less. Hence, the links are not falling into the critical path of the overall NoC.

Energy consumption in the victim line for all possible transitions in the wires of a three-wire model can be calculated using *HSPICE*. Table 4.3 shows a look-up table where the first and last rows for every state indicate the energy consumption (in Joules) per transition in the middle wire of lengths 1.25 and 2.5 mm, respectively. This look-up table is used to calculate the link energy from the NoC simulator for 200,000 cycles. It can be observed from Table 4.3 that energy consumption in the middle line is negative for some of the transitions (e.g., 000 → 101 in 2.5-mm wire), as also observed and explained in the work of Sotiriadis and Chandrakasan (2002). Due to the capacitive coupling for some specific state transitions, the current flows back to the power supply through the middle wire.

4.2 Traffic Modeling

In this chapter, traffic injected by the IP cores follows self-similar distribution. In the following, a precise way of self-similar traffic generation has been presented. It has been shown that modeling of self-similar traffic can be obtained by aggregating a large number of ON–OFF message sources (Park and Willinger 2000). The length of time each message spends in either

TABLE 4.3

Energy (in Joules) on the Middle Line of Three-Wire Model Obtained from HSPICE

Present State	Length (mm)	Next State							
		000	001	010	011	100	101	110	111
000	1.25	5.25E-18	1.04E-13	7.70E-14	1.72E-13	1.04E-13	2.12E-13	1.72E-13	2.76E-13
	2.5	3.41E-18	-9.61E-14	2.94E-13	2.04E-13	-9.61E-14	-1.93E-13	2.04E-13	1.13E-13
001	1.25	1.07E-13	4.16E-18	2.74E-13	1.60E-13	2.14E-13	1.11E-13	3.65E-13	2.68E-13
	2.5	9.56E-14	3.41E-18	3.84E-13	2.94E-13	-2.62E-16	-9.62E-14	2.91E-13	2.04E-13
010	1.25	1.80E-13	3.47E-13	3.77E-18	1.91E-13	3.47E-13	5.13E-13	1.91E-13	3.82E-13
	2.5	2.76E-13	2.92E-13	2.65E-18	9.97E-14	2.92E-13	1.90E-13	9.97E-14	1.99E-13
011	1.25	1.22E-13	9.93E-14	2.07E-14	2.69E-18	3.14E-13	2.92E-13	2.15E-13	1.98E-13
	2.5	1.88E-13	2.75E-13	-9.98E-14	2.65E-18	2.66E-13	2.92E-13	5.30E-16	9.96E-14
100	1.25	1.07E-13	2.14E-13	2.74E-13	3.65E-15	4.16E-18	1.11E-13	1.60E-13	2.68E-13
	2.5	9.57E-14	-2.62E-16	3.84E-13	2.92E-13	3.41E-18	-9.62E-14	2.94E-13	2.04E-13
101	1.25	2.01E-13	9.93E-14	4.45E-13	3.34E-13	9.93E-14	3.07E-18	3.34E-13	2.43E-13
	2.5	1.91E-13	9.57E-14	4.48E-13	3.84E-13	9.57E-14	3.41E-18	3.84E-13	2.94E-13
110	1.25	1.22E-13	3.14E-13	2.07E-14	2.15E-13	9.93E-14	2.92E-13	2.69E-18	1.98E-13
	2.5	1.88E-13	2.66E-13	-9.98E-14	5.30E-16	2.75E-13	2.92E-13	2.65E-18	9.96E-14
111	1.25	4.21E-14	2.75E-14	3.27E-14	1.40E-14	2.75E-14	1.86E-14	1.40E-14	1.60E-18
	2.5	9.26E-14	1.88E-13	-1.99E-13	-9.97E-14	1.88E-13	2.75E-13	-9.97E-14	2.65E-18

the ON or the OFF state should be selected according to a distribution that exhibits long-range dependence. The *Pareto distribution* $[F(x) = 1 - x^{-\alpha}$, with $1 < \alpha < 2]$ is found to fit well to this kind of traffic. The duration of each ON–OFF period is assumed to be a random variable T_i $(i \in \{ON, OFF\})$. The degree of self-similarity is expressed using only a single parameter, namely, *Hurst parameter* (HP). The value α at ON slot is related to HP as given below:

$$HP = \frac{3 - \alpha_{ON}}{2}, \text{where } 0.5 < HP < 1 \tag{4.5}$$

If the network utilization parameter is given by ρ, the α_{OFF} parameter is obtained as

$$\alpha_{OFF} = \frac{(1-\rho)\alpha_{ON}}{(1-\rho)\alpha_{ON} - \rho(\alpha_{ON} - 1)} \tag{4.6}$$

For a random variable U with uniform distribution on $[0, 1]$, the following transformation can be used to generate the random number P of time slots during active and idle periods:

$$P_i = \text{round}\left[U^{\frac{-1}{\alpha_i}}\right], i \in \{ON, OFF\} \tag{4.7}$$

The active and idle periods in each iteration is calculated as

$$T_i = \frac{P_i}{IR} \tag{4.8}$$

where IR is the packet injection rate within the ON slot. The pseudocode of the algorithm for generating a self-similar traffic is shown in Figure 4.5.

Algorithm: *Generation of Self-Similar Traffic*

1. Set IR, HP, ρ, and time to 0.
2. Calculate α_{ON} from HP by using (Eq. 5.5).
3. Calculate α_{OFF} from ρ and α_{ON} by using (Eq. 5.6)
4. While time $\leq$ SIMULATION_TIME do
   4.1 Generate a random number U between 0 and 1.
   4.2 Calculate P_{ON} and P_{OFF} by using (Eq. 5.7).
   4.3 Calculate T_{ON} and T_{OFF} by using (Eq. 5.8).
   4.4 For $j \leftarrow 0$ to P_{ON} do
           Generate packet for the destination.
   4.5 *time* $\leftarrow$ *time* $+ T_{ON} + T_{OFF}$
5. Stop

FIGURE 4.5
Pseudocode of the algorithm for generating self-similar traffic.

In the simulation process, the parameters ρ and HP have been set to 0.3 and 0.75, respectively. Hence, from Equations 4.5 and 4.6, the values of α_{ON} and α_{OFF} are calculated to be 1.5 and 1.21, respectively. The packet injection rate within the ON slot is varying as 0.002, 0.004, 0.006, and so on. The user may also choose between uniform and localized traffic patterns. In our simulation, we have fixed the packet length to be of 64 flits, as in the work of Pande et al. (2005). The packet injection is continued for a simulation time of 200,000 cycles of the routers' clock including 10,000 cycles to make the network stable from initial transient effects. For accurate estimation of the energy consumption, we have assumed that each traffic generator module injects traffic into the network with a switching activity of 0.9 as it is expected to introduce a large number of transitions, and thus energy consumption in the network.

4.3 Selection of Channel Width and Flit Size

To design a NoC-based system, selection of flit size and channel width plays a crucial role in the overall system performance and cost. Although there are several works reported in the literature that use wider flits [e.g., 64 bits (Soteriou et al. 2007), 96 bits (Rijpkema et al. 2003), 128 bits (Millberg et al. 2004), and 256 bits (Chi and Chen 2004)], Salminen et al. (2008) recommended the usage of 32-bit flit size and the same link width for mesh topology. Tilera's 64-core-based *TILE64* processor chip, implemented in 90-nm technology, uses 32-bit-wide two unidirectional inter-router links where the routers are connected in an 8 × 8 two-dimensional (2D) mesh fashion (Wentzlaff et al. 2007). Intel's 80-core-based *Teraflops* research chip (Vangal et al. 2008) also uses 32-bit flit size and 39-bit wider links (including some handshaking signals) in 65-nm technology where the routers are connected in an 8 × 10 mesh network. *XPIPES* (Bertozzi and Benini 2004) and *Dy-AD* mesh (Hu and Marculescu 2004) also use 32-bit link width in 100- and 160-nm technologies, respectively. Considering all these examples, this chapter also uses both flit size and link width equal to 32 bits for all the networks under consideration, keeping the wire dimension unchanged.

4.4 Simulation Results and Analysis of MoT Network with WH Router

In this section, the evaluation methodology mentioned is applied to find out the performance and energy consumption of MoT-based network consisting of 32 IP cores by varying the offered load and locality factor in a self-similar traffic. In the WH router architecture, FIFOs of depth 6 are used at both input and output channels.

4.4.1 Accepted Traffic versus Offered Load

The accepted traffic depends on the rate at which the IP blocks inject traffic into the network. Ideally, accepted traffic should increase in direct proportion to the increasing offered load. However, due to the limitation of routing and arbitration strategy and the unavailability of enough buffer space within the WH router (FIFO depth being much lesser than the size of the packet), the network suffers from contention. Therefore, the accepted traffic saturates after a certain value of the offered load. Figure 4.6 depicts this scenario for uniformly distributed traffic in a MoT-based network. The maximum accepted traffic where the network is saturated is termed as throughput and it relates to the maximum sustainable data rate by the network.

4.4.2 Throughput versus Locality Factor

Locality factor is defined as the ratio of local traffic to total traffic and its value is zero for uniformly distributed traffic. For a 4×4 MoT, as shown in Figure 4.7, the possible distances (d) of the destinations from any source are $d = 0, 2, 4, 6,$ and 8. There is only one destination core at the nearest cluster ($d = 0$). For example, if the locality factor is 0.5, 50% of traffic is targeted at the nearest cluster from source. The rest of the traffic is distributed according to their distances from the source, such that a destination at the nearer cluster gets more traffic compared to one at farther cluster. In a 4×4 MoT, for a locality factor of 0.5, the distribution of the traffic is as follows:

- Fifty percent of the total traffic goes to the cluster having $d = 0$.
- Rest 50% of the traffic is distributed as follows:
- $(8 * 50)/(2 + 4 + 6 + 8) = 20\%$ of the total traffic goes to the cluster having $d = 2$

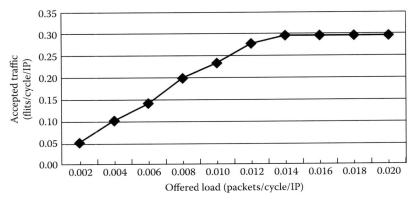

FIGURE 4.6
Accepted traffic with varying offered load at uniformly distributed traffic.

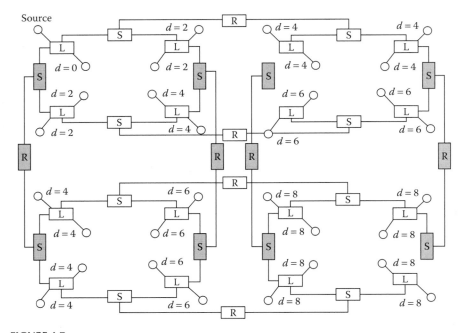

FIGURE 4.7
Distances of the destination cores from any source in 4 × 4 MoT. L, S, and R represent Leaf, Stem, and Root routers respectively.

- (6 * 50)/(2 + 4 + 6 + 8) = 15% of the total traffic goes to the cluster having $d = 4$
- (4 * 50)/(2 + 4 + 6 + 8) = 10% of the total traffic goes to the cluster having $d = 6$
- (2 * 50)/(2 + 4 + 6 + 8) = 5% of the total traffic goes to the cluster having $d = 8$

Now, as there are more than one destination cores in some clusters ($d = 2, 4, 6,$ and 8), the traffic gets randomly distributed among them.

The effect of traffic spatial localization on the throughput of the MoT-based network is shown in Figure 4.8. It can be observed that network throughput increases with increasing locality factor. This is due to the fact that as the locality factor increases, more traffic is destined for their local clusters, thus traversing lesser number of hops, which in turn increases throughput.

4.4.3 Average Overall Latency at Different Locality Factors

The average overall latency of any network depends on both the offered load and the locality factor. Figure 4.9 shows the average overall latency profile with an offered load under uniformly distributed and localized traffic in a

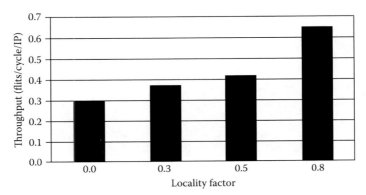

FIGURE 4.8
Variation of throughput with locality factor in MoT having 32 cores.

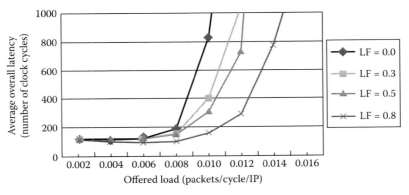

FIGURE 4.9
Latency profile of MoT with offered load at different locality factors.

4 × 4 MoT topology. It shows that at lower traffic, the latency variation is not significant. This is due to the fact that at lower traffic, the contention in the network is less, but it increases as the offered load increases, which in turn increases the latency. The simulation result shows that as the offered load increases toward the network saturation point, the latency increases exponentially, which signifies that the packets take much more time to reach their destinations. Therefore, it is always desirable to operate the network below its saturation point. The effect of spatial localization of traffic on the average overall latency in a MoT based network is also shown in Figure 4.9, where locality factors are represented by LF. It can be observed that localization of traffic has significant impact on the latency, which decreases with increasing locality factor. As the locality factor increases, more traffic goes to local cluster and hence traverses lesser number of hops. Moreover, this causes lesser contention in the network. Therefore, the network can carry more traffic, before going to saturation, which in turn enhances the operating point of the network.

4.4.4 Energy Consumption at Different Locality Factors

Total energy consumption in NoC is the summation of energy consumed by the routers and communication links. Both the factors are network topology dependent. The total energy consumption of a MoT-based network for uniformly distributed self-similar traffic is shown in Figure 4.10 (simulation for 200,000 clock cycles with a 666-ps clock period taken as evaluation parameter). It can be observed that the network energy consumption increases linearly with the offered load but saturates as the offered load increases to the throughput limit. Beyond saturation, no additional packets can be injected successfully into the network and, consequently, no additional energy is consumed.

Figure 4.11 depicts the component-wise energy consumption of the network at saturation. It can be observed that the energy consumption of all the FIFOs is 60% of the total network energy consumption, whereas all the links consume only 30% of it. The combined energy consumption of all the routing logic, arbiters, and control logics is about 10% of the same. Thus, from Figure 4.11, this can be concluded that FIFOs are the most energy hungry component in NoC.

The average energy consumption per cycle of a MoT-based NoC at saturation with uniformly distributed and localized traffic is shown in Figure 4.12. With increasing locality factor, packets traverse lesser hops to reach their destinations. Although the energy consumption of the local link increases with increasing locality factor, the *stem* and *root* routers and the inter-router links consume lesser energy due to lesser switching. From Figure 4.12, it can be noticed that the average energy consumption of the overall network decreases as the locality factor increases.

To get an idea about the energy spent per packet, the average packet energy is computed. This is another important attribute for characterizing NoC structures. Figure 4.13 shows the average packet energy consumed at different locality factors at saturation. As the energy consumption decreases

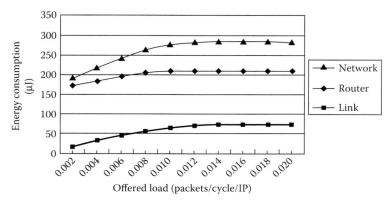

FIGURE 4.10
Energy consumption in MoT network with uniformly distributed load.

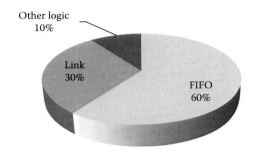

FIGURE 4.11
Percentage energy consumption by FIFOs, links, and other logics. Other logic includes routing logic, arbitration, and control logic.

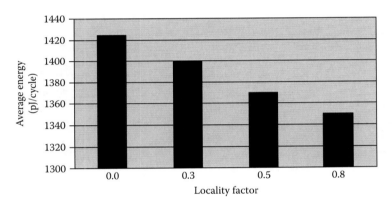

FIGURE 4.12
Average cycle energy at saturation with varying locality factors.

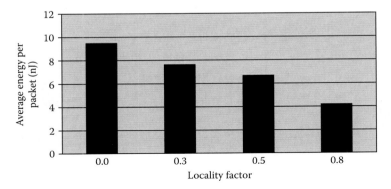

FIGURE 4.13
Average packet energy at saturation with varying locality factors.

and the number of accepted traffic increases with increasing locality factor, the average energy consumption per packet also decreases.

4.5 Impact of FIFO Size and Placement in Energy and Performance of a Network

In Section 4.4, we have seen that FIFO is the most energy-consuming element in the network. The WH router uses both input and output buffering. There are considerable amount of works in the literature (Pande et al. 2003; Wentzlaff et al. 2007; Zhang et al. 2006) that follow similar methodology of placing the FIFO buffers at both input and output channels of a router. Researchers have also proposed alternative WH routers that use only input buffering (Kumar et al. 2007; Zeferino and Susin 2003). Since both the schemes are widely used in literature, this section evaluates and compares the two. We will proceed with the better one in remaining analysis. Moreover, it has also been reported in literature (Marculescu et al. 2009; Ogras et al. 2005) that the performance and energy consumption of NoC are largely dependent on the depth of the FIFO buffer. Experimentation has been carried out on this issue as well. Table 4.4 reports the percentage performance degradation and network energy saving in MoT at saturation by varying FIFO size at different locality factors. *FIFO Depth i–j* in the second column of Table 4.4 signifies

TABLE 4.4

Energy and Performance Variation in MoT at Saturation by Varying FIFO Depths

	FIFO Depth	Locality Factor			
		0	0.3	0.5	0.8
Throughput	4–6	7.183	5.763	4.574	3.504
degradation (%)	4–4	15.024	11.318	7.944	6.261
	6–0	**1.192**	**1.391**	**0.229**	**0.156**
	4–0	12.674	8.642	7.029	6.439
Latency increment (%)	4–6	44.94	58.539	53.831	102.942
	4–4	62.889	107.06	103.456	146.595
	6–0	**13.641**	**12.187**	**11.716**	**14.317**
	4–0	71.101	130.541	131.809	163.581
Energy saving (%)	4–6	14.425	13.623	12.403	13.668
	4–4	23.93	23.206	21.516	22.235
	6–0	**37.063**	**36.899**	**36.321**	**35.443**
	4–0	41.043	42.06	41.228	41.779

Note: Bold values signifies that FIFO depth 6–0 is the optimum solution in terms of throughput, latency, and energy consumption.

that FIFO at the input channel has depth *i* and that at the output channel has depth *j*, where *j* = 0 signifies that no FIFO is present at the output channel. In this study, *FIFO Depth* 6–6 (FIFO depths at both the input and output channels are 6) has been taken as the reference. When contention occurs, FIFO at the input channel allows some more data flits to make progress even though the output channel is not available. However, FIFO at the output channel allows a few flits to cross the crossbar even when the FIFO of the next router's input channel is full. Thus, decreasing the depth of any of the FIFOs will generate *FULL* signal more frequently and thus have a negative impact on performance. This is readily reflected in Table 4.4, as performance degrades with decreasing FIFO depth.

Table 4.4 shows that the elimination of output channel FIFO in the MoT router has very little impact on throughput and latency. In this case, the number of flits stored in the router data path is lesser than that in reference case (*FIFO Depth* 6–6) by 6. However, after getting the grant signal from any output channel, header passes directly to the next router, saving one cycle delay of FIFO at the output channel. This diminishes the negative impact of eliminating FIFO at the output channel to some extent.

Network energy consumption also decreases with reducing the total FIFO size of the router. From Table 4.4, it can be concluded that the elimination of the output buffer reduces the energy consumption significantly at the cost marginal performance degradation. Therefore, to optimize the performance and energy consumption, the WH router is modified by eliminating FIFO at the output channel (*FIFO Depth* 6–0). This modified router architecture is used in rest of the chapter. The modified router architecture has only two-cycle latency, one cycle each in FIFO at the input channel and arbiter. The energy consumption profile of the MoT-based network after elimination of FIFO at the output channel with uniformly distributed offered load is shown in Figure 4.14.

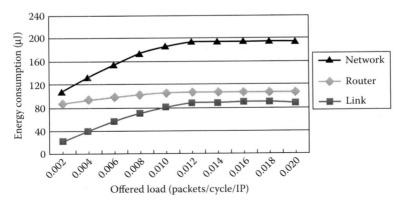

FIGURE 4.14
Network energy consumption in MoT with modified router under uniformly distributed offered load.

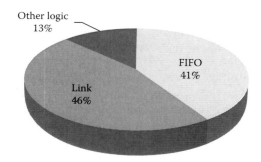

FIGURE 4.15
Component-wise energy consumption with modified router at saturation. Other logic includes routing logic, arbitration, and control logic.

The component-wise energy consumption of the network with the modified router under uniformly distributed offered load at saturation is shown in Figure 4.15. It can be observed that the energy consumption of all the FIFOs is about 41% of the total network energy consumption, whereas all the links consume only about 46% energy. The combined energy consumption of all the routing logic, arbiters, and control logic is about 13% of the same. Thus, even after the architecture modification, FIFOs are the most energy-consuming elements of a router.

The dynamic power reported by *Synopsys Prime Power* tool consists of switching power and internal power; the internal power is the summation of short-circuit power and internal node switching power. As the internal node switching power is dominated by the transition of clock in any sequential design (Synopsys Prime Power Manual 2006), the capacitances internal to the cell are either charging or discharging with every transition of clock, and hence consume very high internal power. It can be observed from Figure 4.14 that the change in the total router energy consumption is not significant with variation in the offered load. The routers consume a significant amount of energy even at very low traffic, though switching of input data is low. This appears as the internal power consumption due to free-running clock dominates over the switching and leakage power. Therefore, to reduce the internal power, it is essential to stop the free-running clock, when the network is idle. The implementation of clock gating in FIFO is presented in Chapter 6. The total energy consumption of a MoT-based network after clock gating in FIFO with uniformly distributed self-similar traffic is shown in Figure 4.16. It is noticeable that the link energy consumption exceeds the router energy consumption after clock gating in FIFO. It can be observed from Figure 4.17 that the energy consumption of all the FIFOs is 27% of the total energy, whereas all the inter-router links consume 55% of it. The combined energy consumption of all the routing logic, arbiters, and control logic is about 18% of the same.

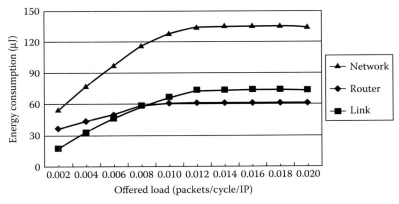

FIGURE 4.16
Network energy consumption in MoT after clock gating in FIFO.

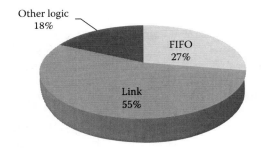

FIGURE 4.17
Component-wise energy consumption of the MoT network after clock gating in FIFO at saturation. Other logic includes routing logic, arbitration, and control logic.

4.6 Performance and Cost Comparison of MoT with Other NoC Structures Having WH Router under Self-Similar Traffic

This section compares the proposed MoT-based network with other network topologies under consideration consisting of 32 IP cores. For deterministic routing in mesh networks, XY routing (Duato et al. 2003) is used, whereas a least common ancestor (LCA) routing (Pande et al. 2003) can be adopted for BFT-based networks. In the BFT-based network, localized traffic is constrained within a cluster of four cores. In case of mesh networks having single core with each router, the number of destination cores in the local cluster is 2, 3, or 4 depending on the position of the source. Figure 4.18 depicts the two scenarios for BFT and mesh. In the mesh-based network having two cores with each router, such as MoT, there is only a single core in the local cluster. The performance and cost of all the networks are evaluated and compared

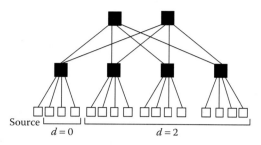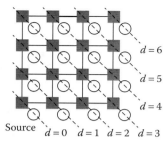

FIGURE 4.18
Distribution of destination cores from any source in BFT and mesh networks.

after eliminating FIFO at the output channel from the routers and applying clock gating to FIFO at the input channel.

4.6.1 Network Area Estimation

Table 4.5 shows the silicon area required by each type of router to implement the networks under consideration taking a 32-bit flit size. For fair comparison of the networks, this section revisits all the networks having 32 cores as shown in Figures 4.1 through 4.4. Although each tile is taken as a square of side 2.5 mm, inter-tile spacing varies significantly with the underlying topology due to varying sizes of routers, which in turn causes variations in estimated chip dimensions. It can also be noticed that the number of links running through inter-tile spaces are varying in different networks. While a mesh structure (both Mesh-1 and Mesh-2) has a uniform wiring density, MoT and BFT have nonuniform wire densities and use flyover links over the top of another router. Thus, to compare the chip dimensions of different topologies, this work has taken a uniform channel width of 32 bits for all the networks. The width of each wire and inter-wire spacing is taken to be 0.25 µm as mentioned earlier. The dimension of each router is assumed to be a perfect square.

The dimension of a mesh-based network can be estimated as follows: The routers in each row are placed between two row-wise adjacent tiles such that

TABLE 4.5

Silicon Area Required for Each Type of Router after Clock Gating

Networks	Type-1 Router		Type-2 Router		Type-3 Router	
	Position	Area (mm²)	Position	Area (mm²)	Position	Area (mm²)
Mesh-1	Center	0.079	Edge	0.061	Corner	0.045
Mesh-2	Center	0.115	Edge	0.079	Corner	0.057
BFT	Leaf	0.115	Stem	0.115	Root	0.022
MoT	Leaf	0.057	Stem	0.041	Root	0.022

the breadth of the chip increases just because of the channel width. The length of the chip will increase by the length of all *center* routers and two *edge* routers of a row. Thus, a better estimation of the chip dimension of a 4 × 8 Mesh-1 network considering unidirectional opposite links increases the area from 20 mm × 10 mm to 22.18 mm × 10.128 mm, and that of a 4 × 4 Mesh-2 network becomes 21.24 mm × 10.128 mm. For the BFT-based network, the length of the chip increases because of the *leaf*, *stem*, and *root* routers, whereas its breadth increases just because of the channel width. The estimated dimension of the BFT network becomes 22.33 mm × 10.128 mm. For the MoT-based network, all routers and repeaters of a row tree can be placed between row-wise adjacent tiles such that they do not increase the breadth of the chip. It can be noted that only the *stem* routers of the column trees of each 2 × 2 MoT subnetwork will increase the breadth of the chip. The length of the chip will be increased because of the routers and repeaters of the row tree. Taking all these facts into account, the estimated chip dimension of the 4 × 4 MoT increases to 21.5 mm × 10.4 mm. Assuming each router to be a perfect square, the length of each side of the *stem* router is about 200 μm, which is almost 6 times wider than the cross section of two opposite unidirectional 32-bit links. Thus, unlike mesh and BFT networks, MoT network connects up to 16 cores in a single row tree; the channel width will not increase the breadth of the chip.

Table 4.6 depicts the total area and area overheads due to underlying networks for 32-core- and 256-core-based systems. It may be noted that for a 32-core system, the 4 × 8 Mesh-1 network needs more area than the 4 × 4 MoT. In the same way, for a 256-core system where 16 cores are placed in each row, the areas of 16 × 16 Mesh-1 and 16 × 8 MoT networks are incremented from 40 mm × 40 mm to 44.42 mm × 40.512 mm and 43.272 mm × 41.6 mm, respectively. Table 4.6 shows that the 16 × 16 Mesh-1 network requires almost the same area as that of the 16 × 8 MoT network. In both the cases, Mesh-2 network requires the least area, whereas BFT needs the maximum area. Although there exists a possibility of trading-off this additional area for energy/performance benefits, in this book, we do not explore this avenue as it goes deep into the physical design issues of systems involving these NoC topologies.

TABLE 4.6

Area Overhead for 32- and 256-Core-Based NoCs

Networks	32 Cores		256 Cores	
	Overall Area (mm²)	Overhead (%)	Overall Area (mm²)	Overhead (%)
Mesh-1	224.64	12.32	1799.87	12.49
Mesh-2	215.12	7.56	1725.65	7.85
BFT	226.16	13.08	1890.82	18.17
MoT	223.6	11.8	1800.12	12.5

4.6.2 Network Aspect Ratio

Besides the channel width and flit size, the network aspect ratio also has an important role in determining the overall performance and cost of NoC. In general, for an $M \times \lceil C / M \rceil$ Mesh-1 network (M is the number of nodes in each row and C is the total number of cores attached in the network), Equations 4.9 and 4.10 give the average distance (D) and the number of directed edges (E) as mentioned in the work of Pavlidis and Friedman (2007):

$$D = \frac{M + \lceil C / M \rceil}{3} \tag{4.9}$$

$$E = 2 \times \left\{ M \times \left(\left\lceil \frac{C}{M} \right\rceil - 1 \right) + \left\lceil \frac{C}{M} \right\rceil \times (M - 1) \right\} \tag{4.10}$$

$$\frac{E}{D} = 2 \times \left(\frac{6C}{M + \lceil C / M \rceil} - 3 \right) \tag{4.11}$$

The ratio of E to D is shown in the above equation. The E/D ratio is a good indicator of network throughput (Decina et al. 1991). It can be shown that for the Mesh-1 network, the value of E/D reaches its maximum and the value of D reaches its minimum when the condition $M = \lceil C / M \rceil$ holds. This signifies that a square mesh network with an equal number of rows and columns is expected to show the best performance. The performance will degrade as the network becomes more and more rectangular in nature. Thus, to connect 2^n cores, where n is odd, a mesh network that connects single core to each router may not be the ideal choice to the NoC designers due to its rectangular shape. This statement is also true for a Mesh-2 network for 2^n cores, where n is even.

Table 4.7 shows different topological parameters such as diameter, average distance in hops (D), number of directed edges (E), number of destination cores in the local cluster, and bisection width of all the networks under consideration for connecting 32 IP cores. It may be observed that although Mesh-1 has the highest E/D ratio, its diameter is also the highest among all.

TABLE 4.7

Topological Parameters of Different Networks with 32 Cores

Networks	Number of Destination Core in Local Cluster	Number of Edges (E)	Average Distance (D)	Edges/ Avg. Dist.	Bisection Width	Diameter
Mesh-1	2, 3, 4	104	4.000	26.00	4	10
Mesh-2	1	48	2.645	18.15	4	6
BFT	3	40	2.840	14.08	4	4
MoT	1	96	5.161	18.60	4	8

Thus, it is difficult to comment on the superiority of any network based on the values noted in Table 4.7 alone. It is necessary to perform a detailed simulation to compare the topologies.

4.6.3 Performance Comparison

4.6.3.1 Accepted Traffic versus Offered Load

Figure 4.19 compares the accepted traffic of all the networks under consideration by applying uniformly distributed self-similar traffic. Although in a contention-free environment, the *E/D* ratio is a good indicator of network throughput, in actual traffic condition, the network dimension has a significant role to play. Due to the rectangular structure of a Mesh-1 network, packets are expected to traverse more hops in the horizontal direction under uniform distribution. Thus, the network suffers more contention. In a Mesh-2 network, due to its square structure for 32 cores, traffic movement is identical in both horizontal and vertical directions. But due to lesser number of directed edges (Table 4.7), it also experiences contention compared to the Mesh-1 network. In BFT, due to lesser number of edges, packets suffer more contention as they traverse toward the root. For a 4×4 MoT, as the number of row trees and column trees are same, packet traversal through the row trees and column trees is identical. Moreover, MoT networks have more number of edges and the connectivity of the routers is also the least (Table 4.1) among all the networks considered here. Thus, the packets are expected to experience lesser contention as they traverse toward the roots of the trees. In our simulation, it can be observed that the throughput of the MoT-based network is higher than other topologies considered here under uniformly distributed traffic, as shown in Figure 4.19.

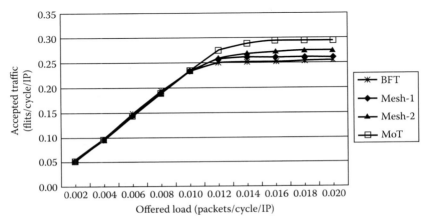

FIGURE 4.19
Accepted traffic with uniformly distributed offered load in different networks under consideration.

4.6.3.2 Throughput versus Locality Factor

The effect of spatial localization of traffic on the network throughput is shown in Figure 4.20. It can be observed that localization of traffic has significant impact on all the networks as it enhances the throughput. As the locality factor increases, more traffic is directed toward their local clusters, thus traversing lesser number of hops and increasing the throughput.

In BFT, localized traffic is constrained within a cluster consisting of a single subtree having four cores. It can be observed that the throughput of the BFT-based network is the least among all the networks under uniformly distributed traffic. For localized traffic, although its throughput increases with increasing locality factor, it always has the minimum value compared to other networks. This is due to the fact that the BFT-based networks are more congested, since there are three destination cores in local cluster and the routers have high connectivity as shown in Table 4.1.

In Mesh-1 network, localized traffic is constrained within the four destination IPs placed at the shortest Manhattan distance, whereas Mesh-2 network has a single core in its local cluster. Although Mesh-2 network has higher connectivity than Mesh-1 network, the former enjoys the advantage of having only a single core in its local cluster. Thus, under localized traffic, Mesh-2 network has higher throughput than Mesh-1 network. In case of highly localized traffic, the benefit of connecting two cores to each router is clearly reflected. At highly localized traffic, as there is a single destination core in the local cluster, more packets reach their destinations, resulting in higher throughput.

The MoT-based networks enjoy the benefits of both small node degree and single destination node in local cluster. As a result, throughput remains the highest among all the networks under localized traffic as well.

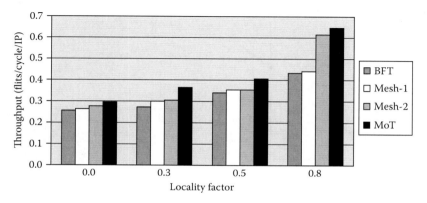

FIGURE 4.20
Variation of throughput of different networks under consideration with locality factor.

TABLE 4.8

Number of Links, FIFOs, and Zero-Load Latency of Network Topologies under Consideration with 32 Cores

Networks	Number of Links			Number of FIFOs	Zero-Load Latency (cycle)
	Few Micrometers	1.25 mm	2.5 mm		
Mesh-1	24	64	80	136	10
Mesh-2	8	64	64	80	8.45
BFT	0	64	96	80	10
MoT	32	64	80	128	12.32

4.6.3.3 Average Overall Latency under Localized Traffic

In a contention-free environment, zero-load latency (in cycles) is a widely used performance metric. Zero-load latency of a network is the latency where only one packet traverses through the network (Pavlidis and Friedman 2007). Table 4.8 shows the zero-load latency of all the networks in terms of cycle delay of a source router. According to the WH router architecture, each router has two-cycle latency (one cycle in each IB and SA unit), whereas single-cycle latency is taken for the *root* routers of MoT and BFT networks. The cycle latency of inter-router link traversal of all the networks is taken from Figures 4.1 through 4.4.

Under an actual traffic scenario, contention of packets being a major challenge, the latency of any network depends on both the offered load and the locality factor. Simulation has been carried out to estimate the average overall latency for all the networks with uniformly distributed and localized load as shown in Figures 4.21 through 4.24. It shows that at lower load, the latency variation is not significant. This is because at lower traffic, contention in the network is less. The contention increases as the offered load increases, which in turn increases the latency. The simulation results show that as the offered load increases toward the network saturation point, the latency increases exponentially. The packets take much longer time to reach their destinations. Therefore, it is always desirable to operate the network below its saturation point.

In determining the network contention, the network structure has an important role to play. It can be observed from Figure 4.21 that under uniformly distributed traffic, the latency profile of the BFT-based network is the worse among all the topologies. This happens as the BFT-based network has the least number of edges. Packets experience more contention as they traverse toward the root of the tree. However, in Mesh-1 network, due to its rectangular structure, more packets are traversing in the horizontal direction. Thus, the network suffers from more contention. Due to the square structure of Mesh-2 network, contention in this network is lesser than in Mesh-1

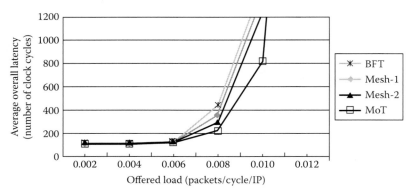

FIGURE 4.21
Latency variation with offered load (locality factor = 0.0).

network. Thus, packets take lesser time to reach their destinations compared to Mesh-1. Hence, this network has better latency profile under uniformly distributed traffic.

Although the MoT-based network has the maximum zero-load latency as shown in Table 4.8, under an actual traffic condition, the 4 × 4 MoT network experiences lesser contention than others. While comparing with Mesh-2 network with the same number of cores in the local cluster, it can be observed that the proposed MoT-based network has better latency profile under uniformly distributed traffic. A MoT network has more edges and lesser connectivity than a Mesh-2 network. Hence, it encounters lesser contention under uniformly distributed offered load.

Next, we have studied the effect of traffic spatial localization on the average overall latency of all the networks under consideration (shown in Figures 4.22 through 4.24). The average overall latency of all the networks decreases with increasing locality factor. As the locality factor increases, more traffic is destined for their local clusters. Hence, packets traverse lesser number of hops and cause lesser contention in the network. With increasing locality factor, simulation results show that the rate of decrease of latency in Mesh-2 and MoT networks is higher compared to others. Due to the presence of only a single core in their local clusters, more packets reach their destinations facing less contention, at medium (locality factor = 0.5) and highly (locality factor = 0.8) localized traffic. Thus, an actual benefit of having single destination core in the local cluster is observed in medium and highly localized traffic. Although Mesh-2 network has higher connectivity and suffers from more contention than MoT-based network under uniform distribution, it has been observed that the latency profile of Mesh-2 network comes closer to that of MoT-based network with increasing locality factor. This is due to the fact that the contention in both the networks becomes almost identical at highly localized traffic as both have single destination core in their local clusters.

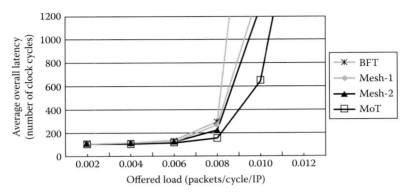

FIGURE 4.22
Latency variation with offered load (locality factor = 0.3).

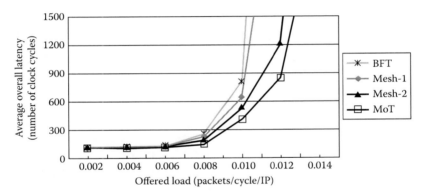

FIGURE 4.23
Latency variation with offered load (locality factor = 0.5).

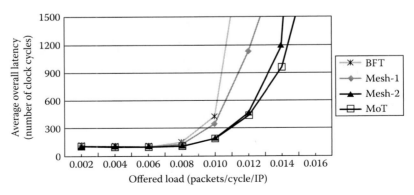

FIGURE 4.24
Latency variation with offered load (locality factor = 0.8).

4.6.4 Comparison of Energy Consumption

This section reports the comparison of energy consumption of all the networks at saturation. A network with more number of channels will definitely have higher network throughput but at the cost of more link energy consumption. Therefore, the average energy consumption per packet reception is a meaningful metric while comparing the energy consumption of various network structures. The number of FIFOs required to build a network is a topology-dependent parameter. Table 4.8 contains information about the number of links along with their lengths and the number of FIFOs required to implement the networks under consideration. It can be observed from Table 4.8 that for connecting 32 IP cores, 64 unidirectional local links of length 1.25 mm are required. There are also some very short links (of the order of few micrometers) that exist in the proposed MoT and both type of mesh networks. They consume very small energy. It has been observed that Mesh-1 and MoT-based networks require 80 links of length 2.5 mm, whereas for BFT and Mesh-2, the number of 2.5-mm links is 96 and 64, respectively.

Table 4.8 shows that the number of FIFOs required to implement the BFT and Mesh-2 networks connecting 32 cores is less compared to the other two networks under consideration. As the throughput of the BFT network is the least, Figure 4.25 shows that at uniformly distributed traffic, aggregation of average energy consumption per cycle of router and repeater is the least in BFT networks. Due to more number of FIFOs in Mesh-1 and MoT networks, they consume higher router energy compared to the other two. It has also been observed that links consume more energy than the routers for all the networks at saturation. With increasing locality factor, as more packets are reaching to their local clusters, energy consumption of the local links increases, whereas that of the inter-router links decreases. The toggling of data in those ports of a router connected to cores also increases, whereas the toggling in other ports

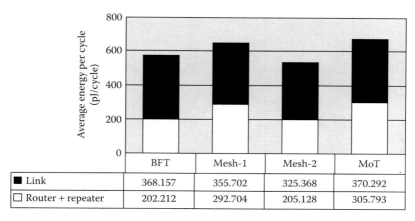

FIGURE 4.25
Average energy consumption per cycle at saturation under uniformly distributed traffic.

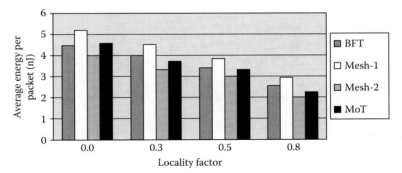

FIGURE 4.26
Average energy consumption per packet received at network saturation under different locality factors.

decreases at high locality factors. It has been observed that the overall energy consumption decreases with increasing locality factors for all the networks.

Due to the relatively lesser number of FIFOs and inter-router links, the energy consumption of Mesh-2 network turns out to be the least for all locality factors. The energy consumption of BFT and Mesh-2 networks are almost the same under uniform distribution, as shown in Figure 4.25. However, due to higher throughput, Mesh-2 network has lesser average energy consumption per packet reception at saturation, as shown in Figure 4.26. The throughput of the BFT network is significantly lesser than that of Mesh-2 and MoT networks at highly localized traffic, which causes higher average energy consumption per packet. As Mesh-1 network has higher number of FIFOs and inter-router links, and its throughput is lesser for any locality factor, simulation result shows that the average energy consumption per packet of the Mesh-1 network is inferior to others.

Although the MoT-based network requires more FIFOs and inter-router links than Mesh-2 networks, at high locality factor, the toggle of the data in *stem* and *root* routers of MoT networks is significantly less. At a high locality factor, throughputs of both the networks are almost the same. As a consequence, the average energy consumption per packet reception in the MoT network is slightly higher than that in Mesh-2 network, but lesser than other networks considered here.

4.7 Simulation Results and Analysis of MoT Network with Virtual Channel Router

The usage of WH router in NoC design leads to performance degradation due to contention in the network. Virtual channel (VC) router, however, mitigates the network contention problem, to some extent, by using multiple FIFOs at every incoming physical channel, at the cost of extra energy consumption and area overhead. The design of VC router is shown in Chapter 3. This section presents a comparison of performance and cost between the MoT networks with WH and VC routers for a 32-core-based system.

4.7.1 Throughput versus Offered Load

Figure 4.27 compares the throughput of VC and WH router-based MoT networks under uniformly distributed self-similar traffic. It can be observed that at lower injection load, the accepted traffic of VC router-based network increases linearly as also in WH router-based network, but saturates at a higher value. The simulation result depicts that almost 24% throughput improvement can be achieved by using VC router-based network over WH-based network under uniformly distributed self-similar traffic.

Figure 4.28 compares the throughput of both the networks under the localized traffic condition. It can be observed that the rate of increment of throughput decreases with increasing locality factor. This can be explained as the locality factor increases, more traffic is going to their local clusters having only a single core. Hence the network suffers less contention. At higher localization of traffic, throughputs of both the networks are almost identical.

4.7.2 Latency versus Offered Load

Figure 4.29 compares the average overall latency of MoT network under uniformly distributed self-similar traffic by using VC- and WH-based routers. At lower offered load, due to lesser contention in the network, the latency of both the cases is identical. At higher value of the offered load, it can be

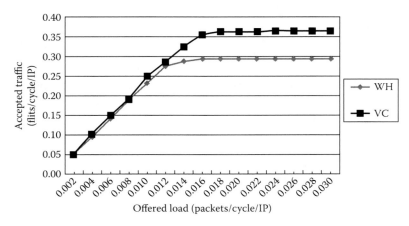

FIGURE 4.27
Comparison of accepted traffic in WH- and VC-based MoT networks under uniformly distributed offered load.

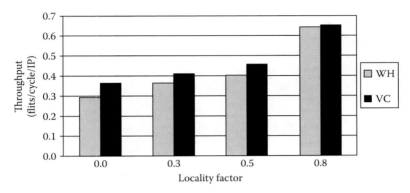

FIGURE 4.28
Throughput comparison of WH- and VC-based MoT networks under uniformly distributed and localized offered load.

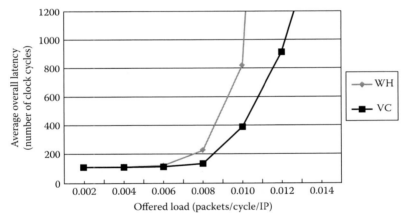

FIGURE 4.29
Latency comparison of WH- and VC-based MoT networks under uniformly distributed offered load.

observed that the MoT network with VC routers has lesser latency than that with WH router-based one.

Similar trend is also observed for localized traffic as shown in Figures 4.30 through 4.32. It can be noticed that for a particular offered load, due to lesser contention in VC-based network, the latency gap between WH- and VC-based networks is gradually decreasing with increasing locality factor.

4.7.3 Energy Consumption

The total energy consumption of MoT network after applying clock gating to all the VCs under uniformly distributed self-similar traffic is shown in Figure 4.33 (simulation performed for 200,000 clock cycles with 666-ps clock

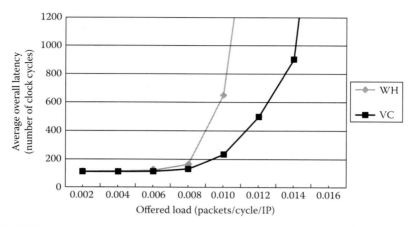

FIGURE 4.30
Latency comparison of WH- and VC-based MoT networks under localized offered load (locality factor = 0.3).

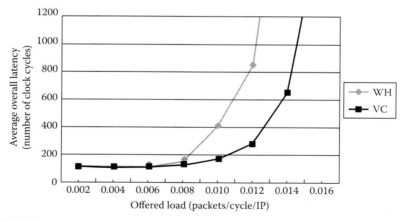

FIGURE 4.31
Latency comparison of WH- and VC-based MoT networks under localized offered load (locality factor = 0.5).

period). The component-wise energy consumption details are shown in Figure 4.34.

While comparing with the WH-based MoT network as shown in Figures 4.16 and 4.17, the energy consumption of VC router is significantly large. This is due to the fact that VC routers are having four FIFOs in each incoming physical channel and more complex round-robin arbiter. As VC-based network has higher throughput, its links consume more energy than the WH-based network, due to higher switching. Hence, VC-based network improves the performance at the cost of higher energy consumption. Figure 4.35 shows the comparison of energy per packet metric in WH and VC-based networks

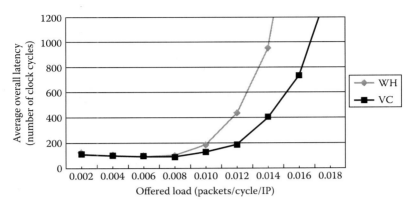

FIGURE 4.32
Latency comparison of WH- and VC-based MoT networks under localized offered load (locality factor = 0.8).

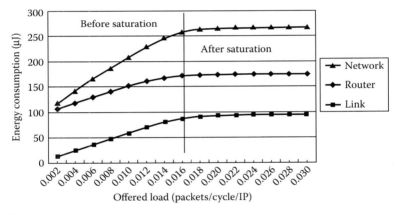

FIGURE 4.33
Network energy consumption in MoT network after clock gating in VC routers.

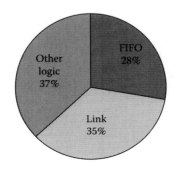

FIGURE 4.34
Component-wise energy consumption after clock gating in all the VC routers. Other logic includes routing logic, arbitration, and control logic.

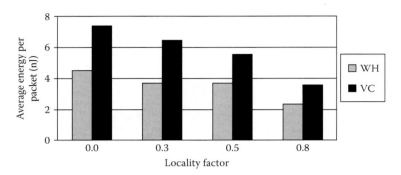

FIGURE 4.35
Comparison of average packet energy at saturation under different locality factors in both WH- and VC-based MoT networks.

TABLE 4.9

Area Required for Each Type of WH and VC Routers for Designing a MoT-Based NoC

	Type-1 Router		Type-2 Router		Type-3 Router	
Router	Position	Area (mm²)	Position	Area (mm²)	Position	Area (mm²)
WH	Leaf	0.057	Stem	0.041	Root	0.022
VC	Leaf	0.183	Stem	0.18	Root	0.08

TABLE 4.10

Area Overhead for 32- and 256-Core MoT-Based NoCs with WH and VC Routers

	32 Cores		256 Cores	
Router	Overall Area (mm²)	Overhead (%)	Overall Area (mm²)	Overhead (%)
WH	223.6	11.8	1800.12	12.5
VC	247.7	23.85	2006.57	25.4

under uniformly distributed and localized traffic. It can be observed that the energy per packet in a VC-based network is higher than in a WH-based network for all locality factors.

4.7.4 Area Required

The silicon area required for *leaf, stem,* and *root* routers of MoT network in both WH and VC schemes is shown in Table 4.9. As described in Section 4.6.1, only the *stem* routers of the column trees of each 2 × 2 MoT subnetwork contribute to increase the breadth of the chip, whereas the length of the chip gets increased due to the routers and repeaters of the row tree. Table 4.10 depicts the total area of the hand layouts and area overheads due to underlying networks for 32- and 256-core-based systems.

4.8 Performance and Cost Comparison of MoT with Other NoC Structures Having VC Router

4.8.1 Accepted Traffic versus Offered Load

Figure 4.36 compares the accepted traffic in all the networks with VC router for uniformly distributed self-similar traffic. It can be observed that the accepted traffic in all the VC router-based networks saturate at higher values compared to their WH counterparts. The relative ranking of the networks in terms of accepted traffic under uniformly distributed offered load is almost identical to that obtained in WH router-based networks. From Figure 4.36, it can be noticed that accepted traffic in 4 × 4 MoT network is more than in other networks taken here into consideration.

4.8.2 Throughput versus Locality Factor

The effect of traffic spatial localization on throughput in different VC router-based networks is shown in Figure 4.37. Like the WH router-based network, throughput of the VC router-based network also increases with locality factor. As VC can mitigate the traffic contention, throughput in all the networks is higher than in WH case at low locality factors. In higher localization of traffic, as the traffic traverses toward its local clusters, the contention is less.

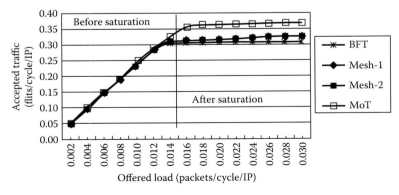

FIGURE 4.36
Accepted traffic with uniformly distributed offered load in VC router-based networks.

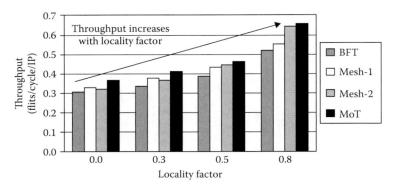

FIGURE 4.37
Throughput variation with locality factor in different VC router-based networks under consideration.

Hence, a marginal improvement in throughput is observed compared to its WH counterpart. From Figure 4.37, it can be noticed that the throughput in 4 × 4 MoT network is more than in other networks taken here into consideration under localized traffic condition.

4.8.3 Average Overall Latency under Localized Traffic

For a MoT network, Figures 4.29 through 4.32 show that introducing VC router in building the network improves the average overall latency compared to its WH counterpart under both uniformly distributed and localized traffic conditions. A similar trend has been observed for all other networks under consideration. The relative position of the VC router-based networks in this comparative study of average latency is found to be similar to that of WH case as shown in Figures 4.38 through 4.41.

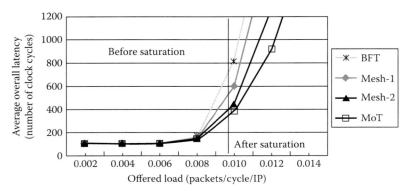

FIGURE 4.38
Latency variation in different VC router-based networks under consideration with uniformly distributed offered load.

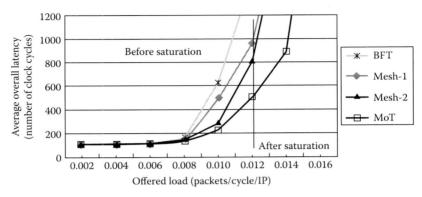

FIGURE 4.39
Latency variation in different VC router-based networks under consideration with offered load (locality factor = 0.3).

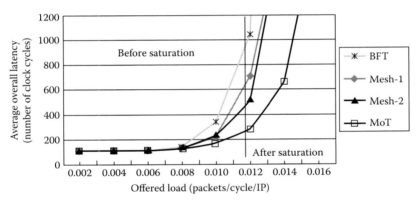

FIGURE 4.40
Latency variation in different VC router based networks under consideration with offered load (locality factor = 0.5).

The 4 × 4 MoT network shows the best latency profile, whereas the BFT network shows the worst under uniform distribution and localized traffic conditions. Due to the square network structure and having a single destination core in the local cluster, Mesh-2 network shows better latency profile than BFT and Mesh-1 networks. As both MoT and Mesh-2 networks have a single destination core in their local clusters, at highly localized traffic, the contention in both the networks becomes almost identical, and hence their latency profile is close to each other as shown in Figure 4.41.

4.8.4 Energy Consumption

Table 4.11 depicts the clock frequencies of different types of VC routers used in implementing the networks under consideration after applying clock

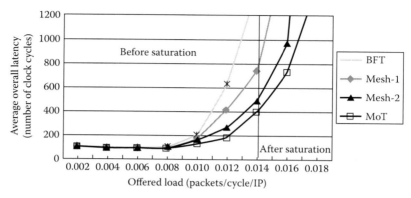

FIGURE 4.41
Latency variation in different VC router-based networks under consideration with offered load
(locality factor = 0.8).

TABLE 4.11

Frequency of Different Types of VC Routers after Clock Gating in FIFO to
Implement the Networks under Consideration for Connecting 32 Cores

	Type-1 Router		Type-2 Router		Type-3 Router	
Networks	Position	Frequency (GHz)	Position	Frequency (GHz)	Position	Frequency (GHz)
Mesh-1	Center	1.56	Edge	1.60	Corner	1.66
Mesh-2	Center	1.52	Edge	1.56	Corner	1.60
BFT	Leaf	1.52	Stem	1.52	Root	1.90
MoT	Leaf	1.66	Stem	1.70	Root	1.90

gating to the FIFO. Though MoT has the highest minimum frequency, in
this work, to support mesochronous clocking and to provide a consistent
comparison with other networks, all the routers are driven at 1.5-GHz clock.

Figure 4.42 shows the average energy consumption per cycle at saturation
under uniformly distributed traffic in VC router-based networks. It can be
observed that the router energy is higher than the link energy. Due to lesser
number of routers in BFT and Mesh-2 networks, the aggregation of router
and repeater energies of these two networks is lesser than that of Mesh-1
and MoT networks. Figure 4.43 shows the average energy consumption per
packet in all the networks under consideration with a VC-based router. Due
to higher energy consumption in Mesh-1 network, its average energy consumption per packet is the highest among all in any traffic condition. The
Mesh-2 network, due to its least energy consumption, shows the least energy
per packet. Although MoT consumes higher energy than BFT, due to higher
throughput of MoT networks, it shows almost similar average energy consumption per packet as BFT.

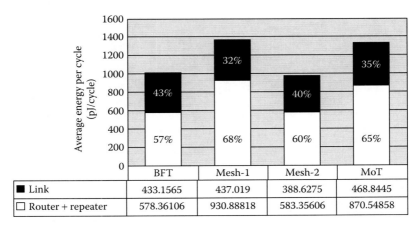

FIGURE 4.42
Average energy consumption per cycle at saturation under uniformly distributed traffic in different VC router-based networks under consideration.

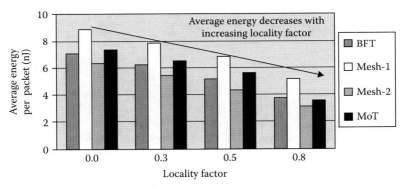

FIGURE 4.43
Average energy consumption per packet received at saturation under different locality factors in VC router-based networks.

4.8.5 Area Overhead

The area occupied by each type of router in the VC-based networks under consideration is shown in Table 4.12. For a fair estimation of network area, this section once again revisits all the networks having 32 cores as shown in Figures 4.1 through 4.4. By placing the routers in the similar way as in WH case, the dimension of the 4 × 8 Mesh-1 network increases from 20 mm × 10 mm to 24.434 mm × 10.128 mm, and that of the 4 × 4 Mesh-2 network becomes 22.392 mm × 10.128 mm. The dimensions of BFT and MoT networks become 23.872 mm × 10.128 mm and 22.839 mm × 10.848 mm, respectively. For a 256-core system, where 16 cores are placed in each row, the areas of 16 × 16 Mesh-1, 16 × 8 Mesh-2, BFT, and 16 × 8 MoT networks are incremented

TABLE 4.12

Silicon Area Required for Each Type of Router after Clock Gating

	Type-1 Router		Type-2 Router		Type-3 Router	
Networks	**Position**	**Area (mm²)**	**Position**	**Area (mm²)**	**Position**	**Area (mm²)**
Mesh-1	Center	0.331	Edge	0.242	Corner	0.220
Mesh-2	Center	0.400	Edge	0.318	Corner	0.263
BFT	Leaf	0.290	Stem	0.330	Root	0.080
MoT	Leaf	0.183	Stem	0.180	Root	0.080

TABLE 4.13

Area Overhead for 32- and 256-Core-Based NoCs

	32 Cores		256 Cores	
Networks	**Overall Area (mm²)**	**Overhead (%)**	**Overall Area (mm²)**	**Overhead (%)**
Mesh-1	247.47	23.74	1986.47	24.15
Mesh-2	226.79	13.40	1819.80	13.74
BFT	241.78	20.89	2018.02	26.13
MoT	247.7	23.85	2006.57	25.4

from 40 mm × 40 mm to 49.034 mm × 40.512 mm, 44.92 mm × 40.512 mm, 49.5 mm × 40.512 mm, and 46.243 mm × 43.392 mm, respectively. Table 4.13 shows the area required by each network having 32- and 256-core-based systems. It can be observed that the area overhead of Mesh-2 network is lesser than that of Mesh-1 network and the area overhead of MoT network is slightly more than that of Mesh-1 network.

4.9 Limitations of Tree-Based Topologies

Although simulation results show the performance and energy consumption benefits of MoT-based NoC over Mesh-1 network having 32 cores under the same bisection width constraint, the tree-based topologies may not be a good choice for NoC designers while attempting large number of cores. This is because the lengths of the edges of tree-based topologies increase with increasing network size, whereas that of the mesh structure does not vary. In general, the longest edge of MoT topology is the connection between *stem* and *root* routers and its length can be estimated as max(l_1, l_2)/4, where l_1 and l_2 are the length and the breadth of the hand layout, respectively. As the wire delay increases with its length, it is essential to pipeline the links after a certain length such that its delay does not fall into a critical path of the design.

TABLE 4.14

Number of Unidirectional Opposite Edges of Different Lengths for BFT and MoT Networks with Increasing Number of Cores

		BFT				MoT			
	Layout	Number of Edges of Different Length (mm)				Number of Edges of Different Length (mm)			
Number of Cores	Dimension (mm × mm)	2.5	5	10	20	2.5	5	10	20
16	10 × 10	–	16	–	–	24	–	–	–
32	20 × 10	–	48	–	–	48	16	–	–
64	20 × 20	–	64	32	–	96	48	–	–
128	40 × 20	–	128	96	–	192	96	32	–
256	40 × 40	–	256	128	64	384	192	96	–

Therefore, the cycle latency of the packets will increase while traversing through those links, which, in turn, will enhance the cycle latency of the overall system. Moreover, the feedback signals will come back after multiple clock cycles that can be overcome by deploying deeper FIFO. The above limitation is also true for other tree-based topologies such as fat tree, flattened butterfly, and BFT. For a BFT-based network, the length of the longest edge is $\max(l_1, l_2)/2$.

Table 4.14 depicts the number of unidirectional opposite edges of different lengths in BFT and MoT networks with increasing number of cores considering the tile size of 2.5 mm × 2.5 mm. It can be observed that BFT requires more number of long edges than MoT. Thus, packets in the BFT network experience more pipeline stages and hence more cycle latency compared to those in MoT. The above problem of having long edges in any tree-based topology can be addressed in the following two ways: current-mode signaling (Bashirullah et al. 2003) in NoC link and integrating the cores in three-dimesnional (3D) integrated circuit (IC) with multiple silicon layers in a stack (Feero and Pande 2009).

4.10 Summary

This chapter presents a thorough performance and cost evaluation of MoT topology by applying self-similar traffic and compares a well-known tree-based topology, BFT, and two variants of mesh topology connecting single/two cores to each router under same bisection width constraint. It is customary to say that if this constraint does not hold, the network with more bisection width is expected to perform better under uniformly distributed and low localized traffic. The simulation results place MoT to be the best in terms of performance for both WH and VC router-based NoCs. This evaluation corresponds to self-similar traffic with varying locality factors. In terms of per packet energy consumption and area overhead, the WH router-based MoT network ranks second in the list, next to the mesh structure that connects two cores to each router. With VC router, MoT consumes lesser average packet energy than the mesh network that connects single core to each router and occupies almost similar area such as mesh network connecting single core to each router. The comparative study shows that MoT-based NoC also works fine under application-specific traffic. On the architecture front of WH and VC routers, due to lesser connectivity of MoT routers, synthesis result shows that they can be operated at a higher frequency compared to other network structures, thus increasing the speed of the overall network. However, for a system having large number of cores, it can be predicted that MoTs will suffer in both energy and latency fronts mainly due to the large number of pipelining stages required for the longer edges. In BFT networks, this problem is much more severe as they require more number of long edges as shown in Table 4.14. The upcoming trends such as current mode signaling in NoC link and 3D NoC are expected to alleviate this bottleneck, making MoT a more acceptable topology for larger core-based NoC design.

Although in investigating the promise of any topology in a NoC paradigm, applying self-similar traffic is expected to produce the average behavior of the network, Chapter 5 will focus on a different application mapping algorithm in NoC paradigm and also evaluate the performance and cost of each network under consideration under a set of real benchmark applications.

References

Bashirullah, R., Liu, W., and Cavin, R. K. 2003. Current-mode signaling in deep submicrometer global interconnects. *IEEE Transactions on Very Large Scale Integration (VLSI) Systems*, vol. 11, no. 3, pp. 406–417.

Bertozzi, D. and Benini, L. 2004. Xpipes: A network-on-chip architecture for gigascale systems-on-chip. *IEEE Circuits and Systems Magazine*, vol. 4, no. 2, pp. 18–31.

Chi, H. C. and Chen, J. H. 2004. Design and implementation of a routing switch for on-chip interconnection networks. *Proceedings of Asia-Pacific Conference on Advanced System Integrated Circuits*, pp. 392–395, August 4–5, Fukuoka, Japan.

Decina, M., Trecordi, V., and Zanolini, G. 1991. Throughput and packet loss in deflection routing multichannel-metropolitan area networks. *IEEE GLOBECOM*, pp. 1200–1208, December 2–5, Phoenix, AZ.

Duato, J., Yalamanchili, S., and Ni, L. 2003. *Interconnection Networks: An Engineering Approach*. Morgan Kaufmann Publishers, San Francisco, CA.

Feero, B. S. and Pande, P. P. 2009. Networks-on-chip in a three dimensional environment: A performance evaluation. *IEEE Transactions on Computers*, vol. 58, no. 1, pp. 32–45.

Hu, J. and Marculescu, R. 2004. DyAD-smart routing for networks-on-chip. *Proceedings of Design and Automation Conference*, pp. 260–263, July 7–11, San Diego, CA.

Kumar, A., Kundu, P., Singh, A. P., Peh, L. S., and Jha, N. K. 2007. A 4.6Tbits/s 3.6GHz single-cycle NoC router with a novel switch allocator in 65nm CMOS. *Proceedings of IEEE International Conference on Computer Design*, pp. 63–70, October 7–10, Lake Tahoe, CA.

Kundu, S., Soumya, J., and Chattopadhyay, S. 2012. Design and evaluation of mesh-of-tree based network-on-chip using virtual channel router. *Microprocessors and Microsystems Journal*, vol. 36, pp. 471–488.

Marculescu, R., Ogras, U. Y., Peh, L. S., Jerger, N. E., and Hoskote, Y. 2009. Outstanding research problems in NoC design: Systems, microarchitecture, and circuit perspectives. *IEEE Transactions on Computer-Aided Design of Integrated Circuits and Systems*, vol. 28, no. 1, pp. 3–21.

Millberg, M., Nilsson, E., Thid, R., and Jantsch, A. 2004. Guaranteed bandwidth using looped containers in temporally disjoint networks within the Nostrum network on chip. *Proceedings of IEEE Design, Automation, and Test in Europe*, pp. 890–895, February 16–20, Paris, France.

Ogras, U. Y., Hu, J., and Marculescu, R. 2005. Key research problems in NoC design: A holistic perspective. *Proceedings of IEEE/ACM/IFIP International Conference on Hardware/Software Codesign and System Synthesis*, pp. 69–74, September, Jersey City, NJ.

Pande, P. P., Grecu, C., Ivanov, A., and Saleh, R. 2003. High-throughput switch-based interconnect for future SoCs. *Proceedings of IEEE International Workshop on System-on-Chip for Real Time Applications*, pp. 304–310, June 30-July 2, Alberta, Canada.

Pande, P. P., Grecu, C., Jones, M., Ivanov, A., and Saleh, R. 2005. Performance evaluation and design trade-offs for MP-SOC interconnect architectures. *IEEE Transactions on Computers*, vol. 54, no. 8, pp. 1025–1040, August.

Park, K. and Willinger, W. 2000. *Self-Similar Network Traffic and Performance Evaluation*. Wiley, New York.

Pavlidis, V. F. and Friedman, E. G. 2007. 3-D Topologies for networks-on-chip. *IEEE Transactions on VLSI Systems*, vol. 15, no. 10, pp. 1081–1090.

Rijpkema, E., Goossens, K. G. W., and Radulescu, A. 2003. Trade offs in the design of a router with both guaranteed and best-effort services for network on chip (extended version). *IEE Proceedings of Computers and Digital Techniques*, vol. 150, no. 5, pp. 294–302.

Salminen, E., Kulmala, A., and Hamalainen, T. D. 2008. Survey of network-on-chip proposals. *White Paper*, © OCP-IP.

Soteriou, V., Eisley, N., Wang, H., Li, B., and Peh, L. S. 2007. Polaris: A system-level roadmap for on-chip interconnection networks. *IEEE Transactions on Very Large Scale Integration (VLSI) Systems*, vol. 15, no. 8, pp. 855–868.

Sotiriadis, P. P. and Chandrakasan, A. P. 2002. A bus energy model for deep submicron technology. *IEEE Transactions on Very Large Scale Integration (VLSI) Systems*, vol. 10, no. 3, pp. 341–350.

Synopsys. 2006. *Synopsys Prime Power Manual*. Version Y-2006.06. http://www.synopsys .com/Support/LI/Installation/Documents/Archive/iugux_y2006_06.pdf.

Vangal, S. R., Howard, J., Ruhl, G., Dighe, S., Wilson, H., Tschanz, J., Finan, D. et al. 2008. An 80-tile sub-100-W TeraFLOPS processor in 65-nm CMOS. *IEEE Journal of Solid-State Circuits*, vol. 43, no. 1, pp. 29–41.

Wentzlaff, D., Griffin, P., Hoffmann, H., Bao, L., Edwards, B., Ramey, C., Mattina, M., Miao, C. C., Brown, J. F., and Agarwal, A. 2007. On-chip interconnection architecture of the TILE processor. *IEEE Micro*, vol. 27, no. 5, pp. 15–31.

Zeferino, C. A. and Susin, A. A. 2003. SoCIN: A parametric and scalable network-on-chip. *Proceedings of IEEE Symposium on Integrated Circuits and Systems Design*, pp. 169–175, September 8–11, São Paulo, Brazil.

Zhang, Y. P., Jeong, T., Chen, F., and Wu, H. 2006. A study of the on-chip interconnection network for the IBM Cyclops64 multi-core architecture. *Proceedings of IEEE International Parallel and Distributed Symposium*, April 25–29, Rhodes Island.

5

Application Mapping on Network-on-Chip

5.1 Introduction

In chapters 2, 3, and 4, we have seen how to design a router fabric following a specific topology (such as mesh, mesh-of-tree, and butterfly fat tree). However, a complete network-on-chip (NoC) system consists of the router fabric and the intellectual property (IP) cores. The router network is utilized to facilitate communication between these cores. Looking at the NoC design problem from a system perspective, the whole process starts with the specification of the system. The specification is refined into a set of interactive tasks that accomplish the system functionality. For example, Figure 5.1 shows a task-level decomposition of a *video object plane decoder* (*VOPD*) application. It consists of several tasks, such as *down sampler* and *run length decoder*. The tasks need to communicate between themselves at the rates specified as edge labels. A label represents the required bandwidth between the tasks in megabytes per second. The corresponding graph-based representation is shown in Figure 5.2a. The router-based communication backbone is utilized for these communications. Assuming that each task is realized by an IP core, it is desirable that the cores corresponding to two highly communicating tasks be placed close to each other. If only one core is attached to a router, such cores should be attached to two neighboring routers, whereas cores having little bandwidth requirement between them can be placed relatively far apart, without affecting the system performance significantly. Once the communication infrastructure of a NoC is finalized, a major challenge in the overall system design is to associate the IP cores implementing tasks of an application with the routers. This problem of *application mapping* has a very significant role to play in determining the performance of the overall system, as it directly influences the communication time, the required link bandwidth, and the admissible delay in the routers. For example, Figure 5.2b shows a possible mapping of the application shown in Figure 5.2a onto a mesh topology. The mapping problem is NP-hard (Pop and Kumar 2004). In this chapter, we will look

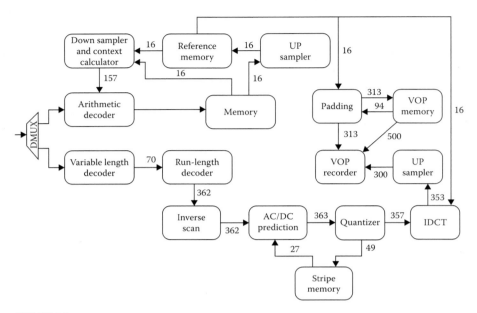

FIGURE 5.1
Block diagram of a VOPD, IDCT, Inverse Discrete Cosine Transform; DMUX, Demultiplexer.

into the various strategies adopted to solve the problem. To start with, we will look into the mathematical definition of the problem.

5.2 Mapping Problem

Given an application consisting of a set of communicating tasks, the first step toward NoC realization is to identify the cores to carry out the tasks. After the cores participating in an application have been decided, the application can be represented in the form of a *core graph*, which is defined as follows:

Definition 5.1

The *core graph* for an application is a directed graph, $G(C, E)$ with each vertex $c_i \in C$ representing a core and the directed edge $e_{i,j} \in E$ representing the communication between the cores c_i and c_j. The weight of the edge $e_{i,j}$, denoted by $comm_{i,j}$, represents the bandwidth requirement of the communication from c_i to c_j.

However, the given NoC topology can be represented in the form of a *topology graph*.

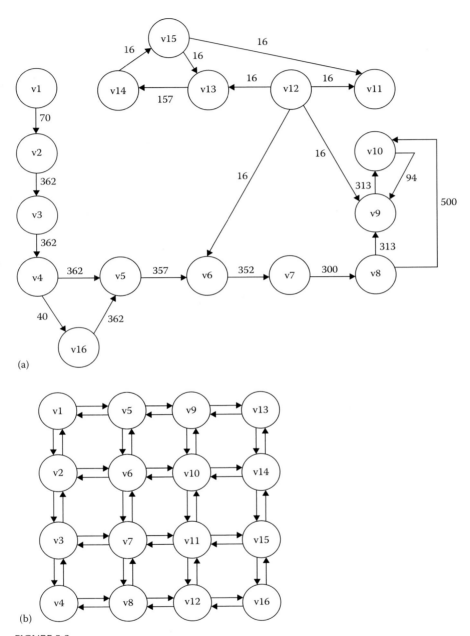

FIGURE 5.2
(a) Application graph for VOPD; (b) a possible mapping onto mesh topology.

Definition 5.2

The NoC *topology graph* is a directed graph $P(U, F)$, with each vertex $u_i \in U$ representing a node in the topology and the directed edge $f_{i,j} \in F$ representing a direct communication between the vertices u_i and u_j. The weight of the edge $f_{i,j}$, denoted as $bw_{i,j}$, represents the bandwidth available across the edge $f_{i,j}$.

A mapping of the core graph $G(C, E)$ onto the topology graph $P(U, F)$ is defined by the function, *map*: $C \rightarrow U$, such that $c_i \in C$, $u_j \in U$, and $map(c_i) = u_j$. The function associates core c_i to router u_j. Mapping is defined only when $|C| \leq |U|$, assuming that at most a single core is associated with a router. The quality of such a mapping is defined in terms of the total *communication cost* of the application under this mapping. The communication between each pair of cores can be treated as flow of a single commodity, d^k, where $k = 1, 2, \ldots, |E|$. The value of commodity d^k, corresponding to the communication between cores c_i and c_j, is equal to $comm_{i,j}$, the bandwidth requirement. If c_i is mapped to the router $map(c_i)$ and c_j is mapped to $map(c_j)$, the set of all commodities $D = \{d^k\}$ is defined as follows:

$$D = \{d^k \mid value(d^k) = comm_{i,j}, \text{ for } k = 1, 2, \ldots, |E| \text{ and } e_{i,j} \in E\}$$

$$source(d^k) = map(c_i) \text{ and } dest(d^k) = map(c_j)$$

The link between two individual routers u_i and u_j of the topology has a maximum bandwidth of $bw_{i,j}$. The total commodity flowing through such a link should not exceed this bandwidth. The quantity $x_{i,j}^k$ indicating the value of commodity d^k flowing through the link (u_i, u_j) is given by

$$x_{i,j}^k = \begin{cases} value(d^k), & \text{if } link(u_i, u_j) \in Path(source(d^k), dest(d^k)) \\ 0, & \text{otherwise} \end{cases}$$

where $Path(a, b)$ indicates the deterministic routing path between the mesh nodes a and b in the topology.

Satisfaction of bandwidth limitations of individual links must be ensured. That is, all mapping solutions should satisfy the following relation:

$$\sum_{k=1}^{|E|} x_{i,j}^k \leq bw_{i,j}, \quad \forall i, j \in \{1, 2, \ldots |U|\}$$

If all bandwidth constraints are satisfied, the communication cost of a mapping solution is given by

$$T = \sum_{k=1}^{|E|} value(d^k) \times hopcount(source(d^k), dest(d^k))$$

where *hopcount*(*a, b*) is the number of hops between the topology nodes *a* and *b*, assuming that all routers take the same number of clock cycles to pass a packet from an input port to an output port.

Otherwise, the hopcount metric needs to be replaced by the number of router cycles involved. For a deterministic shortest path routing, *hopcount* corresponds to the minimum number of hops between the constituent nodes. Since the communication cost is very much dependent on the mapping solution, the overall mapping problem is to optimize the communication cost, ensuring that the bandwidth constraints of all individual links are satisfied. The communication cost affects the performance of the overall system and its energy consumption, as both of these factors are directly proportional to the total *hopcount*. The application mapping problem is to determine the *map* function of an application to be mapped onto a given topology graph such that the overall communication cost *T* is minimized.

Several strategies have been proposed to solve the mapping problem. The techniques may broadly be classified belonging to one or more of the following categories:

- Exact mapping strategies, such as integer linear programming (ILP)
- Constructive heuristics with/without iterative improvement
- Evolutionary techniques, such as genetic algorithms (GAs), ant colony optimization (ACO), and particle swarm optimization (PSO)

Some proposed techniques from each of the categories will be discussed in Sections 5.3 through 5.6.

5.3 ILP Formulation

ILP is an exact technique to solve optimization problems. It attempts to assign values to the unknowns (and thus constructs a solution) satisfying a set of constraints and optimizing the given objective function. However, as the problem size grows, it takes exponential time to arrive at the optimum result. Due to its exact nature, ILP is often used to judge the quality of other nonexact approaches (such as heuristics and meta-search techniques). Often approximations are introduced into the ILP formulation to arrive at fast heuristic techniques. In the following, we will discuss about an ILP formulation of the application mapping problem. We have used the following variables to express the constraints and the objective function.

- U: The set of routers $\{u_1, u_2, \ldots\}$ of the topology graph
- C: The set of cores $\{c_1, c_2, \ldots\}$ of the application

- $m_{c_i}^{u_s}$: The mapping result taking values $\{0,1\}$. The variable is set to 1 if core c_i is mapped onto router u_s.
- $P_{c_i c_j}^{u_s u_t}$: The communication path result $\in \{0,1\}$. The variable is set to 1 if the communication path exists between routers u_s and u_t to which the cores c_i and c_j have been mapped.
- $BW_{c_i c_j}$: The bandwidth requirement between the cores c_i and c_j
- $MD_{u_s u_t}$: The Manhattan distance between the routers u_s and u_t

With this, we proceed to define the constraints for mapping.

1. *One-to-one mapping*: Each core be mapped to a router and each router may have at most one core attached to it.

$$\forall u_s \in U, \quad \sum_{c_i \in C} m_{c_i}^{u_s} \leq 1 \tag{5.1}$$

$$\forall c_i \in C, \quad \sum_{u_s \in U} m_{c_i}^{u_s} = 1 \tag{5.2}$$

Equation 5.1 implies that any router has at most one core mapped onto it. Equation 5.2 means each core has to be mapped onto only one router.

2. *Communication path*: For any two communicating cores c_i and c_j, a communication path is needed between the routers to which they are mapped. That is, for E being the set of edges of the application graph,

$$\forall (c_i, c_j) \in E, \quad P_{c_i c_j}^{u_s u_t} = \begin{cases} 1, & \text{if}(m_{c_i}^{u_s}{=}1) \text{ and } (m_{c_j}^{u_t}{=}1) \\ 0, & \text{otherwise} \end{cases} \tag{5.3}$$

It can be rewritten as

$$m_{c_i}^{u_s} + m_{c_j}^{u_t} - 1 \leq P_{c_i c_j}^{u_s u_t} \leq \frac{m_{c_i}^{u_s} + m_{c_j}^{u_t}}{2} \tag{5.4}$$

$$0 \leq P_{c_i c_j}^{u_s u_t} \leq 1 \tag{5.5}$$

If it is assumed that the links of the topology graph do not impose any constraint on the amount of traffic they can carry, the objective function is given by

$$\min \left\{ \sum_{(c_i c_j) \in E} \left[BW_{c_i c_j} \times \left(\sum_{(u_s u_t) \in U} MD_{u_s u_t} \times P_{c_i c_j}^{u_s u_t} \right) \right] \right\} \tag{5.6}$$

This equation computes the total communication cost for all the edges of the application graph. It may be noted that for regular topologies such as mesh and tree, computing the Manhattan distance is easy. However, for irregular topologies, computing the distance may not be that trivial. If cores c_i and c_j are mapped to routers u_s and u_t, respectively, $P_{c_i c_j}^{u_s u_t}$ is equal to 1. In such case, the Manhattan distance between the routers u_s and u_t is multiplied by the bandwidth requirement of the communication between the cores. Summing this product over all the edges of the application graph gives the total communication cost of the mapping solution produced.

However, if we need to consider the link bandwidth limitation as well, two issues are to be resolved. First, for each link the total communication scheduled through it should not exceed the limit. Second, the distance between the cores is not simply the Manhattan distance, but it depends on the route through which the communication takes place. To incorporate these considerations into the ILP formulation, we need to introduce a few more variables.

- $L(u_s, u_t)$: It is the set of links forming the path from u_s to u_t.
- $N(u_s, u_t)$: It is the set of nodes in the path from u_s to u_t.
- $l_{u_i u_j}^{u_s u_t}$: It marks whether the link from u_i to u_j forms a part of the path from u_s to u_t. It is equal to 1 if (u_i, u_j) is a part of the path, otherwise 0.
- $n_i^{u_s u_t}$: It marks if n_i is a node on the path from u_s to u_t, otherwise 0.
- Lc: It is the link capacity expressed as Mbits/s.
- $D_{u_s u_t}$: It is the distance between the routers u_s and u_t measured in terms of the number of routers in the path.

While considering the bandwidth constraints for individual links, the shortest path between two router nodes is not fixed because of the limiting link capacity. Hence, it is necessary to determine the path that satisfies the link capacities for individual links between the routers. For such a path, we can compute the distance (number of hops or routers) between the source and destination routers.

1. *Link capacity constraint*: It limits the bandwidth of individual links, so that the traffic in each link remains within the maximum given link capacity (Lc).

$$\forall (u_i, u_j) \in U \left[\sum_{(c_i, c_j) \in E} BW_{c_i c_j} \times \left(\sum_{u_s, u_t \in U} l_{u_i u_j}^{u_s u_t} \times P_{c_i c_j}^{u_s u_t} \right) \leq Lc \right] \quad (5.7)$$

In the above equation, if $l_{u_i u_j}^{u_s u_t}$ is 1, it implies that the link (u_i, u_j) is a part of the path from router u_s to u_t. The link can be a part of other paths also, between other pairs of routers. Hence, we need to add all those bandwidths and ensure that the total bandwidth requirement is less than the link capacity (Lc). Otherwise, some or all paths should be modified till all the link capacity constraints are satisfied.

2. *Source and destination constraints*: The source and destination routers must belong to the path between themselves.

$$\forall (c_i, c_j) \in E, \forall (u_s, u_t) \in U \left(P_{c_i c_j}^{u_s u_t} - n_s^{u_s u_t} = 0 \right) \tag{5.8}$$

3. *Starting link constraint*: Once the source node is fixed, in the topology graph it will have a number of neighbors. Out of these, only one neighbor is to be chosen for continuation of the path, which is specified in Equation 5.9. It will make the variable corresponding to only one link as 1 and the rest will be 0.

$$\forall (c_i, c_j) \in E, \forall (u_s, u_t) \in U \left[P_{c_i c_j}^{u_s u_t} - \sum_{u_k \in \text{neighbor}(u_s)} l_{u_s u_k}^{u_s u_t} = 0 \right] \tag{5.9}$$

4. *Intermediary node constraint*: If a link is a part of the path, the source and destination nodes of the link must also be included in the path. This is captured in the following equation:

$$\forall (u_s, u_t) \in U, \forall (u_i, u_j) \in U \left(2 \times l_{u_i u_j}^{u_s u_t} - n_i^{u_s u_t} - n_j^{u_s u_t} \leq 0 \right) \tag{5.10}$$

Excepting the start and end nodes of a path, each other node must have an incoming edge to it and an outgoing edge from it in the path.

$$\forall (u_s, u_t) \in U, \forall u_i \in U, u_i \neq u_s, u_i \neq u_t \left[2 \times n_i^{u_s u_t} - \sum_{u_j \in \text{neighbor}(u_i)} l_{u_i u_j}^{u_s u_t} = 0 \right] \tag{5.11}$$

This will form the shortest path under the link capacity constraint.

5. *Shortest path*: To compute $D_{u_s u_t}$, the distance between the routers u_s and u_t, we need to count the number of routers in the path from u_s to u_t, such that after mapping all link capacities are satisfied, which is specified in Equation 5.12. $D_{u_s u_t}$ can take integer values in the range from 0 to the total number of routers.

$$\forall (u_s, u_t) \in U \left[D_{u_s u_t} - \sum_{(u_i, u_j) \in U} l_{u_i u_j}^{u_s u_t} = 0 \right] \tag{5.12}$$

Once $D_{u_s u_t}$ is determined, the total communication cost can be determined and the objective function to be optimized can be constructed as follows:

$$\min \left[\sum_{(c_i, c_j) \in E} BW_{c_i, c_j} \left(\sum_{(u_s, u_t) \in U} D_{u_s u_t} \times P_{c_i c_j}^{u_s u_t} \right) \right]$$

Solving this optimization problem can produce a solution to the mapping problem.

5.3.1 Other ILP Formulations

A mixed ILP (MILP)-based task mapping for heterogeneous multiprocessor systems is reported in the work of Bender (1996). In this heterogeneous multiprocessor, some processors are programmable, whereas others are application specific. The model determines the optimization trade-off between the execution time, the processor (general-purpose or application-specific processor), and the communication cost. This is a hardware/software codesign process that runs iteratively until the design goal is met. An MILP formulation for mapping cores onto NoC while considering the choice of core placements, switches for each core, and network interfaces (NIs) for communication has been proposed by Rhee et al. (2004). It is reported that the energy consumption is much less compared to other mapping techniques for some real, as well as, random benchmarks. An integrated approach for mapping of cores onto heterogeneous processor/memory-based NoC topologies and physical planning has been presented by Murali et al. (2005), where the position and size of the cores and network components are computed. For initial mapping, they followed a greedy mapping of cores onto the specified topology, and then in the improvement phase, the relative core positions are fixed by *Tabu search*. An MILP-based physical planning algorithm has been formulated to improve the area and power of the final design and also to guarantee the quality of service (QoS) for the application. Srinivasan et al. (2006) presented an MILP formulation for synthesis of custom NoC architectures. Here the optimization objective is to minimize the power consumption, subject to the performance constraints. In case of linear programming (LP), the main bottleneck is runtime. To reduce runtime, they partitioned the application task graph into a number of clusters. The MILP formulation for topology design is then utilized and partial solutions are generated. At the end, the final mapped custom topology is generated by adding physical links between the ports of neighboring routers of the clusters.

The *network processors* incorporate features such as symmetric multiprocessing (SMP), block multithreading, and multiple memory elements to support high-performance networking applications. Mapping an application onto a complex multiprocessor, multithreaded network processor is a difficult task. Ostler and Chatha (2007) presented a two-stage ILP formulation for process allocation and data mapping on SMP and block multithreading-based network processor. They compared the normalized results from their models, such as without optimizations, multithreading-aware data mapping, process transformation, and multithreading-aware data mapping with process transformation. Power/energy control is a very important issue in case of NoC-based chip multiprocessors (CMPs). Ozturk et al. (2007) attempted to minimize the energy by shutting down certain communication links in such architectures. This formulation can be used for selecting the links in use, their voltage and frequency values. The problem of minimizing energy consumption during application execution while satisfying the performance constraint may be a combination of some subproblems, such as mapping of application tasks to IPs, mapping of IPs to the routers of NoC architecture, assignment of operating voltages to IPs, and routing. Different operating voltages are assigned to IPs if they are operating at multiple voltages. A unified approach of energy-efficient application mapping that utilizes MILP formulation of the problem has been presented by Ghosh et al. (2009), taking care of all the subproblems, such as application mapping, operating voltage assignment, and routing. In the work of Huang et al. (2011), the existing ILP (Ghosh et al. 2009) is extended to find a trade-off between computation and communication energy. In the work of Chou et al. (2008), factors that produce network contention are analyzed. An ILP formulation for contention-aware application mapping algorithm in a tile-based NoC is proposed to minimize inter-tile network contention. In NoC-based design, the global wires are replaced by a network of shared links and the routers exchange data packets simultaneously through the links. Therefore, there is traffic congestion within the links, which significantly degrades the system performance. The network contention may be source based, destination based, and path based. The result shows that there is a significant reduction of packet latency by reducing the network contention, but the loss of communication energy is high. Tosun et al. (2009) presented an ILP formulation for application mapping onto a mesh-based NoC to minimize energy consumption for different benchmarks. However, the formulation does not include bandwidth constraints. The CPU time for different benchmarks reported in this work is also quite high. To overcome the high CPU time, a clustering-based relaxation for ILP formulation has been proposed by Tosun (2011a). The tasks of the application graph are clustered suitably, as in Srinivasan et al. (2006). Based on the number of clusters, the mesh architecture is divided into smaller sized meshes. The ILP-based formulation of Tosun et al. (2009) is used to map the clusters onto the corresponding sub-meshes. At the end, it merges all such sub-meshes to determine the final solution. It is noted that the CPU time gets improved with a sacrifice in the communication cost of the mapping solution.

5.4 Constructive Heuristics for Application Mapping

In constructive heuristics, partial solutions are generated sequentially, and at the end the final mapping solution is obtained. Some of the techniques perform an additional iterative improvement phase after getting the initial solution. In this section, we look into one of the constructive techniques, known as *binomial mapping (BMAP)* algorithm (Shen et al. 2007). It works in the context of mesh topology. The on-chip network (OCN) design flow in BMAP is shown in Figure 5.3. Given a system-on-chip (SoC) application to be implemented, first the designer chooses the NI to be used by the cores. It is assumed that the target NoC will utilize the wormhole architecture for the routers, and thus, every packet consists of a header flit, a number of body flits, and a tail flit. From the given SoC application, a traffic model is extracted. The OCN synthesis process in BMAP can be divided into the following three stages:

- *Mapping*: It performs a mapping of the application task graph onto the topology graph.
- *Optimization*: The mapped network is optimized at this stage.
- *Simulation*: A cycle-accurate simulator can be used to evaluate the performance of the synthesized NoC.

Out of these three stages, the mapping and the optimization phases constitute the BMAP algorithm.

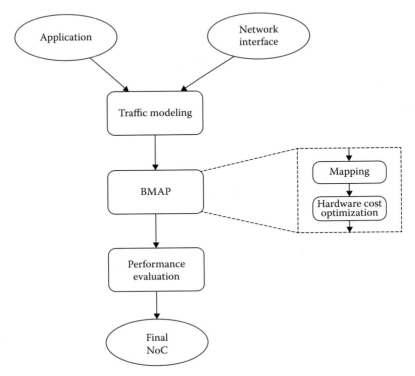

FIGURE 5.3
OCN design flow in BMAP.

A flowchart depicting the mapping and optimization stages of the BMAP algorithm is shown in Figure 5.4. It mainly consists of three major operations: binomial merging iteration, topology mapping and traffic surface creation, and hardware cost optimization.

5.4.1 Binomial Merging Iteration

This iterative step finds the relative positions of the IP cores in the topology. For this, the IP cores are put into sets, called *IP sets*. Initially, individual IPs form single-member sets. Now, depending upon the traffic between the IPs, the grouping process starts by forming sets with two IPs. The process continues to iterate merging smaller sized sets into larger sized ones. Each iteration consists of three steps: *calculate IP ranking, merging IP sets,* and *refreshing IP sets.*

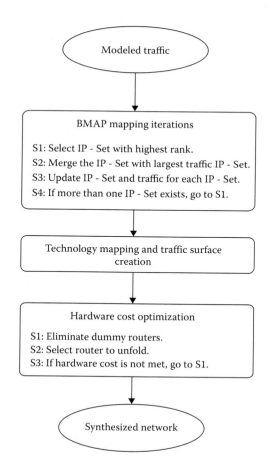

FIGURE 5.4
BMAP flowchart.

1. *Calculate IP ranking*: Ranking of an individual IP core can be calculated by summing up the traffic from the core to other IPs and from other IPs to the core. For IP core i, rank is computed as follows:

$$\text{ranking}(i) = \sum_{j=1}^{N} \text{communication}(i,j) + \text{communication}(j,i)$$

In the above equation, communication(i,j) indicates the bandwidth requirement for communication from IP core i to IP core j. As noted earlier, these communication requirements are extracted from the traffic load of the IPs obtained via simulation of the original SoC.

2. *Merging IP sets*: Based upon the IP ranking, the IP sets are merged. To start with, each IP set consists of a single IP. Thus, in the first step of iteration, two-element sets are formed. Since the IPs are ranked, IP sets corresponding to two highest ranked IPs are merged to form a two-core IP set. Next two IP sets are merged in a similar fashion creating two-core sets. Thus, at the end of the first iteration, each IP set contains two cores. In general, at the end of ith iteration, each IP set consists of 2^{i-1} cores, which is shown in Figure 5.5. At any stage, an IP set corresponds to a sub-mesh. All such sub-meshes are of equal dimension, say $d_1 \times d_2$. If $d_1 = d_2$, there are 16 possible ways to merge two such sub-meshes (as shown in Figure 5.6a). However, if $d_1 \neq d_2$, there are four possible merges. It may be noted that in Figure 5.6b, the edges A1, A2, D1, and D2 are not aligned as this affects the aspect ratio of the resultant mesh. A square-shaped mesh has less average distance than a rectangular mesh. Among all these 4 or 16 contact options between the boundaries of sub-meshes for the IP sets, the best one is chosen that minimizes the traffic load.

3. *Refreshing IP sets*: The new requirements of the merged IP sets are recalculated. If the merging of IP sets i and j creates the IP set k, the ranking of k is obtained as follows:

$$\text{ranking}(k) = \text{ranking}(i) + \text{ranking}(j) - \text{communication}(i,j) \\ - \text{communication}(j,i)$$

5.4.2 Topology Mapping and Traffic Surface Creation

At the end of BMAP, the IP cores are mapped to individual routers. The accumulated traffic at each router is calculated by considering the traffic flow through the entire network. Since the underlying topology is mesh, a minimal path XY routing is used. The process shows the traffic distribution at various routers. The information so produced is used in the next stage to optimize the hardware. It helps in selecting proper routers from the given library of hardware models.

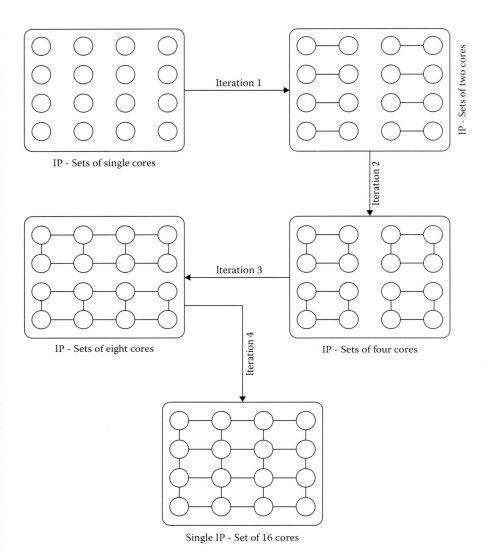

FIGURE 5.5
Example binomial merging.

5.4.3 Hardware Cost Optimization

Since the routers are the dominant hardware components in an OCN, BMAP algorithm attempts to optimize the same, using the traffic loading information generated at the previous stage. The routers, specially the buffers, consume a good amount of area and power in the network. This hardware cost is optimized using the strategies as follows:

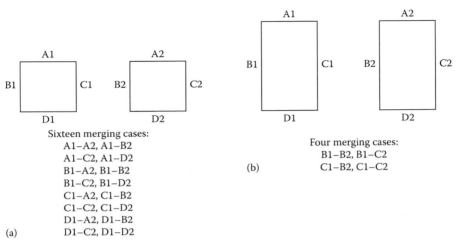

FIGURE 5.6
Merging of IP sets: (a) 16 cases in $d_1 = d_2$; (b) 4 cases in $d_1 \neq d_2$. (Data from Shen, T., et al., A new binomial mapping and optimization algorithm for reduced-complexity mesh-based on-chip network, *Proceedings of NOCS'07*, 317–322, 2007.)

1. *Dummy router elimination*: In the binomial merging process of BMAP algorithm, it is assumed that the mesh network is of the dimension $2^n \times 2^n$ such that the number of IP cores $k \leq 2^n \times 2^n$. If the number of cores is less than that of possible routers, in the final network there will be a few routers to which no core is attached. Such routers constitute dummy routers. Communication between such dummy cores or routers is zero. Hence, after the iterative merging process, such routers are put to the boundary of the mesh. These dummy routers can be removed without sacrificing the network speed. This saves the area, as well as the power, since the idle routers also consume static power.

2. *Router selection*: In an OCN development environment, several alternative router designs may be available, with varying cost, in terms of area, performance, power, and so on. For example, a router has a number of channels. Since the buffers are the most costly components (in terms of area and power consumption) of a router, several trade-offs can be made. For fastest operation, each of the individual input–output channels may have their own buffer. However, a number of channels can be made to share their buffer space, and therefore reduce the cost of the router. Buffer depths can also vary. Routers having low bandwidth cores attached to them may use a less costly router, whereas the routers having high traffic passing through them may have separate buffers (with possibly higher depth) in each channel.

3. *Unfolding*: Some of the routers and links in the network may need to sustain very high traffic load, more than their capacity. This may happen as the communication requirements are decided by the application. Unfolding technique uses duplicate resources (i.e., duplicate routers and links) so that the extra traffic can be carried through the network.

5.5 Constructive Heuristics with Iterative Improvement

These methods attempt to solve the mapping problem by first constructing a candidate solution that satisfies all the bandwidth requirements. The solution is then improved using an iterative approach to obtain solutions with less overall communication cost. One of the very prominent works in this category is the *NMAP* algorithm proposed by Murali and Micheli (2004a). The algorithm, though presented for mesh topology, can also be extended to other topologies. The algorithm has three phases as follows:

- Initialization phase: It computes an initial mapping.
- Minimum path computation phase: It identifies the minimum cost available path between two mapped cores.
- Iterative improvement phase: It invokes the second phase for each pair-wise swapping of mapped core positions.

5.5.1 Initialization Phase

This phase constructs an initial mapping solution for the application on a mesh topology. To start with, the cores are sorted based on their communication demands. The core with the highest communication demand is placed at one of the mesh nodes having the maximum number of neighbors. Let the core having the highest communication be c_1 and let it be associated with router u_j. Next, the core having maximum communication with c_1 is identified; let it be c_2. The core c_2 is placed at one of the neighbors of u_j, so that the communication cost is the minimum. Since, at this time, only one core has been placed, c_2 can get mapped to any of the neighbors of u_j. However, in general, at any stage of the algorithm, let R be the set of cores already mapped and W be the set of corresponding router positions. Let c_k be an unmapped core with the maximum communication requirement with the cores in R. Then, c_k is chosen as the next candidate for mapping. The core c_k can be mapped to any of the available router positions. Associating c_k with router r_m will incur a certain communication cost. For all available mapping positions, the communication costs are evaluated. The core is mapped to the router position resulting in the minimum communication cost. The procedure is repeated until all the cores are mapped.

Procedure Initialize(G, P)

Input: $G(V,E)$—the task graph, $P(U,F)$—the topology graph
Output: A mapping function **map: V→U**, such that *map(c)* gives the mapping of core c onto a router
Begin

   $Placed = NULL$ /* Initialize the set *Placed* to a null set */
   $maxs$ = Core in V with maximum communication
   $maxt$ = Node with maximum neighbors in U
   $map(maxs) = maxt$ /*Map core $maxs$ to the router with maximum
         neighbors */
   $U = U - \{maxt\}$
   $V = V - \{maxs\}$
   $Placed = Placed \cup \{maxt\}$
   While ($|V| > 0$) do
   begin
      $nexts$ = Core in V having maximum communication with
        cores corresponding to routers in *Placed*
      for all router position $u_j \in U$ do
        $commcost_j = 0$
        for all cores c_k corresponding to routers in *Placed* do
          u_k = router corresponding to c_k
          $commcost_j = commcost_j + comm(nexts, ck)$ *
          $hopcount(u_j, u_k)$
      $nextt$ = Router position u_j with minimum *commcost*
      $map(nexts) = nextt$
      $U = U - \{nextt\}$
      $V = V - \{nexts\}$
      $Placed = Placed \cup \{nextt\}$
  end
  return *map, Placed*
End

5.5.2 Shortest Path Computation

This phase identifies the shortest paths for communication between the cores placed at different routers. The set *Placed* is used to identify the routers having cores associated with them. To start with, if two cores are mapped to adjacent routers, the corresponding edge weight is set to be equal to the total bandwidth requirement between them. The communications of the application graph are sorted in descending order of bandwidth requirement. The first such communication is picked up. The minimum path between the corresponding router pair is identified. The weights of all edges in the topology graph belonging to the path are incremented by the bandwidth requirement.

If, at the end, the bandwidth constraints are satisfied by all the edges of the topology graph, the mapping is successful. In such case, the overall communication cost is computed and returned; otherwise a very high-constant *MAXVALUE* is returned, indicating that the mapping violates the bandwidth constraints of at least one edge. In a mesh topology, the shortest path between two nodes always lies between the minimum quadrant involving the nodes. Hence the algorithm restricts the minimum path search within the quadrant only.

Procedure shortestpath(*Placed*)

Input: A set of router positions having cores, that is, the *Placed* set computed earlier
Output: Total communication cost if bandwidths are satisfied, *MAXVALUE* otherwise
Begin
> Initialize edge weights of *Placed* with total communication bandwidth for adjacent nodes and *MAXVALUE* for others
> Sort communications in core graph with decreasing communication cost
> For each communication *d* do
> begin
>> Make quadrant graph *Q* with *source(d)* and *dest(d)* as the corner vertices
>> *Path = Minpath(Q)*
>> Increase edge weights for edges in Path by the bandwidth requirement of the communication
> end
> If all bandwidth constraints are satisfied
>> *Cost* = Total communication cost
> Else
>> *Cost = MAXVALUE*
> Return *Cost*
End

5.5.3 Iterative Improvement Phase

This phase attempts to improve upon the result obtained in the shortest path computation phase. For this, it tries to swap the position of two cores (i.e., their router allotment) and check whether it leads to a better solution or not. In case it results in a better solution, the new mapping and the corresponding cost are remembered.

Procedure iterative_improvement(*G*, *P*)

Input: The application core graph *G* and the topology graph *P*
Output: Best communication cost and mapping via swapping
Begin

> *Bestcommcost = shortestpath(Placed)* /* Find the communication cost for current placement */

> *Bestmapping = Placed*
> For *i* = 1 to *number of nodes in topology graph* do
> > For *j* = *i* + 1 to *number of nodes in topology graph* do
> > begin

> > > *Ptemp = Placed*
> > > Swap nodes w_i and w_j in *Ptemp*
> > > *commcost = shortestpath(Ptemp)*
> > > If (*commcost < bestcommcost*)
> > > begin

> > > > *Bestmapping = Ptemp*
> > > > *Bestcommcost = commcost*

> > > end

> > end

> end
> Return *Bestcommcost, Bestmapping*

End.

5.5.4 Other Constructive Strategies

PMAP, a two-phase mapping algorithm for placing clusters onto processors, was presented by Koziris et al. (2000), where highly communicating clusters are placed on adjacent nodes of the processor network. Each cluster contains all tasks that are to be executed in the same processor having zero interconnection overhead to increase parallelism. A tool, SUNMAP, was presented by Murali and Micheli (2004b) to automatically select the best standard topology for a given application and produce a mapping of cores onto that topology. It minimizes the average communication delay, area, power dissipation subject to bandwidth and area constraints. MOCA is a two-phase heuristic for low-energy mesh-based on-chip interconnection architecture proposed by Srinivasan and Chatha (2005) to reduce the communication energy considering the bandwidth and latency constraints. In the first phase, the cores are mapped to different routers of the mesh by invoking a bipartitioning-based slicing tree generation technique. In the second phase, it attempts to find a minimal path from source to destination for each traffic trace. It does not give good solution when latency constraints are considered. All the mapping techniques proposed earlier use the communication weighted model (CWM) to account for the overall communication volume of each channel. It does not consider communication timing. To capture both timing of application communication and communication volume, communication dependence and computation model (CDCM) was proposed by Marcon et al. (2005a, 2005b), which maps applications on regular NoC under bandwidth constraint and minimizes average communication delay. Marcon et al. (2008) compared different algorithms for obtaining low-energy mappings onto NoCs using a CWM. They also proposed two heuristics, largest communication first (LCF) and greedy incremental (GI) for low-energy mapping using CWM. UMARS, a unified mapping, routing, and slot allocation algorithm presented by Hansson et al. (2005), couples mapping, path allocation, and time slot allocation to minimize communication energy. This technique maps cores onto the NoC topology, routes the communication, and allocates time division multiplexed access (TDMA) time slots on network channels so that application constraints are met. SMAP (Saeidi et al. 2007) is a simulation-based environment, which performs application mapping and task routing for a two-dimensional (2D) mesh-based NoC to minimize the execution time and communication energy. In this technique, the highest priority task is mapped at the center and other tasks are mapped from the mapped tasks spirally to the boundaries of the mesh-based NoC by placing highly communicating cores as close as possible to each other. Spiral is a mapping algorithm proposed by Mehran et al. (2007), which reduces the cumulative energy consumption of communication links and the overall system execution time. In this mapping technique, the high-priority resources are mapped spirally from the center to the boundaries of the mesh-based NoC by placing highly communicating cores as close as possible to each other as in the work of Saeidi et al. (2007). A simulated annealing (SA)-based application mapping technique proposed by Harmanani and Farah (2008) for a 2D mesh-based NoC minimizes the area requirement and the maximum bandwidth. It also proposes an efficient routing algorithm that selects a route among alternative paths based on the network state and occupancy of queues. Cluster-based technique combined with SA was proposed by Lu et al. (2008) for application mapping onto a 2D mesh-based NoC. In this technique, mapping is done cluster-wise, instead of node-wise, to reduce the mapping complexity. Clustering is a technique to partition nodes into groups according to the physical distance among them in the network topology. Clustering exploits the knowledge about the network architecture and communication demand of applications. Therefore, in this mapping technique, first a cluster-based core to node initial mapping is done and then a SA technique is applied upon it to find good mapping solution. Elmiligi et al. (2008a, 2008b, 2009) analyzed different approaches to minimize the total communication energy by inserting some permissible longer links and by-passing some routers of application-specific NoC. In this process, by network partitioning, the area cost is reduced by reducing both the router area and the number of links. Elmiligi et al. (2009) proposed an efficient methodology to choose the most power-efficient application-specific NoC architecture. They compared different topologies taking only one application benchmark and reported the best one, but that topology may not be good for other applications. Topology design is one of the significant factors that affects the net delay and energy consumption of an application-specific NoC. The topology must satisfy the design constraints. For very-high I/O rate streaming type of application mapping, a guaranteed and high-throughput pipelined mechanism for NoC is introduced in the work of Yu et al. (2009). They proposed a pipeline-based high-throughput low-energy mapping algorithm that performs task allocation, pipelined task scheduling, and communication scheduling simultaneously on the heterogeneous NoC and minimizes the energy consumption. Onyx, a new bandwidth-constrained application mapping, was presented by Janidarmian et al. (2009) to minimize the overall communication cost of NoC. In this technique, a core with the highest communication bandwidth is mapped at the center. Then the ranking of other unmapped cores is settled according to the communication volume with mapped cores. The unmapped cores are placed at the nearest possible distance with their related cores by looking the lozenge-shaped path with one or two hop distances and so on till the empty tile is identified. CHMAP (Tavanpour et al. 2009) is a chain-mapping algorithm that produces chains of connected cores in order to introduce a method for application mapping onto a mesh-based NoC. Crinkle, a mapping algorithm, was presented by Saeidi et al. (2009) to reduce the overall communication cost. In this technique, priority lists are prepared depending on the interconnection degree of nodes and communication bandwidth before mapping onto a mesh-based NoC. Depending on the priority lists, the heuristic maps the tasks from the corner of 2D mesh platform and ends on another corner in a zigzag manner. A multiobjective optimization strategy was proposed by Tornero et al. (2009) to determine the pareto optimal NoC configurations to optimize an average delay of the network and routing robustness. In this technique, both the topological mapping and the routing are considered concurrently. Wang et al. (2009, 2010) proposed a power-aware template-based efficient mapping (TEM) algorithm for NoC to generate good mapping solutions with low runtime under bandwidth and latency constraints. CMAP (Chen et al. 2009) is a fast constructive application mapping algorithm that maps tasks onto NoC minimizing the total communication cost and energy. It is a hybrid of two constructive mapping algorithms, link-based mapping (LBMAP), and sort-based mapping (SBMAP). After comparing the results of these two, the better one is taken as output. Citrine is a two-step 2D mesh mapping algorithm proposed by Janidarmian et al. (2010), which uses the mapping technique Onyx (Janidarmian et al. 2009) to retrieve the order of cores, and then a branch-and-bound search tries to search different permutations by a lozenge-shaped rule of Onyx. RMAP is a reliability-aware application mapping technique for a mesh-based NoC proposed by Patooghy et al. (2010). It divides the application graph into two subgraphs, which minimizes the communication traffic between the subgraphs and maximizes the traffic within each subgraph. Then one subgraph is mapped onto upper triangular nodes of the NoC and the other is mapped to lower triangular nodes of the NoC. This technique utilizes the nonuniformity of traffic distribution over the network channels to efficiently route the packets of redundant communications. In the work of Yang et al. (2010b), all the nodes and the interconnections among nodes of a 2D mesh-based NoC are abstracted as a tree. In this tree model, the vertex with highest communication volume is selected as root node. The vertices communicating to the root (node) are the children of that node and so on. During mapping, the root node is placed at the center of the mesh-based NoC, and the traversal is made from the center toward the borders of the NoC. The child nodes are placed by seeing the tree structure and the communication volume of interconnects from the center toward the borders. Yang et al. (2010a) proposed a two-step multiapplication mapping algorithm that maps multiple applications simultaneously onto different regions of NoC to minimize network latency and energy consumption for a set of applications. The algorithm consists of an application mapping phase followed by a task mapping phase. The application mapping phase deals with the multiple applications mapping to optimize the layout of multiple applications on the NoC. After the application mapping phase, the role of task mapping phase is to map the tasks of the application so that the average communication distance is minimized. The task mapping of each application follows the tree model-based mapping as described in the work of Yang et al. (2010b). LMAP is a mapping algorithm proposed by Sahu et al. (2010) to reduce both static and dynamic costs of a mesh-based NoC. In the initial mapping phase, a Kernighan–Lin (K–L) partitioning scheme is used to identify the closeness of cores by analyzing their bandwidth or communication requirements. This bipartitioning is applied (recursively) until the closest two cores are left in one final partition. After initial mapping, an iterative improvement phase is applied to arrive at a final mapping. CastNet is an energy-aware application mapping and routing technique for 2D NoC proposed by Tosun (2011b). Before mapping, a priority list of the tasks is formed based on its total communication with its neighbors. Depending on the priority list, the initial task is selected. For mapping the first task, a set of initial node positions is selected. A set of solutions is generated by this technique for each initial node position of the initial task. The remaining tasks are placed on the nodes of NoC according to the priority list. After each mapping the priority list is also updated. Finally, from the set of solutions, the best one is taken as the solution for mapping of applications onto NoC.

All the application mapping techniques of NoC discussed above are based on the mesh-based network architecture. But it is essential to check the suitability of other network topology when applications are mapped onto that. An energy-aware mapping technique was proposed by Chang et al. (2008), which maps the IPs onto a tree-based NoC architecture such that the total communication energy can be minimized. In this technique, first an energy-aware mapping is formulated, and then a recursive bipartitioning algorithm is used to solve it. An application mapping heuristic was proposed by Majeti et al. (2009) for generating an optimal tree-based topology for multimedia applications to minimize energy consumption while meeting the design constraints. Application mapping techniques were proposed by Sahu et al. (2011a, 2011b) to map applications onto butterfly fat tree- and mesh-of-tree-based NoCs, respectively. In this technique, a K–L partitioning scheme was used by Sahu et al. (2010) to identify the closeness of cores by analyzing their bandwidth or communication requirements. An energy-aware mapping algorithm has been presented in Hu and Merculescu (2007) that computes the network energy in terms of energy consumed per bit transmission through the routers and the links. A bandwidth constrained mapping has been presented in Reshadi et al. (2010).

5.6 Mapping Using Discrete PSO

PSO is a population-based stochastic technique developed by Kennedy and Eberhart (1995), inspired by social behavior of bird flocking or fish schooling. In a PSO system, multiple candidate solutions coexist and collaborate simultaneously. Each solution, called a *particle*, flies in the problem space according to its own experience as well as the experience of neighboring particles. It has been successfully applied in many problem areas. In a PSO, each single solution is a particle in the search space, having a fitness value. The quality of a particle is evaluated by its fitness. Inspired by its success in solving problems in continuous domain, several researchers have attempted to apply it in discrete domain as well (Wang et al. 2003). A well-known problem that was attempted to be solved using discrete PSO (DPSO) technique is the travelling salesman problem (TSP) (Wang et al. 2003). A solution to a TSP problem consists of a sequence of all cities, such that the total distance travelled is minimized. Structurally, the NoC application mapping problem is very much similar to TSP. If the router positions in the topology graph are given unique numbers in the range 0 to *number_of_routers* − 1, the solution associates each core of the application graph to one such router. Thus, the mapping problem can also be viewed as an ordering of the cores. This leads to a DPSO formulation of the application mapping problem.

5.6.1 Particle Structure

In application mapping, a particle corresponds to a possible mapping of cores to the routers. An example of a particle structure is shown in Figure 5.7. The numbers shown within circles in the boxes are the core numbers present in the core graph. The numbers outside the box are the router numbers of the topology graph. The figure shows that core 1 is attached to router 0, core 4 is attached to router 1, and so on. If the number of nodes (routers) present in the topology graph is greater than the number of cores present in the core graph, dummy nodes are added to the core graph to make the two numbers same. Dummy nodes are connected to all core nodes and between

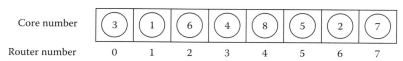

Core number

Router number   0   1   2   3   4   5   6   7

FIGURE 5.7
A particle structure.

themselves. Edges connecting a core node to dummy nodes and the edges between dummy nodes are assigned a cost 0. Let N be the number of cores present in the core graph for mapping cores onto the topology graph, after connecting dummy nodes, if required. For these N cores, there are N node positions in the topology graph. A particle is a permutation of numbers from 1 to N, which shows the placement of cores to the node positions of the topology graph. The overall communication cost is influenced by the position of cores in a particle. The overall communication cost forms the fitness function. Fitness of a particle p_i is equal to the overall communication cost after placement of cores of the core graph to different routers as specified by the particle.

5.6.2 Evolution of Generations

In the general DPSO framework, let the position of a particle (in an n-dimensional space) at kth iteration be $p_k = <p_{k,1}, p_{k,2}, \ldots, p_{k,n}>$. For ith particle, the quantity is denoted as p_k^i. Let $pbest^i$ be the local best solution that particle i has seen so far and $gbest_k$ be the global best particle of iteration k. The new position of particle i is calculated as follows:

$$p_{k+1}^i = [s_1 \times I \oplus s_2 \times (p_k \rightarrow pbest^i) \oplus s_3 \times (p_k \rightarrow gbest_k)] \times p_k^i$$

In the above equation, $a \rightarrow b$ represents the minimum length sequence of swapping to be applied on components of a to transform it to b. For example, if $a = <1, 3, 4, 2>$ and $b = <2, 1, 3, 4>$, $a \rightarrow b = <swap(1, 4), swap(2, 4), swap(3, 4)>$. The operator $\oplus$ is the fusion operator. Applied on two swap sequences, $a \oplus b$ is equal to the sequence in which the sequence of swaps in a is followed by the sequence of swaps in b. The constants s_1, s_2, s_3 are the inertia, self-confidence, and swarm confidence values, respectively. The quantity $s_i \times (a \rightarrow b)$ means that the swaps in the sequence $a \rightarrow b$ will be applied with a probability s_i. I is the sequence of identity swaps, such as $<swap(1, 1), swap(2, 2), \ldots, swap(n, n)>$. It corresponds to the inertia of the particle to maintain its current configuration. The final swap corresponding to $s_1 \times I \oplus s_2 \times (p_k \rightarrow pbest^i) \oplus s_3 \times (p_k \rightarrow gbest_k)$ is applied on particle p_k^i to generate p_{k+1}^i.

In reference to the application mapping problem, for a particle p, the router associated with a core is identified by the position index of the core in p. The indexing of the position takes values between 0 and $N - 1$ (N being the number of routers). The index corresponds to the router number, as shown in Figure 5.7. Let the swap operator be $SO_{j,k}$ (where j and $k = 0, 1, \ldots, N-1$) that swaps the jth and kth positions of particle p to create a new particle p_{new}. For example, consider the particle $p = \{1, 4, 3, 6, 2, 8, 5, 7\}$, where the numbers represent the core numbers of the core graph and the position represents the router numbers in the topology graph. The swap operator $SO_{4,6}$ swaps the cores at positions 4 and 6, which creates a new particle $p_{new} = \{1, 4, 3, 6, 5, 8, 2, 7\}$.

To align a particle p_i with its local best, the swap sequence is identified. Let this be SS_i^{l-best}. Then another swap sequence is identified to align the particle with the global best. Let this be SS_i^{g-best}. Now the swap sequence SS_i^{l-best} is applied on particle p_i with a probability of s_2. Let the modified particle be p_i^{l-best}. Then the swap sequence SS_i^{g-best} is applied on p_i^{l-best} with a probability of s_3. This creates a new particle p_i^{new}. Its fitness is evaluated and the local best is updated for particle i, if it is better than the previous local best for the particle. If the best fitness in a generation is better than the global best of the previous generation, the global best is also updated.

Procedure Compute_Swap_Sequence

Input: Source sequence *Sour_seq* Destination sequence *Dest_seq*
Output: Swap sequence *Swap_seq* to align *Sour_seq* to *Dest_seq*
Begin
      For $i = 1$ to *total number of nodes* in *Sour_seq*
          Swap_seq[i] = Index of *Sour_seq[i]* in *Dest_seq*
      End for
End

Assuming that none of the sequences are sorted, the time complexity of the procedure *Compute_Swap_Sequence* is $O(n^2)$, n being the number of nodes.

5.6.3 Convergence of DPSO

From Guilan et al. (2008), it can be found that the convergence condition for this DPSO is given by

$$\left(1-\sqrt{s_1}\right)^2 \leq s_2 + s_3 \leq (1+s_1)^2$$

Setting the values of $s_1 = 1.0$, $s_2 = 0.04$, and $s_3 = 0.02$ is observed to produce good results for most of the applications we have experimented with. A typical trace of the evolution of a particle with these parameter settings shows that in the process of convergence, it is safe to assume that the particle has converged to its final value if there are no significant improvements in the solution quality for last 100 generations.

5.6.4 Overall PSO Algorithm

The overall PSO algorithm is presented as follows:

Initialization

For each particle
        Initialize particle with random solution
        Evaluate *fitness* value of each particle
        Set *local_best* of each particle to itself
End for
Set *global_best* to the best fit particle
Evolutions
Do
        For each particle p_i

            Identify $SS_i^{l\_best}$ and $SS_i^{g\_best}$
        p_i^{new} = Modify p_i by applying $SS_i^{l\_best}$ with probability s_2 followed
            by $SS_i^{g\_best}$ with probability s_3
        Evaluate fitness of p_i^{new}
        If fitness of p_i^{new} is better than the local best for p_i then
update *local_best* for p_i End for
        Find the particle with the best fitness and update *global_best*
While maximum generation (prespecified) not attained and *global_best* is
not remaining unaltered for a prespecified number of generations

---

5.6.5 Augmentations to the DPSO

The DPSO formulation discussed in Section 5.6.4 can be augmented in the following two ways to achieve better solutions.

5.6.5.1 Multiple PSO

The PSO formulation can be run several times to improve upon the global best solution. Suppose that the ith run of the PSO produces the local best $pbest_i^k$ for each particle k and the global best $gbest_i$. The $(i + 1)$th run of the PSO starts with a new set of particles. However, the global and local best information of the particles is passed from ith to $(i + 1)$th PSO. The number of times for which the PSO is run, that is, the terminating criteria, is decided by the following:

1. There may be a user-specified upper limit. In the work of Sahu et al. (2012), it has been kept at 200 individual PSO runs.
2. The global best cost does not improve in the last 20 PSO runs.

5.6.5.2 Initial Population Generation

For an application with n cores to be mapped onto a mesh topology having n routers in it, the total number of possible mappings is $n!$. Thus, exploration of the promising region of this huge search space depends to a great extent on the initial population with which each PSO starts evolving. To augment the solution quality, in the initial set of particles, some particles are included that are generated via a deterministic mapping technique discussed in this section. For the topology with n routers, exactly n deterministically generated particles are included. The remaining particles are generated randomly. The deterministic particle generation works as follows.

First, the edges of the core graph are sorted on descending communication requirements, as specified in edge label. Let $e = (c_1, c_2)$ be the edge with the maximum bandwidth requirement. Mapping process starts with this edge. For core c_1, the total bandwidth requirement is computed by summing up the labels of all edges of c_1 to its neighbors. The same is done for c_2. Let the value computed for c_1 be higher than that for c_2. The mapping process generates solutions with c_1 mapped to each router position of the topology. For a particular placement of c_1, the remaining cores are mapped judiciously to obtain a good solution. Thus, a set of particles equal to the number of routers gets created. The set forms a subset of particles for the initial population.

Suppose that c_1 is mapped onto router u_1, and in the topology graph, u_1 has neighbors u_2, u_3, and u_4. Since all these routers are one hop away from u_1, all of them are equally suitable for mapping of c_2. In general, at a point during execution of this constructive mapping algorithm, a subset of cores is already mapped onto the routers of the topology graph. Let this set of cores be C' and the corresponding router set be U'. The algorithm now considers those edges of the core graph of which exactly one vertex has already been mapped. It selects such an edge with the highest bandwidth requirement. Let the unmapped core of that edge be c_i. We try out the mapping of c_i to each router placed at a one-hop distance from any router in U' (the set of routers with already assigned cores). For each such mapping, the cost of mapping is evaluated by considering the subgraph consisting of cores in the set $C' \cup \{c_i\}$. If there is a single mapping with the minimum cost, it is accepted for mapping of c_i, and the process continues with the next candidate node selected in a similar fashion. However, if multiple mappings of c_i are of the same cost, let us assume $M = \{m_1, m_2, \ldots, m_k\}$ be the set of k candidate positions for c_i resulting in equal mapping cost for the subgraph with a vertex set $C' \cup \{c_i\}$. To distinguish between these k positions, temporarily select m_1 to be the mapping of c_i. With this, we proceed to find the mapping for the remaining cores in a similar fashion, as noted earlier. That is, for the next core to be mapped, the router positions neighboring to the topology subgraph $U' \cup m_1$ are evaluated. However, in this case we do not distinguish between contending positions with minimum cost values. Instead, we take the first such position and continue with mapping of the remaining cores. When all cores are mapped, the cost of the final mapping solution is taken as the predicted cost of selecting router position m_1 for c_i. Similarly, other $k - 1$ positions m_2, m_3,..., m_k are evaluated and the core c_i is mapped onto the router position with the minimum predicted cost. The process continues by selecting the next core. The following algorithm enumerates the process:

Initial Mapping Algorithm: Map_Graph

Input: Core graph G, Topology graph P
Output: Mapping of G onto P
Begin
    Sort edges of G on descending order of communication cost
    For each router position u of P do
        Mark all cores of G as unmapped
        $Best\_Cost = \infty$
        $Best\_Mapping = \Phi$
        $Mapping = $ Find_Mapping (G, P, u)
        Output $Mapping$ as a particle
    End for
End

Procedure Find_Mapping
Input: Core graph G, Topology graph P,
    $Core$: core to be mapped,
    $Start\_Posn$: Position in P where first core to be mapped
Output: Mapping of all cores of G onto P with the first core mapped to $Start\_Posn$
Begin
    Let (c_1, c_2) be the edge of G with the highest required bandwidth

$$Cost_1 = \sum_{c_i \in neighbour(c_1)} Bandwidth \; requirement \; of \; (c_1, c_i)$$

$$Cost_2 = \sum_{c_i \in neighbour(c_2)} Bandwidth \; requirement \; of \; (c_2, c_i)$$

    If $(Cost_1 > Cost_2)$ then $Core = c_1$ else $Core = c_2$
    $Mapping[Start\_Posn] = Core$
    Mark $Core$ as mapped
    While there exist unmapped cores in G do
    Let (c_i, c_j) be the edge of G with highest bandwidth such that exactly one of c_i and c_j is already mapped
    Let $c = c_i$ if c_j is already mapped else $c = c_j$
    $Positions.$ = set of positions in P with one hop distance from already mapped positions
    Evaluate_Positions($Positions$). $Min\_Positions$ = Set of. $Positions$ with minimum cost

If (cardinality of set *Min_Positions* =.)

    Best_Position = Min_Position[0]

Else

    Best_Position = Predict_Best(Min_positions, G, P, c). Mark *cc* mapped

End while

Return *Mapping*

End

Procedure Predict_Best

Input: Core graph *G*, Topology graph *P*,

    Core: core to be mapped,

    Posn: set of contending position of *P*

Output: Predicted best position of core amongst *Posn*

Begin

    Min_Cost = ∞

    Newly_Marked_Cores = Φ

    For each position *p* in *Posn* do

        Mapping[Core] = p

        Newly_Marked_Cores = Newly_Marked_Cores ∪ *{Core}*

        Mark *Core* mapped

        While there exists unmapped cores in *G* do

            Let (c_i, c_j) be the edge of *G* with the highest bandwidth such that exactly one of c_i and c_j is already mapped

            Let $c = c_i$ if c_j is already mapped else $c = c_j$ *Positions* = Set of positions in *P* one hop distance from already mapped positions

            Evaluate_Positions(*Positions*)

            Best_Position = First *Position* with minimum cost

            Mapping[Best_Position] = c

            Mark *c* mapped

        End while

        Cost = Total communication cost for this mapping

        If *Min_cost* > *Cost* then

            Min_cost = Cost

            Min_posn = Φ

        End If

        Unmark all cores in *Newly_Marked_Cores*

        Newly_Marked_Cores = Φ

    End for

    Return *Min_posn*

End

5.6.6 Other Evolutionary Approaches

A two-step GA for mapping applications onto NoC was proposed by Lei and Kumar (2003), which reduces the overall execution time. In the first step, the tasks are assigned onto different IPs assuming the edge delays to be constant and equal to the average edge delay. In the second step, the IPs are mapped to tiles of NoC taking the actual edge delay, based on the network traffic model, and the total system delay is minimized. In this mapping, some delay factors, such as the message sending probability of cores, the packet length, and the network contention for communication, are not been considered. Zhou et al. (2006) proposed a delay model for application mapping onto NoC considering all these factors. Their proposed GA-based delay model can map the application onto NoC optimally with a minimum average delay. PLBMR, a PSO-based two-phase application mapping algorithm proposed by Zhou et al. (2007), minimizes the NoC communication energy and allocates the routing path for balancing the link load. In the first phase, the PSO maps IP cores onto NoC to minimize the energy consumption, and in the second phase, the routing paths are allocated to every pair to satisfy the link–load balance. Ascia et al. (2004) proposed a pareto-based multiobjective evolutionary computing technique that optimizes the performance and power consumption of mapped NoC. Ascia et al. (2006) used the above technique for application task mapping. For dynamic evaluation, an event-driven trace-based simulator was used to compare their results with a pareto-based branch-and-bound approach and a pareto-based NMAP approach. A multiobjective GA-based application mapping for NoC was presented by Benyamina and Boulet (2007), which targets mapping with a network assignment (NA) for heterogeneous distributed embedded systems to improve the performance and reduce the power consumption and area. This technique first allocates tasks to cores and then maps the cores to different tiles of NoC satisfying communication requirements. The mapping of IP cores onto NoC tiles, together with routing path allocation, is referred to as NA. The NA is usually performed after task mapping to reduce the on-chip intercommunication distance. The GA-based optimization technique, MGAP, proposed by Jena and Sharma (2007) minimizes the power consumption by reducing the number of switches in the communication path between cores and also maximizes the throughput. Although Lei and Kumar (2003) used a similar technique, they considered the dynamic effect of traffic. They also gave a set of solutions using pareto mapping as used in the work of Ascia et al. (2004, 2006). A multiobjective GA (MOGA)-based application mapping technique was proposed by Bhardwaj and Jena (2009), where one–one as well as many–many mapping between switches and tiles were taken into consideration to minimize energy consumption and the required link bandwidth. It is used to find an optimal solution from the pareto optimal solutions as in the work of Jena and Sharma (2007). Darbari et al. (2009a, 2009b) proposed CGMAP, a GA-based application mapping technique that uses the chaotic mapping operator instead of the random processes in GA. Fard et al. (2009)

presented a different one-dimensional chaotic mapping technique onto NoC. GBMAP, an evolutionary approach for mapping cores onto the NoC architecture, was proposed by Tavanpour et al. (2010), which reduces energy consumption and the total bandwidth requirement of NoC. Fekr et al. (2010) proposed a PSO-based application mapping technique for NoC in which merit of the scheme is not clear, as no comparison was made with the existing approaches. A mapping technique based on discrete PSO was presented by Lei and Xiang (2010). However, it only considers improvement over a GA-based method and reports relative improvements only. Benyamina et al. (2010) proposed a hybrid multiobjective algorithm, where Dijkstra's shortest path algorithm is used to find the shortest path among the communicating cores to satisfy the bandwidth constraints and then a multiobjective pareto-based PSO technique is applied upon that to improve performance. GMAR, a GA-based mapping and routing approach proposed by Fen and Ning (2010), addresses a two-phase mapping of IP cores onto the NoC architecture and generates a deterministic deadlock-free minimal routing path for each communication to minimize the total communication energy and maximize the link bandwidth utilization of the NoC architecture. In the first phase, GMAR maps IP cores onto different resource nodes of the mesh-based NoC architecture. In the second phase, it generates deterministic deadlock-free minimal routing path for each communication trace. Jang and Pan (2010) proposed an architecture-aware analytic mapping algorithm (A3MAP) for NoC with homogeneous and heterogeneous cores on regular and irregular mesh or custom architecture. The task mapping problem is solved by two effective heuristics: a successive relaxation algorithm as a fast algorithm and a GA to find better mapping solutions. Choudhary et al. (2010) proposed a GA-based mapping technique for a customized NoC architecture to reduce the communication energy. Choudhary et al. (2011) proposed a GA-based congestion-aware mapping technique for an irregular customized NoC architecture to reduce the communication energy. A multiobjective adaptive immune algorithm (MAIA), based on an evolutionary approach, was proposed by Sepulveda et al. (2009), which maps the application tasks onto NoC to reduce the power consumption and overall network latency. The adaptive immune algorithms integrate a wide set of features that improve local search while preventing the premature convergence by preserving the diversity of solutions in the population. Sepulveda et al. (2011) proposed an improved version of MAIA to solve the multiapplication NoC problem. It produces a set of mapping alternatives by exploring the mapping space. Wang et al. (2011) proposed an ACO-based algorithm for application task mapping onto NoC to minimize the bandwidth requirement. The results were compared with random mapping techniques. Sahu et al. (2011c), proposed PSMAP, a metaheuristic strategy using PSO technique, to reduce both static and dynamic costs of NoC for 2D mesh-based application mapping.

5.7 Summary

Application mapping refers to the problem of determining the router positions to which individual cores of an application be mapped. The major goal of the operation is to minimize the communication cost. Communication cost controls the latency of communication between the cores and the overall power consumption. The mapping problem is NP-hard. In this chapter, various strategies for application mapping have been discussed. While ILP-based approaches produce the best results, the overall computation time is high, restricting its usage to only a few cores in the application graph. The constructive heuristic approaches attempt to construct a solution, which may be followed by an iterative improvement phase. The evolutionary algorithms perform particularly well. Such algorithms work well even for reasonably large number of cores (e.g., 128 cores). The mapping problem can be extended to the power- and thermal-aware strategies.

References

Ascia, G., Catania, V., and Palesi, M. 2004. Multi-objective mapping for Mesh-based NoC architectures. *International Conference on Hardware/Software Codesign and System Synthesis*, ACM, pp. 182–187.

Ascia, G., Catania, V., and Palesi, M. 2006. Multi-objective genetic approach to mapping problem on network-on-chip. *Journal of Universal Computer Science*, vol. 12, no. 4, pp. 370–394.

Bender, A. 1996. MILP based task mapping for heterogeneous multiprocessor systems. *Proceedings of International Conference on Design and Automation*, IEEE Computer Society Press, pp. 190–197.

Benyamina, A. H. and Boulet, P. 2007. Multi-objective mapping for NoC architecture. *Journal of Digital Information Management*, vol. 5, pp. 378–384.

Benyamina, A. H., Boulet, P., Aroul, A., Eltar, S., and Dellal, K. 2010. Mapping real time applications on NoC architecture with hybrid multi-objective algorithm. *International Conference on Metaheuristics and Nature Inspired Computing*, pp. 1–10.

Bhardwaj, K. and Jena, R. K. 2009. Energy and bandwidth aware mapping of IPs onto regular NoC architectures using multi-objective genetic algorithms. *International Symposium on System-on-Chip*, IEEE Press, pp. 27–31.

Chang, Z., Xiong, G., and Sang, N. 2008. Energy-aware mapping for tree-based NoC architecture by recursive bipartitioning. *International Conference on Embedded Software and Systems*, IEEE Press, pp. 105–109.

Chen, Y., Xie, L., and Li, J. 2009. An energy-aware heuristic constructive mapping algorithm for network on chip. *International Conference on ASIC*, IEEE Press, pp. 101–104.

Chou, C. L. and Marculescu, R. 2008. Contention-aware application mapping for network-on-chip communication architectures. *IEEE International Conference on Computer Design*, IEEE Press, pp. 164–169.

Choudhary, N., Gaur, M. S., Laxmi, V., and Singh, V. 2010. Energy aware design methedologies for application specific NoC. *Proceedings of NORCHIP*, IEEE Press, pp. 1–4.

Choudhary, N., Gaur, M. S., Laxmi, V., and Singh, V. 2011. GA based congestion aware topology generation for application specific NoC. *IEEE International Symposium on Electronics Design, Test, and Application*, IEEE Press, pp. 93–98.

Darbari, F. M., Khademzadeh, A., and Fard, G. G. 2009a. Evaluating the performance of a chaos genetic algorithm for solving the network on chip mapping problem. *International Conference on Computational Science and Engineering*, IEEE Press, pp. 366–373.

Darbari, F. M., Khademzadeh, A., and Fard, G. G. 2009b. CGMAP: A new approach to network-on-chip mapping problem. *IEICE Electronics Express*, vol. 6, no. 1, pp. 27–34.

Elmiligi, H., Morgan, A. A., Kharashi, M. W. E., and Gebali, F. 2008a. Power-aware topology optimization for network-on-chips. *IEEE International Symposium on Circuits and Systems*, IEEE Press, pp. 360–363.

Elmiligi, H., Morgan, A. A., Kharashi, M. W. E., and Gebali, F. 2008b. Application-specific networks-on-chip topology customization using network partitioning. *1st International Forum on Next-generation Multicore/manycore Technologies*. ACM.

Elmiligi, H., Morgan, A. A., Kharashi, M. W. E., and Gebali, F. 2009. Power optimization for application-specific networks-on-chips: A topology-based approach. *Journal of Microprocessor and Microsystems*, vol. 33, pp. 343–355.

Fard, G. G., Khademzadeh, A., and Darbari, F. M. 2009. Evaluating the performance of one-dimensional chaotic maps in network-on-chip mapping problem. *IEICE Electronics Express*, vol. 6, no. 12, pp. 811–817.

Fekr, A. R., Khademzadeh, A., Janidarmian, M., and Bokharaei, V. S. 2010. Bandwidth/fault/contention aware application-specific NoC using PSO as a mapping generator. *Proceedings of the World Congress on Engineering*, IAENG, vol. 1, pp. 247–252.

Fen, G. and Ning, W. 2010. Genetic algorithm based mapping and routing approach for network on chip architectures. *Chinese Journal of Electronics*, vol. 19, no. 1, pp. 91–96.

Ghosh, P., Sen, A. Sen, and Hall, A. 2009. Energy efficient application mapping to NoC processing elements operating at multiple voltage levels. *IEEE International Symposiun on Network-on-Chip*, IEEE Press, pp. 80–85.

Guilan, L., Hai, Z., and Chunhe, S. 2008. Convergence analysis of a dynamic discrete PSO algorithm. *International Conference on Intelligent Networks and Intelligent Systems*, IEEE Press, pp. 89–92.

Hansson, A., Goossens, K., and Radulescu, A. 2005. A unified approach to constrained mapping and routing on network-on-chip architectures. *IEEE/ACM International Conference on Hardware/Software Codesign and System Synthesis*, IEEE Press, pp. 75–80.

Harmanani, H. M. and Farah, R. 2008. A method for efficient mapping and reliable routing for NoC architectures with minimum bandwidth and area. *IEEE International Workshop on Circuits and Systems and TAISA Conference*, IEEE Press, pp. 29–32.

Hu, J. and Marculescu, R. 2005. Energy- and performance-aware mapping for regular NoC architectures. *IEEE Transactions on Computer Aided Design of Integrated Circuits and Systems*, vol. 24, no. 4, pp. 551–562.

Huang, J., Buckl, C., Raabe, A., and Knool, A. 2011. Energy-aware task allocation for network-on-chip based heterogeneous multiprocessor systems. *Euromicro International Conference on Parallel, Distributed and Network based Processing*, IEEE Press, pp. 447–454.

Jang, W. and Pan, D. Z. 2010. A3MAP: Architecture-aware analytic mapping for network-on-chip. *Asia and South Pacific Design Automation Conference*, IEEE Press, pp. 523–528.

Janidarmian, M., Khademzadeh, A., Fekr, A. R., and Bokharaei, V. S. 2010. Citrine: A methodology for application-specific network-on-chips design. *Proceedings of World Congress on Engineering and Computer Science*, vol. 1, Springer, pp. 196–202.

Janidarmian, M., Khademzadeh, A., and Tavanpour, M. 2009. Onyx: A new heuristic bandwidth-constrained mapping of cores onto network on chip. *IEICE Electronics Express*, vol. 6, no. 1, pp. 1–7.

Jena, R. K. and Sharma, G. K. 2007. A multi-objective evolutionary algorithm based optimization model for network-on-chip. *IEEE International Conference on Information Technology*, IEEE Press, pp. 977–982.

Kennedy, I. and Eberhart, R. C. 1995. Particle swarm optimization. *Proceedings of IEEE International Conference on Neural Networks*, New Jersey, IEEE Press, pp. 1942–1948.

Koziris, N., Romesis, M., Tsanakas, P., and Papakonstantinou, G. 2000. An efficient algorithm for the physical mapping of clustered task graphs onto multiprocessor architectures. *Proceedings of 8th Euro PDP*, IEEE Press, pp. 406–413.

Lei, T. and Kumar, S. 2003. A two-step genetic algorithm for mapping task graphs to a network on chip architecture. *Proceedings of the Euromicro Symposium on Digital System Design (DSD)*, IEEE Press, pp. 180–187.

Lei, W. and Xiang, L. 2010. Energy- and latency-aware NoC mapping based on discrete particle swarm optimization. *Proceedings of IEEE International Conference on Communications and Mobile Computing*, IEEE Press, pp. 263–268.

Lu, Z., Xia, L., and Jantsch, A. 2008. Cluster-based simulated annealing for mapping cores onto 2D mesh networks on chip. *Proceedings of Design and Diagnostics of Electronic Circuits and Systems*, IEEE Press, pp. 1–6.

Majeti, D., Pasalapudi, A., and Yalamanchili, K. 2009. Low energy tree based network on chip architectures using homogeneous routers for bandwidth and latency constrained multimedia applications. *International Conference on Emerging Trends in Engineering and Technology*, IEEE Press, pp. 358–363.

Marcon, C., Borin, A., Susin, A., Carro, L., and Wagner, F. 2005a. Time and energy efficient mapping of embeded applications onto NoCs. *Proceedings of Asia and South Pacific Design Automation Conference*, IEEE Press, vol. 1, pp. 33–38.

Marcon, C., Calazans, N., Moraes, F., Susin, A., Reis, I., and Hessel, F. 2005b. Exploring NoC mapping strategies: An energy and timing aware technique. *Proceedings of Design, Automation and Test in Europe Conference and Exhibition*, IEEE Press, vol. 1, pp. 502–507.

Marcon, C., Moreno, E. I., Calazans, N. L. V., and Moraes, F. G. 2008. Comparison of network-on-chip mapping algorithms targeting low energy consumption. *IET Computer & Digital Technique*, vol. 2, no. 6, pp. 471–482.

Mehran, R., Saeidi, S., Khademzadeh, A., and Kusha, A. A. 2007. Spiral: A heuristic mapping algorithm for network on chip. *IEICE Electronics Express*, vol. 4, no. 15, pp. 478–484.

Murali, S., Benini, L., and Micheli, G. De. 2005. Mapping and physical planning of networks-on-chip architectures with quality-of-service guarantees. *Asia and South Pacific Design Automation Conference*, IEEE Press, pp. 27–32.

Murali, S. and Micheli, G. De. 2004a. Bandwidth constrained mapping of cores onto NoC architectures. *Proceedings of Design, Automation and Test in Europe Conference and Exhibition*, IEEE Press, vol. 2, pp. 896–901.

Murali, S. and Micheli, G. De. 2004b. SUNMAP: A tool for automatic topology selection and generation for NoCs. *Proceedings of 41st Design Automation Conference,* IEEE Press, pp. 914–919.

Ostler, C. and Chatha, K. S. 2007. An ILP formulation for system-level application mapping on network processor architecture. *Proceedings of Design, Automation and Test in Europe,* IEEE Press, pp. 1–6.

Ozturk, O., Kandemir, M., and Son, S. W. 2007. An ILP based approach to reducing energy consumption in NoC based CMPs. *IEEE International Symposiun on Low Power Electronics and Design,* IEEE Press, pp. 411–414.

Patooghy, A., Tabkhi, A., and Miremadi, S. G. 2010. RMAP: A reliability-aware application mapping for network-on-chips. *International Conference on Dependability,* IEEE Press, pp. 112–117.

Pop, R. and Kumar, S. 2004. A survey of techniques for mapping and scheduling applications to network on chip systems. ISSN 1404–0018, Research Report 04:4, School of Engineering, Jönköping University, Sweden.

Reshadi, M., Khademzadeh, A., and Reza, A. 2010. Elixir: A new bandwidth-constrained mapping for networks-on-chip. *IEICE Electronics Express,* vol. 7, no. 2, pp. 73–79.

Rhee, C., Jeong, H., and Ha, S. 2004. Many-to-many core-switch mapping in 2-D mesh NoC architectures. *IEEE International Conference on Computer Design: VLSI in Computers and Processors,* IEEE Press, pp. 438–443.

Saeidi, S., Khademzadeh, A., and Mehran, A. 2007. SMAP: An intelligent mapping tool for network on chip. *International Symposium on Signals, Circuits and Systems,* IEEE Press, pp. 1–4.

Saeidi, S., Khademzadeh, A., and Vardi, F. 2009. Crinkle: A heuristic mapping algorithm for network on chip. *IEICE Electronics Express,* vol. 6, no. 24, pp. 1737–1744.

Sahu, P. K., Shah, N., Manna, K., and Chattopadhyay, S. 2010. A new application mapping algorithm for mesh based network-on-chip design. *IEEE International Conference,* IEEE Press, pp. 1–4.

Sahu, P. K., Shah, N., Manna, K., and Chattopadhyay, S. 2011a. An application mapping technique for butterfly-fat-tree network-on-chip. *IEEE International Conference on Emerging Applications and Information Technology,* IEEE Press, pp. 383–386.

Sahu, P. K., Shah, N., Manna, K., and Chattopadhyay, S. 2011b. A new application mapping strategy for mesh-of-tree based network-on-chip. *IEEE International Conference on Emerging Trends in Electrical and Computer Technology,* IEEE Press, pp. 518–523.

Sahu, P. K., Shah, T., and Chattopadhyay, S. 2012. Application mapping onto mesh based network-on-chip using discrete particle swarm optimization. *IEEE Transactions on VLSI* vol. 22, no. 2, pp. 300–312.

Sahu, P. K., Venkatesh, P., Gollapalli, S., and Chattopadhyay, S. 2011c. Application mapping onto mesh structured network-on-chip using particle swarm optimization. *IEEE International Symposium on VLSI,* IEEE Press, pp. 335–336.

Sepulveda, M. J., Strum, M., and Chau, W. J. 2009. A multi-objective adaptive immune algorithm for NoC mapping. *International Conference on Very Large Scale Integration,* IEEE Press, pp. 193–196.

Sepulveda, M. J., Strum, M., Chau, W. J., and Gogniat, G. 2011. A multi-objective approach for multi-application NoC mapping. *IEEE Latin American Symposium on Circuits and Systems,* IEEE Press, pp. 1–4.

Shen, T., Chao, C. H., Lien, Y. K., and Wu, A. Y. 2007. A new binomial mapping and optimization algorithm for reduced-complexity mesh-based on-chip network. *Proceedings of NOCS'07,* IEEE Press, pp. 317–322.

Srinivasan, K., and Chatha, K. S. 2005. A technique for low energy mapping and routing in network-on-chip architecture. *IEEE International Symposiun on Low Power Electronics and Design*, IEEE Press, pp. 387–392.

Srinivasan, K., Chatha, K. S., and Konjevod, G. 2006. Linear-programming-based techniques for synthesis of network-on-chip architectures. *IEEE Transactions on Very Large Scale Integration (VLSI) Systems*, vol. 14, no. 4, pp. 407–420.

Tavanpour, M., Khademzadeh, A., and Janidarmian, M. 2009. Chain-mapping for mesh based network-on-chip architecture. *IEICE Electronics Express*, vol. 6, no. 22, pp. 1535–1541.

Tavanpour, M., Khademzadeh, A., Pourkiani, S., and Yaghobi, M. 2010. GBMAP: An evolutionary approach to mapping cores onto a mesh-based NoC architecture. *Journal of Communication and Computer*, vol. 7, no. 3, pp. 1–7.

Tornero, R., Sterrantino, V., Palesi, M., and Orduna, J. 2009. A multi-objective strategy for concurrent mapping and routing in networks on chip. *IEEE International Symposium on Parallel and Distributed Processing*, IEEE Press, pp. 1–8.

Tosun, S. 2011a. Clustered-based application mapping method for network-on-chip. *Journal of Advances in Engineering Software* vol. 42, no. 10, pp. 868–874.

Tosun, S. 2011b. New heuristic algorithm for energy aware application mapping and routing on mesh-based NoCs. *Journal of System Architecture*, vol. 57, pp. 69–78.

Tosun, S., Ozturk, O., and Ozen, M. 2009. An ILP formulation for application mapping onto network-on-chips. *International Conference on Application of Information and Communication Technologies*, IEEE Press, pp. 1–5.

Wang, J., Li, Y., Chai, S., and Peng, Q. 2011. Bandwidth-aware application mapping for NoC-based MPSoCs. *Journal of Computational Information Systems*, vol. 7, no. 1, pp. 152–159.

Wang, K., Huang, L., Zhou, C., and Pang, W. 2003. Particle swarm optimization for traveling salesman problem. *Proceedings of the Second International Conference on Machine Learning and Cybernetics*, IEEE Press, pp. 1583–1585.

Wang, X., Yang, M., Jiang, Y., and Liu, P. 2009. Power-aware mapping for network-on-chip architectures under bandwidth and latency constraints. *International Conference on Embedded and Multimedia Computing*, IEEE Press, pp. 1–6.

Wang, X., Yang, M., Jiang, Y., and Liu, P. 2010. Power-aware mapping approach to map IP cores onto NoCs under bandwidth and latency constraints. *ACM Transactions on Architecture and Code Optimization*, vol. 7, no. 1, pp. 1–30.

Yang, B., Guang, L., Xu, T. C., Santti, T., and Plosila, J. 2010a. Multi-application mapping algorithm for network-on-chip platforms. *IEEE 26th Convention of Electrical and Electronics Engineers in Israel*, IEEE Press, pp. 540–544.

Yang, B., Xu, T. C., Santti, T., and Plosila, J. 2010b. Tree-model based mapping for energy-efficient and low-latency network-on-chip. *International Symposium on Design and Diagnostics of Electronics Circuits and Systems*, IEEE Press, pp. 189–192.

Yu, M. Y., Li, M., Song, J. J., Fu, F. F., and Bai, Y. X. 2009. Pipelining-based high throughput low energy mapping on network-on-chip. *Euromicro International Conference on Digital System Design/Architectures, Methods and Tools*, IEEE Press, pp. 427–432.

Zhou, W., Zhang, Y., and Mao, Z. 2006. An application specific NoC mapping for optimized delay. *IEEE International Conference on Design and Test of Integrated Systems in Nanoscale*, IEEE Press, pp. 184–188.

Zhou, W., Zhang, Y., and Mao, Z. 2007. Link-load balance aware mapping and routing for NoC. *WSEAS Transactions on Circuits and Systems*, vol. 6, no. 11, pp. 583–591.

6

Low-Power Techniques
for Network-on-Chip

6.1 Introduction

As the number of processing elements keeps on increasing in network-on-chip (NoC), power consumption is one of the major concerns in designing such systems since it affects their battery life and packaging costs for heat dissipation. The increasing power density not only raises packaging and cooling challenges, but also enhances reliability problems as the *mean time between failures* (MTBF) decreases exponentially with temperature. In addition, timing requirement degrades and leakage current increases with temperature. Since last decade power consumption has not been a primary concern in chip design, while the cost, area, and timing issues were mostly being addressed by the designers. Today, in ultra-deep submicron (UDSM) technology, the power budget is one of the important goals for most system-on-chip (SoC) designs. Exceeding the power budget will increase the packaging cost, thermal design, and regulator design, and will also affect the timing, reliability, and battery life.

As power minimization is one of the major design challenges, this chapter addresses different low-power techniques that have been adopted in the NoC paradigm. The rest of the chapter has been organized as follows: this section briefly discusses about the different power components. Sections 6.2 and 6.3 discuss about the standard low-power methods for NoC routers and links, respectively. Section 6.4 describes the different system-level power reduction techniques such as dynamic voltage scaling (DVS), dynamic frequency scaling (DFS), voltage–frequency island (VFI) partitioning, and runtime power gating. Finally, Section 6.5 summarizes the chapter.

Total chip power consumption can be split into dynamic power, leakage power, interconnect power, and IO power. The two components that constitute dynamic power consumption are switching power and internal power. The switching power consumption can be described as

$$P_{sw} = \alpha \times C_L \times V_{swing} \times V_{DD} \times f_{clock} \tag{6.1}$$

where:

α is the switching activity factor
C_L is the load capacitance
V_{swing} is the voltage swing on the output node
V_{DD} is the supply voltage of the gate
f_{clock} is the clock frequency

Switching activity factor (α) is the probability that a clock event results in a $0 \rightarrow 1$ event at the output of the gate and its maximum value is 1 (for clock buffer). Load capacitance (C_L) is the summation of the output capacitance of the driver, the wiring capacitance, and the input capacitance of the loading gate.

Internal power, however, consists of a short-circuit power that occurs when both n-type metal oxide semiconductor and p-type metal oxide semiconductor transistors are ON, and an internal node switching power that is required to charge the internal capacitance of the cell. The internal power consumption can be described as

$$P_{int} = t_{sc} \times V_{DD} \times I_{peak} \times f_{clock} \tag{6.2}$$

where:

t_{sc} is the time duration of the short-circuit current
I_{peak} is the total internal switching current (the short-circuit current + the current required to charge the internal capacitance)

Leakage power of a chip is consumed when the device is powered ON but no signals are changing value. There are four main sources of leakage current in a complementary metal oxide semiconductor (CMOS) gate as follows:

1. *Subthreshold leakage current* (I_{sub}) flows from drain to source of a transistor operating in the weak inversion region where the gate is not completely turned off. I_{sub} increases exponentially with increasing temperature and decreasing threshold voltage (V_t) of the transistor. The expression for I_{sub} can be written as follows:

$$I_{sub} = \mu \times C_{ox} \times V_{th}^2 \times \left(\frac{W}{L}\right) \times e^{(V_{GS} - V_t)/nV_{th}} \tag{6.3}$$

where:

W and L are the channel width and length of a transistor, respectively
V_{th} is the thermal voltage given by kT/q (25.9 mV at room temperature)
The parameter n is a function of the device fabrication process and ranges from 1.0 to 2.5
The parameter μ represents the career mobility
C_{ox} denotes the gate oxide capacitance per unit area
k denotes Boltzman constant = 1.3806×10^{-23} J/K

T is absolute temperature in Kelvin

q is electrical charge of electron $= 1.6 \times 10^{-19}$ Coulomb

2. *Gate-induced drain leakage current* flows from drain to substrate induced by high field effect in the drain caused by a high V_{DG}. Gate-induced drain leakage current increases exponentially with increasing temperature and decreasing oxide thickness.

3. *Gate leakage current* flows directly from the gate through the oxide to the substrate due to gate oxide tunneling and hot carrier injection. It increases exponentially with decreasing oxide thickness. In previous technology nodes (130 nm and above), gate leakage current was negligible, but starting from the 90-nm technology node, gate leakage can be comparable with subthreshold leakage current. In future technology nodes (28 nm and below), high-k dielectric materials will be needed to keep gate leakage under control.

4. *Reverse-bias junction leakage current* flows due to drift of minority carriers and generation of electron–hole pairs in the depletion regions.

Leakage power is greatly influenced by process, voltage, and temperature. One of the well-known facts about the leakage current is its significantly large variability due to manufacturing conditions and environmental variations. For the above reasons, leakage power can vary by orders of magnitude for different chips manufactured with same design, and hence, a statistical leakage model is necessary in the UDSM technology (Lu and Agarwal 2007).

In Chapter 4, we have shown that interconnect power consumes a very significant portion of the total chip power. The interconnect power mostly depends on the voltage swing, the driver size, the parasitic capacitance of the wire per unit length, and the length of the wire, whereas IO power is mostly considered as analog power.

The most effective way to reduce the dynamic power is to reduce V_{DD}, but the trouble is that it tends to reduce the drain-to-source current (I_{DS}) and hence slower the speed. If we ignore velocity saturation and other subtle effects that are observed in below 90-nm technology, I_{DS} can be expressed as follows:

$$I_{DS} = \mu \times C_{ox} \times \left(\frac{W}{L}\right) \times \frac{(V_{GS} - V_t)^2}{2} \tag{6.4}$$

From the above equation, it is clear that to maintain the I_{DS} to achieve performance target, V_t has to be reduced as we reduce V_{DD} (and hence V_{GS}). But lowering V_t will cause exponential increase of I_{sub} as mentioned above. Hence there is a conflict between dynamic and leakage power, which needs to be addressed in any low-power design in DSM technology.

In the subsequent sections, we will focus the above issues of low-power design. First, in Section 6.2, we will address the standard low-power techniques that have been adopted to reduce the power consumption of NoC routers.

6.2 Standard Low-Power Methods for NoC Routers

There are a number of power reduction techniques that have been widely used in Very Large Scale Integration (VLSI) design and also adopted in NoC router. This section gives an overview of the following methods: (1) clock gating, (2) gate-level power optimization, (3) multi-V_{DD}, (4) multi-V_T, and (5) power gating.

6.2.1 Clock Gating

The internal node switching power is dominated by the transition of clock in any sequential design. With every clock transition, the capacitances internal to the cell are either charging or discharging, and hence consume very high internal power. It can be observed from Figure 4.14 that the change in the total router energy consumption is not significant with variation in the offered load. The routers consume a significant amount of energy even at very low traffic, though switching of input data is low. This appears as the internal power consumption due to free-running clock dominates over the switching and leakage power. Therefore, to reduce the internal power, it is essential to stop the free-running clock, when the network is idle.

As the write clock (*wr-clk*) of first-in first-out (FIFO) is connected to all the write registers in a stack, gating *wr-clk* will reduce a significant amount of internal power. The clock gating in register for power minimization was well described in the work of Benini et al. (1994). Figure 6.1 shows the clock gating applied to gate the *wr_clk* in each write register of a FIFO. A falling

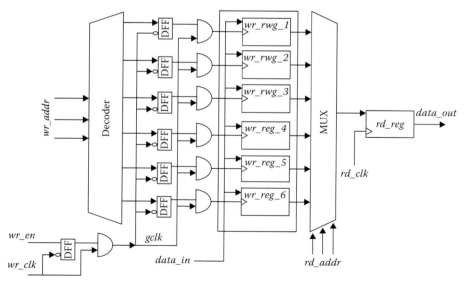

FIGURE 6.1
Gating the write clock of FIFO.

edge-triggered D flip-flop is used for this purpose. The input data to FIFO (*data_in*) and write enable (*wr_en*) signals are synchronous with the rising edge of *wr_clk*. When the network is idle, the *data_in* signal becomes invalid and the *wr_en* signal is at logic 0. Hence, the gated write clock (*gclk*) becomes active low. This *gclk* signal is again gated with the individual write enable signals (selected by the decoder) of the registers to perform register-level clock gating inside the FIFO. It is obvious that the clock gating in the write registers does not introduce any additional cycle latency in the FIFO.

Mullins (2006) proposed a router-level clock gating solution when the routers are idle by inserting a clock gating cell toward the root of the clock tree. The clock enable signal is constrained in a time of $T_{clk} - T_{insertion}$, where T_{clk} is the clock period and $T_{insertion}$ is the clock tree insertion delay. To address this problem, he generated an early-valid signal in each router for each of its outputs as shown in Figure 6.2. These signals are generated quickly and simply determine if it is possible that a particular output port will be used. These signals are then communicated to the router at the end of each output channel, serving as an indication of whether new data will be sent in the current clock cycle or not. In contrast to the actual network data, these signals arrive early enough in the clock cycle to be used in the generation of a router's clock enable signal.

6.2.2 Gate Level Power Optimization

Gate-level power optimization can be of different types and explained briefly as follows:

1. *Reordering of inputs*: The high-activity inputs should always be nearer to the output of the gate.

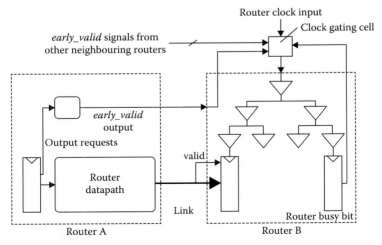

FIGURE 6.2
Router-level clock gating approach.

2. *Buffer insertion*: If any driver drives a long net, it is good to break the net by inserting buffer. This will improve both link delay and link power consumption.

3. *Cell sizing*: Proper adjustment of cell size will reduce the delay and also help to reduce dynamic power consumption.

4. *Logic restructuring*: It is explained here with an example. In Figure 6.3a, if the output of the AND gate is a high active net, it can be redesigned as Figure 6.3b such that the high active net is now inside the cell and hence less capacitance will cause less power consumption.

In any VLSI design, this gate-level optimization is taken care by the Computer Aided Design (CAD) tools. Mullins (2006) showed that using gate-level optimization the dynamic power of a NoC router can be reduced by 28% approximately.

6.2.3 Multivoltage Design

Multiple supply voltage (MSV) is the most frequently used in low-power design. This scheme has the advantage that the gates that are in noncritical paths operate at the low supply voltage, V_{DDL}, whereas the gates that are in critical paths operate at the high supply voltage, V_{DDH}. A multivoltage design can be categorized as follows:

1. *Static voltage scaling (SVS)*: Different blocks or subsystems are given different but fixed supply voltage.

2. *Multilevel voltage scaling (MVS)*: This is an extension of SVS where a block or a subsystem is switched between two or more voltage levels. Only a few, fixed, discrete levels are supported for different operating modes.

3. *Dynamic voltage and frequency scaling (DVFS)*: In this category, a large number of voltage levels are dynamically switched based on changing workloads.

4. *Adaptive voltage scaling (AVS)*: This is an extension of DVFS where a control loop is used to adjust the voltage.

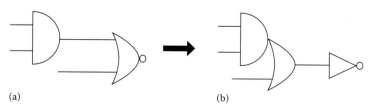

(a)   (b)

FIGURE 6.3
Logic restructuring. (a) Original logic with high active net at the output of AND gate; (b) Restructured logic where high active net is inside the AND–OR cell.

6.2.3.1 Challenges in Multivoltage Design

When a signal traverses from a low-voltage domain (V_{DDL}) to a high-voltage domain (V_{DDH}) or vice versa, the circuit designers face several challenges as follows:

6.2.3.1.1 Short-Circuit Current Flow

Figure 6.4 shows a CMOS logic circuit consisting of first and second CMOS inverters that are directly connected to each other. The first CMOS inverter operates on a lower supply voltage V_{DDL} and the second CMOS inverter on a higher supply voltage V_{DDH}. If $V_{DDL} < V_{DDH} - |V_{thp}|$, the MP2 (is the PMOS transistor of second inverter in Figure 6.4) is incompletely turned off and the short-circuit current flows from a power supply of the higher supply voltage V_{DDH} toward a ground through the second inverter. However, while traversing from a high-voltage domain to a low-voltage domain, the transistor will be overstressed and will cause potential unreliability due to high-voltage input. If V_{GS} or V_{GD} of a transistor exceeds a certain voltage value, the transistor will be overstressed.

To address these issues, inserting a *level shifter* in the voltage domain crossing is utmost necessary. Here we will describe both types of level shifters briefly.

6.2.3.1.1.1 High-to-Low Voltage Level Shifter High-to-low voltage level shifter design has essentially two inverters in series, so it introduces only a single buffer delay. Therefore, the impact of timing is small. Figure 6.5 depicts this scenario.

6.2.3.1.1.2 Low-to-High Voltage Level Shifter Figure 6.6 shows a conventional level shifter, named *dual cascode voltage switch* (DCVS), inserted between gates operating at low- and high-voltage domain. Assume that nodes A and B are initialized at low and high voltages, respectively. When there is a high-to-low transition in input signal X, both MP3 and MN2 are turned on, whereas MP2

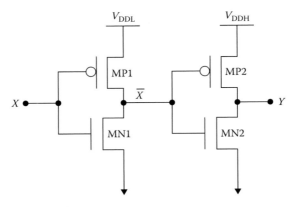

FIGURE 6.4
Low-to-high voltage crossing.

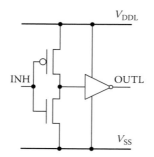

FIGURE 6.5
High-to-low voltage level shifter. INH, input at high voltage; OUTL, output at low voltage.

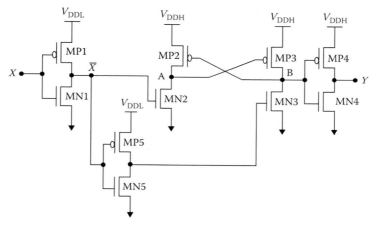

FIGURE 6.6
Conventional low-to-high voltage level shifter.

and MN3 are turned off. Thus, there is no direct path from V_{DDH} to V_{SS}, and hence it prevents the short-circuit current and reduces power consumption. Although the level shifter reduces the short-circuit current, it consumes relatively large dynamic power when it carries out a switching operation. Furthermore, this level shifter has relatively higher delay because it relies on contention between different transistors on the level conversion path.

Yuan and Cheng (2005) proposed an improved circuit as shown in Figure 6.7 to reduce the contention problem so as to achieve high-speed and low-power consumption. In this circuit, the level shifter circuit converts a signal X on the lower supply voltage side into a signal Y on the higher supply voltage side. The signal $\bar{X}$ is transmitted to the gates of transistors MN3 (MN3 is the NMOS transistor of inverter as shown in Figures 6.6 and 6.7), MN6, and MP6. A signal x for the inverted phase, which is generated by an inverter constituted by transistors MP1 and MN1, is transmitted to the gates of transistors MN2, MN5, and MP5. The respective gates of transistors MP2 and MP3

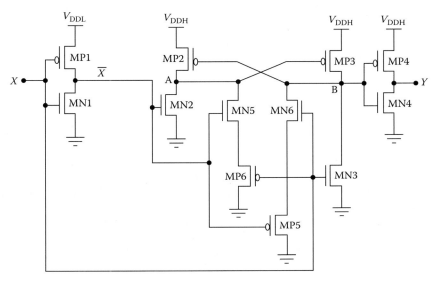

FIGURE 6.7
New low-to-high voltage level shifter.

are cross-connected to the drains of transistors MP3 and MP2, whereas the sources of both transistors are connected to the higher supply voltage V_{DDH}. A node B is connected to an output buffer circuit, constituted by transistors MP4 and MN4, which is connected to the higher supply voltage V_{DDH}.

The operation of the new level converter circuit is explained here. When the voltage level of the input signal swings high to low, the output voltage level of the input inverter becomes the lower supply voltage V_{DDL}. Therefore, MN2, MN5, and MP6 are turned on. As a result, node A is then discharged to a reference voltage V_{SS}. Thus, MP3 becomes on, and then the voltage level of node B becomes a higher supply voltage V_{DDH}. In this case, MP2, MN3, MN6, and MP5 are turned off, and therefore, it is possible to prevent a short-circuit current from flowing between the higher power supply voltage V_{DDH} and the reference voltage V_{SS}. However, when the voltage level of the input signal is switched to logic high, the output voltage level of the input inverter becomes a reference voltage V_{SS}. Therefore, MN3, MN6, and MP5 are turned on. As a result, node B is then discharged to a reference voltage V_{SS}. In this case, MP2 is turned on, and therefore, MP3 becomes off. Moreover, MN2, MN5, and MP6 are also turned off, and thus, it is possible to prevent a short-circuit current from flowing between the higher power supply voltage V_{DDH} and the reference voltage V_{SS}. It is clear that there are three paths to speed up the output level transition in each input signal condition. These results provide faster output transitions as well as an efficient voltage level conversion. As such, it can reduce the contention problem on nodes A and B. As a result, the propagation delay time of the circuit itself becomes short. Moreover, no short-circuit current flows; therefore, it is possible to reduce power consumption.

6.2.3.1.2 Placement of Level Shifter

Multivoltage designs present significant challenges in placement. Figure 6.8 depicts such scenario where two voltage domains are embedded in a third voltage domain. For example, when a signal traversing from a 0.9 V domain to a 1.2 V domain through a 1.1 V domain, power routing will be a challenge, no matter where the level shifter is placed. As the low-to-high voltage level shifter requires both rails, at least one of the rails will have to be routed from another domain. Since the output driver requires more current than the input stage, it is better to place the level shifter in the 1.2 V domain.

If the distance between 0.9 and 1.2 V domains is adequate and an additional buffer is needed to be placed in the 1.1 V domain, it uses the power rail of 0.9 V domain. In this case, 0.9 V rail must be routed in the 1.1 V domain as a signal wire as shown in Figure 6.8. This type of complex power routing is one of the key challenges in automating the implementation of multivoltage design.

For high-to-low voltage level shifter, it is recommended to place the level shifter in the lower voltage domain. If an additional buffer is needed to place in the third voltage domain, 1.2 V rail has to be routed in the 1.1 V domain as a signal wire, which becomes a power routing challenge as mentioned above.

6.2.4 Multi-V_T Design

With the shrinking technology, leakage power is one of the major design challenges. Multi-V_T design is very useful to reduce the leakage power. Many libraries today offer multiple versions of their cells: ultra-low V_T, low V_T, standard V_T, high V_T, and so on. The cell delay increases and the leakage power decreases with the rising V_T. Moreover, for each V_T type, there exist two to three types of channel length. The leakage power increases and the cell delay decreases with decreasing channel length. For a design where performance is the foremost criterion and power reduction is a secondary issue, the CAD tools choose lower V_T cells with lesser channel length during synthesis. Once

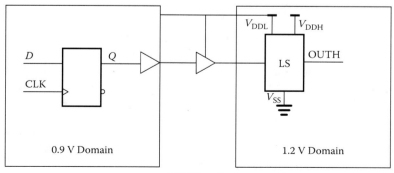

FIGURE 6.8
Placement of low-to-high voltage level shifter.

timing is met, if some positive slack exists in the critical path, optimization tools try to replace some of the lower V_T cells with higher V_T cells and/or with cells having longer channel length. In case of noncritical path, optimization tools also replace the same until they become the critical one.

6.2.5 Power Gating

The principle of power gating is to selectively powering down certain blocks in the chips while keeping other blocks powered up. The goal of power gating is to minimize leakage current by temporarily switching some blocks to the power-down mode that are not required to be in the active mode while minimizing the impact on performance. Power gating is more persistent than clock gating that it affects interblock interface communication and adds significant time delays to safely enter and exit the power-down mode. Power gating to some portions of the design can be controlled by software as a part of device drivers or initiated in hardware by timers or system-level power management controllers. Architectural trade-offs in any power-gated design are as follows:

1. The amount of possible leakage power savings
2. The energy dissipated during entering and leaving such leakage saving modes
3. Frequency of entering into the power gating and active modes
4. Performance penalty during entry and exit times

Figure 6.9 shows a typical example of leakage power saving in a clock gated design due to power gating. During the active state, the circuit consumes both dynamic and leakage power, whereas during the idle state, it consumes only leakage power. During the *sleep* mode, the transition to a power-down state is not instantaneous. It takes several cycles to enter into that state. Similarly,

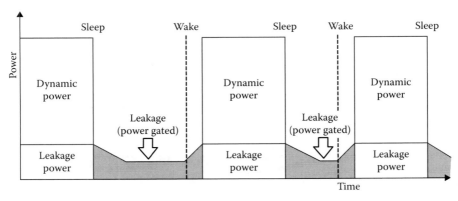

FIGURE 6.9
Leakage power-saving profile using power gating.

after *wake* time, it needs some cycles to go to the active state, which causes performance penalty. Power gating can be applied to circuit blocks with various granularities. Depending on the granularity of target circuit blocks (i.e., power domains), the power gating is classified into coarse-grained and fine-grained approaches.

In fine-grain power gating, the switch is placed locally inside each standard cell. Since the switch must supply the worst-case current required by the cell, it has to be quite large in order not to impact performance. This approach has received a lot of attention in recent years because of its flexibility and short wake-up latency. In coarse-grain power gating, a block of gates has its power switched by a collection of switch cells. The sizing of a coarse-grain switch network is more difficult than that of a fine-grain switch network as the exact switching activity of the logic it supplies is not known and can only be estimated. But coarse-grain gating designs have significantly less area penalty than fine-grain gating designs. Each target circuit block is surrounded by a power/ground ring. Power switches are inserted between the core ring and the power/ground IO cells. The power supply to the circuit block can be controlled by the power switches. Since the power supply to all cells inside the core ring is controlled at one time, this approach is well suited to the IP- or module-level power management. The coarse-grained approach has been popularly used, since its IP- or module-level power management is straightforward and easy to control. However, it typically imposes a microsecond order wake-up latency.

The implementation of power gating presents certain challenges to the designer. which include the following:

1. Design of power switching fabric
2. Design of power gating controller
3. Selection and use of retention registers and isolation cells
4. Minimization of the impact of power gating on timing and area
5. The functional control of clocks and resets
6. Interface isolation
7. Constraint development for implementation and analysis

6.3 Standard Low-Power Methods for NoC Links

There are a number of power reduction techniques for interconnects that have been widely used in VLSI design and also adopted in NoC. This section gives an overview of the following methods: (1) low-power coding (LPC), (2) on-chip serialization, and (3) low-power signaling. First, we will start with bus energy model as described in Section 6.3.1.

6.3.1 Bus Energy Model

In on-chip interconnect, lines are assumed to be distributed, lossy, and capacitively and inductively coupled. The energy model of such interconnect is described in the work of Sotiriadis and Chandrakasan (2002). The effect of inductance (L) can be neglected if $f \ll R/(2\pi L)$ (where f is the frequency and R denotes the bus resistance), which is true in most NoC interconnects. Thus, NoC interconnect in DSM era can be modeled as resistance-capacitance network (Benini and Micheli 2006). The model of an n-wire interconnect in parallel is shown in Figure 6.10. In the figure, C_s and C_c are the substrate and coupling capacitances, respectively; R_i represents the on–off resistance of the ith driver; V_j^i and V_j^f denote the initial and final voltages, respectively, in the jth interconnect. The ratio of coupling capacitance to substrate capacitance is denoted as λ ($\lambda = C_c/C_s$). It is a technology parameter and increases with shrinking technology feature size.

Sotiriadis and Chandrakasan (2000) proposed a three-wire bus energy model. The proposed bus energy model is presented below. Using Kirchoff's current law, the current (I) equation of each line is

$$C_s \times \left[(1+\lambda) \times \frac{dV_1}{dt} - \lambda \times \frac{dV_2}{dt} \right] = \frac{V_1^f}{R_1} - \frac{V_1^i}{R_1}$$

$$C_s \times \left[-\lambda \times \frac{dV_{k-1}}{dt} + (1+2\lambda) \times \frac{dV_k}{dt} - \lambda \times \frac{dV_{k+1}}{dt} \right]$$

$$= \frac{V_k^f}{R_k} - \frac{V_k^i}{R_k}, \text{ where } k = 2,3,4,\ldots, (n-1)$$

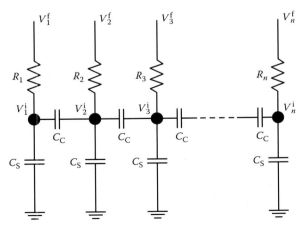

FIGURE 6.10
DSM model of n interconnects.

$$C_s \times \left[-\lambda \times \frac{\mathrm{d}V_{n-1}}{\mathrm{d}t} + (1+\lambda) \times \frac{\mathrm{d}V_n}{\mathrm{d}t} \right] = \frac{V_n^{\mathrm{f}}}{R_k} - \frac{V_n^{\mathrm{i}}}{R_k} \tag{6.5}$$

The energy consumed (or deposited) by each line considering the substrate capacitance and the effect of coupling capacitance of adjacent lines only is given below. Both substrate and coupling capacitances depend on the length of the interconnect wire. Thus, energy will also depend on the length of the wire (Sotiriadis and Chandrakasan 2002).

$$E_1 = C_s \times \left[(1+\lambda) \times \left(V_1^{\mathrm{f}} - V_1^{\mathrm{i}} \right) - \lambda \times \left(V_2^{\mathrm{f}} - V_2^{\mathrm{i}} \right) \right] \times V_1^{\mathrm{f}}$$

$$E_k = C_s \times \left[-\lambda \times \left(V_{k-1}^{\mathrm{f}} - V_{k-1}^{\mathrm{i}} \right) + (1+2\lambda) \times \left(V_k^{\mathrm{f}} - V_k^{\mathrm{i}} \right) - \lambda \times \left(V_{k+1}^{\mathrm{f}} - V_{k+1}^{\mathrm{i}} \right) \right] \times V_k^{\mathrm{f}},$$

where $k = 2, 3, \ldots, (n-1)$

$$E_n = C_s \times \left[-\lambda \times \left(V_{n-1}^{\mathrm{f}} - V_{n-1}^{\mathrm{i}} \right) + (1+\lambda) \times \left(V_n^{\mathrm{f}} - V_n^{\mathrm{i}} \right) \right] \times V_n^{\mathrm{f}} \tag{6.6}$$

6.3.2 Low-Power Coding

Dynamic power dissipation in the bus depends on the number of transitions per time slot. Codes that reduce the average transition activity are referred to as low-power codes (LPCs). In general, transitions in data and address buses are different. For example, transitions on a typical data bus are random in nature. A simple but effective LPC for a data bus is *bus-invert* (BI) code (Stan and Burleson 1995) in which the data are inverted and an invert bit is sent to the decoder if the current data word differs from the previous data word in more than half the number of bits. BI coding is not efficient for buses of higher width. For wide buses, the bus is partitioned into several sub-buses each with its own invert bit (Yoo and Choi 1999). The BI method generates a code to reduce the maximum number of transitions per time slot from n to $n/2$; thus, the average and peak power dissipation of the bus can be reduced by half. Figure 6.11 describes the hardware of BI coding. The overall methodology is shown in Table 6.1 with an example. The coding methodology is explained as follows:

1. Compute the Hamming distance between the previous data value and the present one.
2. If the Hamming distance is larger than $n/2$, set *invert* = 1 and make the present bus value equal to the invert of the present data value.
3. Otherwise, set *invert* = 0 and make the present bus value equal to the present data value.
4. At the receiver side, depending on the status of the *invert* line, the contents of the bus is conditionally inverted.

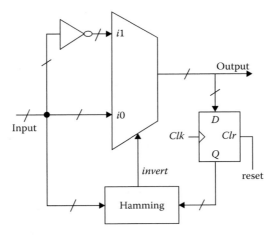

FIGURE 6.11
Hardware of BI coding.

TABLE 6.1

BI Coding Scheme with an Example of 8-bit Data Bus

	Time→																Time→													
D0	1	0	0	0	0	1	0	0	1	1	0	1	1	0	0	D0	1	0	0	0	0	0	0	1	0	0	1	1	0	1	0
D1	1	0	0	0	0	1	0	1	0	1	1	0	1	1	0	D1	1	0	0	0	0	0	0	0	1	0	0	0	0	0	0
D2	0	1	1	0	0	1	0	1	0	0	0	1	0	0	1	D2	0	1	1	0	0	0	0	0	1	1	1	1	1	1	1
D3	1	1	1	1	0	0	0	0	1	1	0	0	0	0	1	D3	1	1	1	1	0	1	0	1	0	0	1	0	1	1	1
D4	0	0	0	1	1	0	0	0	0	1	1	1	0	0	1	D4	0	0	0	1	1	1	0	1	1	0	0	1	1	1	1
D5	0	1	0	1	0	1	0	1	1	0	0	1	1	0	0	D5	0	1	0	1	0	0	0	0	0	1	1	1	0	1	0
D6	1	1	0	0	1	1	1	0	0	0	1	0	1	0	0	D6	1	1	0	0	1	0	1	1	1	1	0	0	0	1	0
D7	1	1	0	0	0	1	0	1	1	0	0	1	0	0	1	D7	1	1	0	0	0	0	0	0	0	1	1	1	1	1	1
																inv	0	0	0	0	0	1	0	1	1	1	1	0	1	1	0
	(Input)																**(Output)**													

BI coding can minimize only self-transition of individual wires and it is non-linear. Sotiriadis (2002) showed that linear codes do not minimize the transition activity. BI has no impact on coupling between two adjacent wires. In DSM buses, both self-transitions and coupling transitions contribute to the power dissipation. In Chapter 7, we have shown a joint coding scheme to reduce both self-transitions and coupling transitions. Ghoneima and Ismail (2004), Kim et al. (2000), and Zhang et al. (2002) proposed another LPC scheme that reduces both self-transitions and coupling transitions by conditionally inverting the bus based on a metric that accounts for both the transitions at the price of increased complexity. While the above-mentioned works describe the methods of eliminating cross talk and/or reducing dynamic power,

TABLE 6.2

Binary and Gray Coding Scheme for 4-bit Address Bus

	Time→																	Time→															
D0	0	1	0	1	0	1	0	1	0	1	0	1	0	1	0	1	D0	0	1	1	0	0	1	1	0	0	1	1	0	0	1	1	0
D1	0	0	1	1	0	0	1	1	0	0	1	1	0	0	1	1	D1	0	0	1	1	1	1	0	0	0	0	1	1	1	1	0	0
D2	0	0	0	0	1	1	1	1	0	0	0	0	1	1	1	1	D2	0	0	0	0	1	1	1	1	1	1	1	1	0	0	0	0
D3	0	0	0	0	0	0	0	0	1	1	1	1	1	1	1	1	D3	0	0	0	0	0	0	0	0	1	1	1	1	1	1	1	1
	(Binary Sequence)																	**(Gray Sequence)**															

Deogun et al. (2004) proposed an encoding scheme that also tackles the rising runtime leakage power levels in such buses along with cross talk and dynamic power. They introduced a new buffer design approach with selective use of high-threshold voltage transistors and coupled this buffer design with a novel bus encoding scheme. For any LPC scheme, a trade-off analysis has to be performed between bus energy reduction and the amount of extra energy consumed by the codec. For technologies such as 90 nm or above, Sridhara and Shanbhag (2005) showed that codec overhead is more in BI scheme than in bus energy saving, but this trade-off will be increasingly favorable in future technologies.

Generally, an address bus will tend to have a sequential behavior; hence, a gray coding scheme is perfect for an address bus where only one transition is occurred per time slot. In actual design, it is always advisable to implement the gray counter from its finite state machine. Table 6.2 presents both binary and gray sequences for an address bus. Although a sequential value on the address bus is generally too simplistic, for a real system only some percentage of bus addresses are typically sequential with the others being essentially random. In such case, a mixed coding, gray and BI coding, will give the best results for both peak and average power dissipation in the bus.

6.3.3 On-Chip Serialization

Bus encoding techniques enlarge the physical transfer unit in NoC. Large physical transfer unit increases the network area and energy consumption, especially for switching circuit and buffering units in switch fabrics. On-chip serializer and deserializer can be used to reduce the physical transfer unit size and further reduce the area and energy consumption of the switch fabric. It reduces the overall network area and optimizes power consumption, which is well explained in the work of Lee et al. (2004, 2005). The power consumption decreases with the increasing ratio of serializer under low frequency. Unfortunately, with the increasing ratio of serialization under higher frequency, the power consumption increases because of large driver to provide high driving ability. Huang et al. (2008) observed that a 4:1 serializer is an optimized ratio to achieve energy saving. Chuang et al. (2008) implemented the serializer

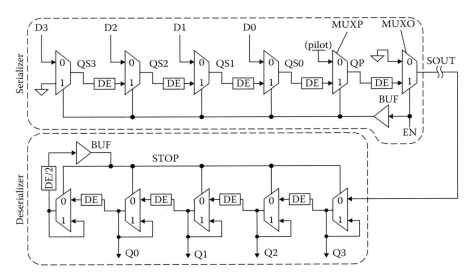

FIGURE 6.12
4:1 Serializer and deserializer.

and deserializer with a digital self-calibrated multiphase delay-locked loop. Lee et al. (2006) proposed another implementation scheme for high-speed and low-overhead 4:1 serializer/deserializer for practical NoCs. The fundamental idea of the implementation scheme is like parallel-to-serial converter (serializer) and serial-to-parallel converter (deserializer) using a shift register. Instead of using D flip-flop in a shift register, the authors used constant delay elements (DEs) such that $T_{DE} < T_{clk}$. The overall scheme is shown in Figure 6.12.

When EN is low, D<3:0> waits at QS<3:0>. The VDD input of MUXP, which is called a pilot signal, is loaded to QP. The GND input of MUXO discharges the serial output (SOUT), while the serializer is disabled. If EN is asserted, QS<3:0> and the pilot signal start to propagate through the serial link wire. Each signal forms a wave front of the SOUT signal, and the timing distance between the wave fronts is the DE and MUX delay which we call a unit delay. The series of wave fronts propagate to the deserializer like a train. When the SOUT signal arrives at the deserializer, it propagates through the deserializer until the pilot signal arrives at the end of the deserializer, or STOP node. As long as the unit delay times of the sender and the receiver are the same, D<3:0> arrives at its exact position when the pilot signal arrives at the STOP node. When the STOP signal is asserted, the MUXs feed back its output to its input, so that the output value is latched.

6.3.4 Low-Swing Signaling

Lowering the swing and driving voltages is the most effective way to reduce the power dissipation on interconnections. Figure 6.13 depicts such scheme

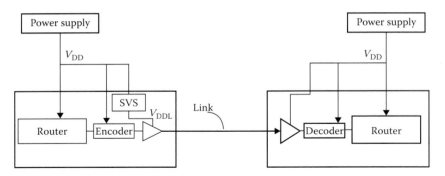

FIGURE 6.13
Low-swing signaling in NoC interconnect.

where SVS is used to supply the lower voltage (V_{DDL}) to the driver. Hence, a high-to-low voltage level shifter is needed in the transmitting router as shown in Section 6.2.3.1. Similarly, in the receiving router side, a low-to-high voltage level shifter is an utmost requirement. The amount of energy saving due to low-swing signaling is shown as follows:

$$E_{saving} = E_{uncoded\_link} - (E_{codec} + E_{coded\_link}) \tag{6.7}$$

6.4 System-Level Power Reduction

The total power consumption in a SoC is the combination of dynamic and leakage power. For dynamic power reduction at system level, DVS/DFS is very well known in VLSI community and also adopted in NoC design.

6.4.1 Dynamic Voltage Scaling

This section highlights the work cited by Shang et al. (2003) on DVS. Dynamic power consumption can be reduced by lowering the supply voltage. This requires reducing the clock frequency accordingly to compensate for the additional gate delay due to the lower voltage. The basic idea is that because of high variance in network traffic, when a link is underutilized, the link can be slowed down without affecting performance.

A variable frequency link consists of the components of a typical high-speed link: a transmitter to convert digital binary signals into electrical signals, a signaling channel, a receiver to convert electrical signals back to digital data, and a clock recovery block to compensate for delay through the signaling channel. In addition, it needs an adaptive power supply regulator that tracks the link frequency, regulates the voltage to the minimum level required, and feeds the regulated supply voltage to multiple links of a network channel, amortizing

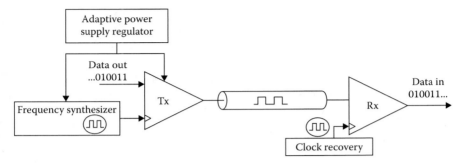

FIGURE 6.14
Components of a DVS link. Rx, receiver; Tx, transmitter.

its area and power costs. To extend variable frequency links to DVS links, an additional frequency synthesizer is needed that supplies the user-controlled frequency to the power supply regulator, as shown in Figure 6.14.

The important characteristics of a DVS link are (1) transition time—how long it takes a link to change from voltage level V_1 to V_2, (2) transition energy—the overhead energy consumed for a transition from V_1 to V_2, (3) transition status—whether the link functions during a transition, and (4) transition step—whether the link supports a continuous range of voltages, or if it only supports a fixed number of voltage levels.

Shang et al. (2003) constructed a multilevel DVS link model that supports 10 discrete frequency levels and their corresponding voltage levels. At each frequency level, the link circuitry can function within a range of voltages. The voltage and frequency transitions between adjacent levels occur separately. When increasing the link speed, the voltage increases first, followed by the frequency. Conversely, when decreasing the link speed, the frequency decreases first, followed by the voltage. The link functions during voltage transition but not during frequency transition. The latency of voltage (frequency) transition between adjacent levels has been assumed to be 10 µs (100 link clock cycles).

Transition energy is derived based on Stratakos's analysis (Stratakos 1998), where the energy overhead when voltage transitions from V_1 to V_2 is calculated with the following first-order estimation equations:

$$\text{Energy}_{\text{overhead}} = (1-\eta) \times C \times \left| V_2^2 - V_1^2 \right|$$

where:
  C is the filter capacitance of the power supply regulator
  η is the power efficiency

Noise is another issue in the design of link circuitry. Different noise sources, power supply noise, cross talk, clock jitter, deviation of parameters, device mismatches, and so on could result in voltage and timing uncertainty. For DVS links, supply voltage reduction magnifies the noise sensitivity of link circuitry. Since a lower link frequency decreases the ratio of timing uncertainty to bit time, frequency reduction improves communication reliability. Bit error rate (BER) is a measure of performance in link circuitry design. Link designs can achieve 10^{-15} BER over a wide range of voltages (0.9–2.5 V) and frequencies (200–700 MHz). In this work, it is assumed that within the range of multiple voltage and frequency levels, link circuitry can always function above the noise margin and achieve low BER. Due to the timing uncertainty, during frequency transition, when the receiver is trying to lock in the input clock, link circuitry is disabled.

6.4.1.1 History-Based DVS

The policy controlling DVS links has to judiciously trade off power and performance, minimizing the power consumption of the network while maintaining high performance. We proposed a distributed history-based DVS policy, where each router port predicts future communication workload based on the analysis of prior traffic, and then dynamically adjusts the frequencies (and corresponding voltages) of its communication links to track network load.

6.4.1.1.1 Communication Traffic Characterization

Communication traffic characteristics can be captured with several potential network measures. An obvious measure for DVS links is link utilization, which is defined as follows:

Link utilization

$$LU = \frac{\sum_{t=1}^{N} A(t)}{N}, \quad 0 \leq LU \leq 1 \tag{6.8}$$

where:

$$A(t) = \begin{cases} 1 & \text{If traffic passes link } i \text{ in cycle } t \\ 0 & \text{If no traffic passes link } i \text{ in cycle } t \end{cases}$$

N is the number of link clock cycles, which is sampled within a history window size H defined in router clock cycles

Link utilization is a direct measure of traffic workload. A higher link utilization reflects that more data are sent to the next router. Assuming that history is predictive, a higher link frequency is needed to meet the performance requirement. Conversely, lower link utilization implies the existence of more idle cycles. Here, decreasing the link frequency can lead to power savings without significantly hurting performance.

To investigate how predictive link utilization is of network load, the utilization of a link within a two-dimensional (2D) 8 × 8 mesh network is traced. At low traffic workloads, contention for buffers and links is rare. In this case, link utilization is bounded by flit arrival rate that is slow at low traffic workloads. As network traffic increases, more flits are relayed between adjacent routers and the utilization of each corresponding link also increases. When the network traffic approaches to the congestion point, resource contention results in flits being stalled in input buffers, since they can be relayed to the next router only if free buffers are available. Limited available buffer space in the succeeding router begins to be a tighter constraint, causing link utilization to decrease. When the network is highly congested, inter-router flit transmission is totally constrained by the availability of free buffers. Link utilization thus starts to dip.

At low network loads, since the flit will not be stalled in the succeeding router, any increase in link delay directly contributes to the overall packet latency. At high network loads, flits will be stalled in the next router for a long time anyway. Getting there faster will not help. In this case, link frequency can be decreased more aggressively with minimal delay overhead. Hence, link utilization alone will not be sufficient for guiding the history-based DVS policy. The other two parameters—input buffer utilization and input buffer age—need to be investigated.

Input buffer utilization

$$
BU = \frac{\sum_{t=1}^{H} [F(t)/B]}{H}, \quad 0 \leq BU \leq 1 \tag{6.9}
$$

where:
$F(t)$ is the number of input buffers that are occupied at time t
B is the input buffer size

Input buffer age

$$
BA = \frac{\sum_{t=1}^{H} \sum_{i=1}^{D(t)} (t_{d_i} - t_{a_i})}{\sum_{t=1}^{H} D(t)} \tag{6.10}
$$

where:
$D(t)$ is the number of flits that leave the input buffer at cycle t of a history interval H
t_d is the departure time of flit i from this buffer
t_a is the arrival time of flit i at the input buffer

Input buffer utilization tracks how many buffers in the succeeding router of the link are occupied. Input buffer age determines how long flits stay in these input buffers before leaving. These measures reflect resource contention in the succeeding router. The input buffers downstream from the same link are tracked. Under low network traffic, resource contention is low and only few buffers are occupied. Flits also do not stay in the input buffers for long. Hence, both input buffer utilization and input buffer age are low. As input traffic increases, more flits are relayed between adjacent routers and resource contention increases, being reflected in higher input buffer utilization and input buffer age. When the network is highly congested, most of the buffers are filled, and flits are stalled within a router for a long time. Both buffer utilization and age thus rise dramatically.

Both input buffer utilization and input buffer age track the network congestion point well. They behave like an indicator function that rises sharply at high network loads. However, compared with link utilization, input buffer utilization and input buffer age are much less sensitive to changes in traffic. Simulation results show that from lightly loaded traffic to high network loads, the average buffer utilization only increases by about 0.1. The average link utilization, however, changes by more than 0.8. Hence, link utilization is much better at tracking nuances in network traffic.

Link utilization and input buffer utilization are selected as the relevant measures for guiding the history-based DVS policy, as input buffer age has similar characteristics as input buffer utilization and is harder to capture. The link utilization is used as the primary indicator, whereas the input buffer utilization is used as a litmus test for detecting network congestion.

6.4.1.1.2 History-Based DVS Policy

Network traffic exhibits two dynamic trends: transient fluctuations and long-term transitions. History-based DVS policy filters out short-term traffic fluctuations and adapts link frequencies and voltages judiciously to long-term traffic transitions. It does this by first sampling link (input buffer) utilization within a predefined history window, and then using exponential weighted average utilization to combine the current and the past utilization history:

$$Par_{predict} = \frac{Weight \times Par_{current} + Par_{past}}{Weight + 1} \tag{6.11}$$

where:

$Par_{predicted}$ is the predicted communication link (input buffer) utilization
$Par_{current}$ is the link (input buffer) utilization in the current history period
Par_{past} is the predicted link (input buffer) utilization in the previous history period

Given the predicted communication link utilization, and input buffer utilization, $BU_{predicted}$, the DVS policy dynamically adapts its voltage scaling to achieve power savings with minimal impact on performance. It prescribes whether to increase the link voltage and frequency to next higher level, decrease the link voltage and frequency to next lower level, or do nothing. Intuitively, when a link is highly utilized, voltage scaling is enabled so that link frequency can be increased to handle the load. Similarly, if a link is mostly idle, voltage scaling is carried out so that the link frequency can drop to save power. Otherwise, voltage scaling is conservatively carried out to minimize the impact on performance. The prescribed action depends on four thresholds, two of which are used when the network is lightly loaded (TH_{high}, TH_{low}) and the other two are used when the network is highly congested (TH_{high}, TH_{low}). In the latter case, since link delay can be hidden, the thresholds prescribe more aggressive power savings. The pseudocode of the proposed DVS policy is shown in Algorithm 1.

Algorithm 1

Dynamic voltage scaling

while (DVS enable) do

$$LU_{predicted} = (W * LU_{current} + LU_{past}) / (W + 1)$$

$$LU_{past} = LU_{predicted}$$

$$BU_{predicted} = (W * BU_{current} + BU_{past}) / (W + 1)$$

$$BU_{past} = BU_{predicted}$$

if ($BU_{predicted} < BU_{congested}$) then

$$T_{low} = TL_{low}, \ T_{high} = TL_{high}$$

else

$$T_{low} = TH_{low}, \ T_{high} = TH_{high}$$

end if

if ($LU_{predicted} < T_{low}$) then

$$NewVol_{link} = Voltage\_table[CurLevel_{link} + 1]$$

$$NewFreq_{link} = Frequency\_table[CurLevel_{link} + 1]$$

else if ($LU_{predicted} > T_{high}$) then

$$NewVol_{link} = Voltage\_table[CurLevel_{link} - 1]$$

$$NewFreq_{link} = Frequency\_table[CurLevel_{link} - 1]$$

else

$$NewVol_{link} = Voltage\_table[CurLevel_{link}]$$

$$NewFreq_{link} = Frequency\_table[CurLevel_{link}]$$

end if

end while

6.4.1.2 Hardware Implementation

The above DVS policy relies only on local link and buffer information. This avoids communication overhead in relaying global information and permits a simple hardware implementation. Figure 6.15 shows the hardware realization of the history-based DVS policy. To measure link utilization, a counter at each output port gathers the total number of cycles that are used to relay flits in each history interval. Another counter captures the ratio between the router and link clocks. A simple Booth multiplier combines these two counters to calculate link utilization. For calculating the exponential weighted average, W is set to 3 so that the division can be implemented as a shift and the numerator as a shift-and-add operation. Two registers store LU_{past} and BU_{past}, which feed the circuit module calculating the exponential weighted average. Finally, some combinational logic performs the threshold comparisons and outputs signals that control the DVS link.

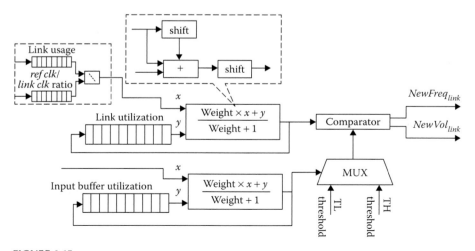

FIGURE 6.15
Hardware implementation of the history-based DVS policy. This circuitry sits at each output port of a router, tracking and controlling the multiple links of that port.

6.4.1.3 Results and Discussions

It has been observed in simulation that history-based DVS increases zero-load latency by 10.8% and average latency before congestion by 15.2%, while decreasing throughput by less than 2.5%. This moderate impact on performance is accompanied by a large power saving of up to 6.3 × (4.6 × average). When the network is saturated, flits are stalled for a long time in input buffers. Such congested routers show high input buffer utilization and low communication link utilization. In such a scenario, the history-based DVS policy will try to dynamically reduce the frequencies of affected links to decrease power consumption. It can be observed that the power consumption of the network increases initially as network throughput increases and dips thereafter as network throughput decreases. This interesting phenomenon is largely due to the very bursty nature of communication workload that results in routers in different parts of the network experiencing widely varying loads over time. When some of the routers are congested, traffic through other routers may still be relatively light. Therefore, as the packet injection rate increases, the overall network throughput may still increase. Link utilization is strongly correlated with network throughput. A higher throughput implies higher average link utilization in the network. The history-based DVS policy only decreases the frequencies and voltages of links that are lightly utilized—those connected to the congested routers. It increases the frequencies and voltages of the heavily used links to meet performance requirements. Hence, only when the entire network becomes highly congested, the overall network throughput starts to decrease, which cause an overall reduction in network power.

By dynamically adjusting the DVS policy to maximize power savings when the network is lightly loaded and minimizing the impact on performance when the network is congested, the policy is able to realize substantial power savings without a significant impact on performance. It should be noted that part of the impact on performance is due to the assumptions for DVS links. First, the link is down during frequency scaling. Second, when increasing voltage and frequency, the voltage increases first, which takes a long time. The frequency is kept at the original low level during voltage transition.

6.4.2 Dynamic Frequency Scaling

As discussed in Section 6.4.1, DVS requires hundreds of clock cycles during transition between voltage levels and additional hardware overhead for each link. The other way to manage power consumption is DFS. DFS only adapts the system clock frequency by setting all links in the network to the same voltage, but it does not always reduce the total energy consumption. For instance, the power consumed by a network can be reduced by reducing the operating clock frequency, but it takes long time to forward the same amount of data and the total energy consumed will be similar. DFS is valid when the target system does not support DVS or the goal is to reduce average power dissipation, indirectly reducing the chip's temperature.

This section highlights the work cited by Lee and Bagherzadeh (2009) based on clock boosting mechanism. The key idea of clock boosting mechanism is the use of different clocks in a head flit and body flits because body flits can continue advancing along the reserved path that is already established by the head flit, while the head flit requires the support of complex logic, increasing critical path. Thus, it reduces the latency and increases the throughput of a router by applying faster clock frequency to a boosting clock in order to forward body flits.

In NoC paradigm, DFS only adapts the system clock frequency by setting all links in the network to the same voltage. In addition, the operating frequency of a system is not limited by the critical path because it only changes clock frequency for the body flit transmission. Thus, this method not only provides variable frequency link but also increases interconnection network performance. Also, fast response time of the clock domain variations makes it possible to use narrow control period for DFS, where clock frequency is adjusted more frequently. Figure 6.16a shows an example of a variable frequency link (Lee and Bagherzadeh 2009). The system has multiple clock frequencies represented by Fi. The link controller selects the clock frequency for the router among the supported clock frequencies by using link utilization level. Figure 6.16b shows the time–space diagram for variable frequency links. In this example, the link supports three different frequencies ($F1$, $F2$, and $F3$). The original clock frequency ($F1$ in this example) is still used for the head flit transmission as well as for idle cycles. Selecting higher frequencies

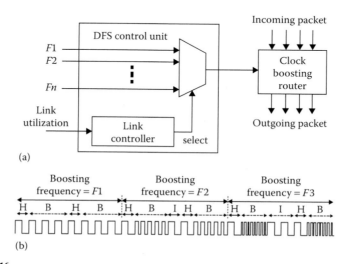

(a)

(b)

FIGURE 6.16
(a) Architecture of a DFS link; (b) time–space diagram showing the clock domain transition in DFS link. B, body flit; H, head flit; I, idle flit.

(*F2* and *F3*) for the body flits ensures the duration of staying in idle condition. Although dynamic power with higher frequency operations is expected to be more than the power with original clock frequency operation, the overall network power may be less because idle operation and head flit transmission have different switching activities even though they operate with the same frequency. Increasing boosting clock frequency results in more idle cycles since the body flits are transmitted with higher frequency.

6.4.2.1 History-Based DFS

Network workload exhibits transient fluctuation and long-term transitions. In order to filter out transient fluctuations from link utilization and to predict future communication workload, a history-based algorithm is used for DFS scheme and is described below. In applying DFS to a system, how to predict future workload with reasonable accuracy is a critical problem. This requires knowing how many packets will traverse a link at any given time. Two issues complicate this problem. First, it is not always possible to accurately predict future traffic activities. Second, a subsystem can be preempted at arbitrary times due to user and I/O device requests, varying traffic beyond what was originally predicted. In order to estimate future workload, link utilization is adopted as an indicator, which is a direct measure of traffic through a link in each unit time. Lower link utilization reflects more idle cycles in a link caused by network congestion with heavy traffic or sparse workload in the incoming port. Conversely, higher link utilization implies that more active cycles in a link pass flits to the destination router. The link utilization is measured by sampling a link at a given time during a pre-defined control period (Tc). The direct link utilization is defined, where k denotes the number of samples in Tc time period:

$$U_L(n) = \frac{\sum_{t=1}^{k} u(t)}{k} \qquad (6.12)$$

where:

$$u(t) = \begin{cases} 1 & \text{If there is link traffic in cycle } t \\ 0 & \text{If there is no link traffic in cycle } t \end{cases}$$

The direct estimator only measures the link utilization whether a link is occupied or not. It does not consider the number of flits traversing through a link during the given time. For instance, even though the link utilizations are the same in time durations Δt_1 and Δt_2, the number of flits passing through the link can be different according to the clock frequency of the router at those times.

Direct estimation can be realized with a counter, reducing the complexity of the estimator and additional hardware overhead caused by the DFS link controller. A counter at each output port gathers the total number of cycles that are used to pass a flit in each control period by counting $u(t)$ with 4 times of the clock frequency for accurate measurement (Figure 6.17a). Function $u(t)$ is assigned to the write enable signal of the router, and the counter value is sampled in each control period to complete the measurement of the link utilization.

History-based link estimator uses exponential weighted average utilization to combine the current $[U_L(n)]$ and the past $[\Psi_L(n-1)]$ link utilization history, smoothing and predicting future link utilization $\Psi_L(n)$ as follows:

$$\Psi_L(n) = \frac{\text{Weight} \times U_L(n) + \Psi_L(n-1)}{\text{Weight} + 1} \tag{6.13}$$

where:

$$\Psi_L(0) = \psi_0, \quad i \in N$$

Weight is the contribution factor of current link utilization level to the history-based link estimator

The hardware overhead is an important factor for the design of the estimator. Soteriou and Peh (2004) realized the history-based estimator with two shifters and an adder by setting weight equal to 3 and reducing additional hardware overhead caused by the prediction mechanism. Figure 6.17b shows the hardware circuit for the exponential weighted average. The result of direct estimator is fed to the exponential weighted average calculator to predict the link utilization. The history-based estimator is a cascade of direct link utilization estimators and an exponential average calculator.

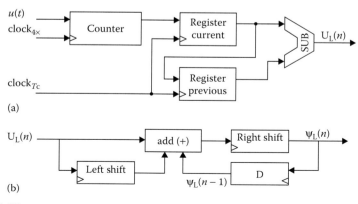

FIGURE 6.17
History-based link utilization estimator: (a) direct link utilization estimator; (b) exponential weighted average calculator.

6.4.2.2 DFS Algorithm

Given the link utilization, the DFS algorithm dynamically adapts its frequency to achieve power savings with minimal impact on performance. It prescribes whether to increase clock frequency to higher level, decrease clock frequency to lower level, or do nothing. Even though the link utilization estimator predicts correctly the workload, determining how fast to run the network is nontrivial. The algorithm controlling DFS link trades off power and performance. Intuitively, if a link utilization is high ($\Psi_L \geq \pi_u$), the clock frequency will be increased. On the contrary, when link utilization falls below the threshold value ($\Psi_L < \pi_l$), the clock frequency will be reduced. The threshold values (π_u and π_l are the threshold values to increase and decrease the frequency, respectively) can be set to a single value for π_u and π_l for the simplest method. Also, multiple thresholds can be set corresponding to each state (three sets of thresholds from π_{l1x} to π_{l4x} and π_{u1x} to π_{u4x}). In addition, threshold values can be predefined in design time or optimized in runtime. A pseudocode of DFS policy is shown in Algorithm 2.

Algorithm 2

Dynamic frequency scaling

    while (DFS enable) do
        $\Psi_L(n) = (W \times U_L(n) + \Psi_L(n-1)) / (W+1)$
        if $\Psi_L(n) \geq \Pi_u$ then
            Increase clock frequency (↑)
        else if $\Psi_L(n) < \Pi_l$ then
            Decrease clock frequency (↓)
        else
            Maintain current clock frequency (–)
        end if
    end while

6.4.2.3 Link Controller

The link controller is implemented with a Moore machine. Each state represents the clock frequency such as f_{1x}, f_{2x}, and f_{4x} with a two-bit value, and the machine output, equal to the state value, is passed on to the clock domain multiplexer. In the link controller, there is no change between f_{1x} and f_{4x}. Clock domain transition occurs only between adjacent clock frequencies. The state values are assigned such that the Hamming distance between state transitions is 1. Clock is the most important and sensitive signal in a system and glitches between clock domain transitions make the system unstable, resulting in erroneous signals. To ensure that constancy of the clock phase during clock domain changes, control period can be set to multiples of the clock period of the original clock frequency, and the link control function is performed in each control period.

6.4.2.4 Results and Discussions

The DFS link characteristics for each boosting clock frequency are obtained by simulation under the given workload (see Table 6.3). The 1× boosting router finishes the entire packet transmission in 24.34 μs, spending more time than 2× and 4× boosting routers. It also has the highest average and peak latency. The 2× boosting router reduces the average latency by about 81% at the expense of 16% more dynamic power in contrast to the 1× boosting router. Similarly, the 4× boosting router is much better compared to the 1× boosting router in terms of latency; however, it consumes 21% more dynamic power, reducing the average latency to 87%. It also reduces the average latency around 32% at the expense of only 5.1% more dynamic power in comparison with the 2× boosting router. These experimental results demonstrate the feasibility of clock boosting router for the DFS link for a power-aware on-chip interconnection network in a NoC platform.

Table 6.4 summarizes the experimental results of the history-based DFS policy varying the control period from 8 to 128 cycles of the 1× clock. DFS policy enables the use of an intermediate value for power consumption between 1× and 2× clock boosting routers. For instance, power consumption of DFS with eight control periods consumes 2.09 mW, whereas 1× and 2×

TABLE 6.3

Characteristic of the DFS Link with the Workload

Boost Clock (MHz)	Peak Latency (ns/flit)	Average Latency (ns/flit)	End Time (μs)	Dynamic Power (mW)	Leakage Power (mW)	Total Power (mW)
100 (1×)	367	81.2	24.34	1.69	0.16	1.85
200 (2×)	97	15.3	24.06	1.96	0.16	2.12
400 (4×)	61	10.4	24.05	2.06	0.16	2.22

TABLE 6.4

Experimental Results of the History-Based DFS Varying Control Period

Control Period (cycle)	Peak Latency (ns/flit)	Average Latency (ns/flit)	End Time (μs)	Dynamic Power (mW)	Leakage Power (mW)	Total Power (mW)
8	116	24.30	24.05	1.93	0.16	2.09
16	146	25.20	24.05	1.91	0.16	2.07
32	187	27.60	24.23	1.90	0.16	2.06
64	242	33.20	24.34	1.88	0.16	2.04
128	265	48.21	24.05	1.91	0.16	2.07

clock boosting routers consume 1.85 and 2.12 mW, respectively, demonstrating the possibility of runtime power management for the given workload. Choosing a wider control period further slows down the adaptation of link frequency for the given traffic, exacerbating latency. While there is a trade-off in power and performance for the control period from 8 to 64 cycles, the history-based DFS with 128 control periods consumes more power. It also increases the latency due to selection of very long control period for the given workload. For on-chip interconnection network, the latency can be a suitable indicator to measure the performance of a network. Trade-off between power consumption and latency depends on the length of control period for the DFS policy. Even though a longer control period saves more power, it suffers from excessive latency. For the given workload, choosing the control period of eight cycles is preferable for the DFS when an application requires tight timing requirements. However, a longer control period might be enough to cope with system requirements, saving more power dissipation. In general, each application has its own power and performance demand to complete an assigned task within the desired time budget. A designer should keep in mind the system requirements in applying DFS for the on-chip interconnection network.

6.4.3 VFI Partitioning

For achieving fine-grain system-level power management, the use of VFIs in the NoC context is likely to provide better power–performance trade-offs than its single-voltage, single-clock frequency counterpart, while taking advantage of the natural partitioning and mapping of applications onto the NoC platform. This section presents the design and optimization of novel NoC architectures partitioned into multiple VFIs that rely on a **globally** asynchronous locally synchronous communication paradigm. In such a system, each voltage island can work at its own speed, while the communication across different voltage islands is achieved through mixed-clock/mixed-voltage FIFOs as shown in Figure 6.18. This provides the flexibility to

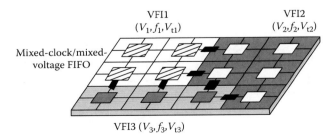

FIGURE 6.18

A sample 2D mesh network with three VFIs. Communication across different islands is achieved through mixed-clock/mixed-voltage FIFOs.

scale the frequency and voltage of various VFIs in order to minimize energy consumption. As a result, the advantages of both NoC and VFI design styles can be exploited simultaneously. This section focuses on the work cited by Ogras et al. (2007) on power management using VFIs.

The design of NoCs with multiple VFIs involves a number of critical steps. First, the granularity (i.e., the number of different VFIs) and chip partitioning into VFIs need to be determined. While a NoC architecture where each processing/storage element constitutes a separate VFI exhibits the largest potential savings for energy consumption, this solution is very costly. Indeed, the associated design complexity increases due to the overhead in implementing the mixed-clock/mixed-voltage FIFOs and voltage converters required for communication across different VFIs, as well as the power distribution network needed to cover multiple VFIs. Additionally, the VFI partitioning needs to be performed together with assigning the supply and threshold voltages and the corresponding clock speeds to each VFI. The energy overhead of adding one additional voltage–frequency island to an already existing design can be written as follows:

$$E_{\text{VFI}} = E_{\text{ClkGen}} + E_{\text{Vconv}} + E_{\text{MixClkFifo}} \qquad (6.14)$$

where:

E_{ClkGen} is the energy overhead of generating additional clock signals
E_{Vconv} denotes the energy consumption of the voltage level converters
$E_{\text{MixClkFifo}}$ is the overhead due to the mixed-clock/mixed-voltage FIFOs used in interfaces

Besides energy, additional VFIs exhibit area and implementation overheads, such as routing multiple power distribution networks. The maximum number of VFIs is assumed to be a constraint. To connect a node in a VFI with another node residing in a different VFI, all data and control signals need to be converted from one frequency/voltage domain to another. For this purpose, a mixed-clock/mixed-voltage interfaces using FIFOs are implemented, which are natural candidates for converting the signals from one VFI to another, as shown in Figure 6.19.

To find the optimum number of VFIs, Ogras et al. (2007) started their experiment with 16 VFIs in a 4 × 4 mesh-based NoC structure. Then, it proceeds by merging the islands until a single island is obtained; as such, it evaluates all possible levels of VFI granularity. Finally, based on different applications, they concluded that two to three VFIs in NoC context provide better power–performance trade-offs than its single-voltage, single-clock frequency counterpart.

6.4.4 Runtime Power Gating

The power consumption is classified into dynamic switching power and static leakage power. The switching power is consumed only when packets

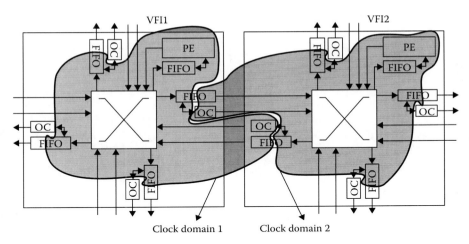

FIGURE 6.19
Illustration of the interface between two different voltage–frequency domains VFI1 and VFI2.

are transferred on a NoC, whereas the leakage power (or static power) is consumed without any packet transfers as long as the NoC is powered on. Since the NoC is the communication infrastructure of chip multiprocessing, it must be always ready for the packet transfers at any workload so as not to increase the communication latency; thus, a runtime power management that dynamically stops the leakage current whenever possible is highly required. This section highlights the work cited by Matustani et al. (2010) on ultra-fine-grained runtime power gating of on-chip router. They partitioned the mesh-based router architecture into several power domains such as virtual channel (VC) buffer for each flit, output latch for each flit, crossbar multiplexer, and VC multiplexer. Isolation cells are inserted to all output ports of the synthesized netlist in order to hold the output values of the domain when the power supply is stopped. The netlist of isolation cells is placed by *Synopsys Astro* tool. They formed the virtual ground (VGND) lines and the power switches are inserted between the VGND and ground (GND) lines by *Synopsys Design Cool Power* tool as shown in Figure 6.20.

In this design, the authors used customized standard cells that have a VGND port in 65-nm technology and modified that according to the cell height. They showed that the area overhead for inserting isolation cells and power switch in the overall design is 4.3%. There is another area overhead of the customized standard cells against the original ones. The total area overhead increases to 15.9%. The authors also assumed that the wake-up latency of each power domain is two, three, and four cycles when the target NoC is operated at 667 MHz, 1 GHz, and 1.33 GHz, respectively. This assumption is a little bit conservative, since the actual wake-up latencies have been observed to be less than 3 ns. Experimental results show a reduction of average router leakage power by 64.6% when applying the runtime power gating technique

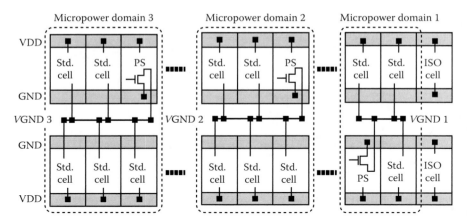

FIGURE 6.20
Fine-grained power gating. ISO, isolation cell; PS, power switch; VGND, virtual ground.

only in VC buffers, assuming that temperature and core voltage are set to 25°C and 1.20 V, respectively. It is also shown that the leakage power reduction can be extended to 78.9% when runtime power gating is applied to VC buffers, VC multiplexers, crossbar multiplexers, and output latches at the expense of 4% performance penalty, assuming that the routers are working at 1 GHz.

6.5 Summary

This chapter provides a clear insight into standard low-power techniques in NoC. Different system-level dynamic and leakage power reduction techniques including power–performance trade-off in NoC platform have also been discussed. In spite of power–performance trade-off, system reliability is another metric to address while lowering the supply voltage. With the decrease in supply voltage, the signal-to-noise ratio reduces, thereby creating an adverse effect on BER. Chapter 7 will focus on the signal integrity and reliability challenges in NoC routers and interconnection links at nanometer regime.

References

Benini, L. and Micheli, G. D. 2006. *Networks on Chips: Technology and Tools*. Morgan Kaufmann Publishers, San Francisco, CA.

Benini, L., Siegel, P., and Micheli, G. D. 1994. Automated synthesis of gated clocks for power reduction in sequential circuits. *IEEE Design and Test of Computers*, pp. 32–41, IEEE.

Chuang, L. P., Chang, M. H., Huang, P. T., Kan, C. H., and Hwang, W. 2008. A 5.2mW all-digital fast-lock self-calibrated multiphase DLL. *Proceedings of International Symposium on Circuits and Systems*, May 18-21, Seattle, WA, IEEE, pp. 3342–3345.

Deogun, H. S., Rao, R. R., Sylvester, D., and Blaauw, D. 2004. Leakage- and crosstalk-aware bus encoding for total power reduction. *Proceedings of Design Automation Conference*, July 7-11, San Diego, CA, pp. 779–782.

Ghoneima, M. and Ismail, Y. 2004. Low power coupling-based encoding for on-chip buses. *Proceedings of IEEE International Symposium on Circuits and Systems*, May 23-26, Vancouver, BC, Canada, IEEE, pp. 325–328.

Huang, P. T., Fang, W. L., Wang, Y. L., and Hwang, W. 2008. Low power and reliable interconnection with self- corrected green coding scheme for network-on-chip. *ACM/IEEE International Symposium on Networks-on-Chip*, April 7-10, Newcastle upon Tyne, IEEE, pp. 77–83.

Kim, K., Baek, K., Shanbhag, N., Liu, C., and Kang, S. 2000. Coupling-driven signal encoding scheme for low-power interface design. *Proceedings of ICCAD*, November 5-9, San Jose, CA, IEEE, pp. 318–321.

Lee, K., Lee, S. J., Kim, S. E., Choi, H. M., Kim, D., Kim, S., Lee, M. W., and Yoo, H. J. 2004. A 51mW 1.6GHz on-chip network for low power heterogeneous SoC platform. *Proceedings of IEEE International Solid-State Circuits Conference*, February 15-19, San Francisco, CA, IEEE.

Lee, S. E. and Bagherzadeh, N. 2009. A variable frequency link for a power-aware network-on-chip. *Integration, the VLSI Journal*, vol. 42, no. 4, pp. 479–485.

Lee, S. J., Kim, K., Kim, H., Cho, N., and Yoo, H. J. 2006. A network-on-chip with 3Gbps/wire serialized on-chip interconnect using adaptive control schemes. *Proceedings of Design Automation and Test in Europe*, pp. 79–80.

Lee, S. J., Lee, K., and Yoo, H. J. 2005. Analysis and implementation of practical, cost-effective networks on chips. *IEEE Design & Test Computers*, vol. 22, no. 5, pp. 422–433.

Lu, Y. and Agarwal, V. D. 2007. Statistical leakage and timing optimization for submicron process variation. *Proceedings of 20th VLSI Design and 6th Embedded Systems*, January 6-10, Bangalore, India, IEEE, pp. 439–444.

Matustani, H., Koibuchi, M., Ikebuchi, D., Usami, K., Nakamura, H., and Amano, H. 2010. Ultra fine-grained run-time power gating of on-chip routers for CMPs. *Proceedings of ACM/IEEE International Symposium on Networks-on-Chip*, May 3-6, Grenoble, France, IEEE, pp. 61–68.

Mullins, R. 2006. Minimising dynamic power consumption in on-chip networks. *Proceedings of International Symposium on System-on-Chip*, November 14–16, Tampere, Finland, IEEE, pp. 1–4.

Ogras, U. Y., Marculescu, R., Choudhary, P., and Marculescu, D. 2007. Voltage-frequency island partitioning for GALS-based networks-on-chip. *Proceedings of Design and Automation Conference*, June 4-8, San Diego, CA, IEEE, pp. 110–115.

Shang, L., Peh, L. S., and Jha, N. K. 2003. Dynamic voltage scaling with links for power optimization of interconnection networks. *Proceedings of IEEE International Conference on High-Performance Computer Architecture*, February 8-12, Anaheim, CA, IEEE, pp. 91–102.

Soteriou, V. and Peh, L.-S. 2004. Design-space exploration of power-aware on/off interconnection networks. *Proceedings of the IEEE International Conference on Computer Design*, October 11-13, San Jose, CA, IEEE, pp. 510–517.

Sotiriadis, P. P. 2002. Interconnect modeling and optimization in deep submicron technologies. Ph.D. Dissertation, Massachusetts Institute of Technology, Cambridge.

Sotiriadis, P. P. and Chandrakasan, A. 2000. Low power bus coding techniques considering inter-wire capacitances. *Proceedings of the IEEE Custom Integrated Circuits Conference*, May 21-24, Orlando, FL, IEEE, pp. 507–510.

Sotiriadis, P. P. and Chandrakasan, A. 2002. A bus energy model for deep submicron technology. *IEEE Transaction on Very Large Scale Integration Systems*, vol. 10, no. 3, pp. 341–350.

Sridhara, S. R. and Shanbhag, N. R. 2005. Coding for system-on-chip networks: A unified framework. *IEEE Transactions on Very Large Scale Integration (VLSI) Systems*, vol. 13, no. 6, pp. 655–667.

Stan, M. R. and Burleson, W. P. 1995. Bus-invert coding for low-power I/O. *IEEE Transactions on Very Large Scale Integration (VLSI) System*, vol. 3, no. 1, pp. 49–58.

Stratakos, A. 1998. High-efficiency low-voltage DC-DC conversion for portable applications. Ph D Thesis, University of California, Berkeley, CA.

Yoo, S. and Choi, K. 1999. Interleaving partial bus-invert Coding for low power reconfiguration of FPGAs. *Proceedings of International Conference on VLSI and CAD*, October 26-27, Seoul, Republic of Korea, IEEE, pp. 549–552.

Yuan, C. P. and Cheng, Y. C. 2005. A voltage level converter circuit design with low power consumption. *Proceedings of International Conference on ASIC*, October 24-27, Shanghai, People's Republic of China, IEEE, pp. 358–359.

Zhang, Y., Lach, J., Skadron, K., and Stan, M. R. 2002. Odd/even bus invert with two-phase transfer for buses with coupling. *Proceedings of International Symposium on Low Power Electronics and Design*, Monterey, CA, IEEE, pp. 80–83.

7

Signal Integrity and Reliability of Network-on-Chip

7.1 Introduction

In deep submicron (DSM) technology, the importance of on-chip wiring (interconnect) becomes significant as it impacts the system performance and cost to a big extent. With the technology shrinking, on-chip interconnect suffers from increased resistance due to decrease in metal cross-sectional area and also suffers from increased capacitance if the metal height is not reduced proportionally with metal spacing. Therefore, resistance-capacitance (RC) parameter in interconnects plays an increasing role in system-on-chip (SoC) performance as feature size scales. In integrated circuit (IC) packaging, usage of multiple interconnection layers is a common trend in order to provide better connectivity. The metal layers are normally divided into three categories: local, semi-global, and global interconnects. In SoC design, several cores [intellectual property (IP), memory, processor, etc.] are integrated on a single silicon die. Local wires are used for connecting intra-core modules. These wires have minimum dimensions and pitch, hence highly resistive. Semi-global wires are used for inter-core communication. These wires are usually longer than local interconnects and have lesser resistance. Global interconnects, however, are used as supply and clock lines. The signals passing through these wires traverse long distance. These wires have higher cross-sectional area and hence lesser resistance. For a typical eight-layerd 90-nm *UMC* process, layers 1–3 are used for local interconnect, layers 4 and 5 for semi-global interconnect, and the top layers are reserved for global interconnect.

Copper is the mostly used material for on-chip interconnects in very-large-scale integration (VLSI) design. It has lesser resistivity compared to aluminum. An on-chip interconnect can be modeled as a distributed RC network or a transmission line with the parasitic resistance, capacitance, and inductance. The resistance of a wire is proportional to its length L and inversely proportional to its cross section A with width (W) and thickness (T). Self-capacitance of a wire is defined as the summation of line-to-substrate capacitance, inter- and intra-layer coupling capacitance, and fringing capacitance from the side walls. The dielectric material used in most ICs is SiO_2. Self- and mutual inductances become prominent for some global signal wires routed in top metal layers due to its lower resistance and faster signal transition time with increasing frequencies. Ismail et al. (1998) proposed a two-sided inequality (Equation 7.1) that determines the range of length of interconnect (L) in which inductance effects are significant and the line can be modeled as a transmission line.

$$\frac{t_r}{2\sqrt{L_s \cdot C_s}} < L < \frac{2}{R}\sqrt{\frac{L_s}{C_s}} \qquad (7.1)$$

In the above relation, the range of interconnect length (L) depends on the parasitic resistance (R), self-inductance (L_s), and self-capacitance (C_s) of the wire per unit length as well as the transition time (t_r) of the signal at the output of its driver. Ismail et al. (1998) also proposed that if the above inequality is nonexistent and Equation 7.2 holds true, the effect of inductance is not important for any length of interconnect and the wire can be modeled as a distributed *RC* line.

$$t_r > 4\left(\frac{L_s}{R}\right) \qquad (7.2)$$

Example 7.1

Consider the following parasitic values for a wire segment: Resistance (R) = 0.4926 MΩ/m, self-inductance (L_s) = 0.3743 µH/m, self-capacitance (C_s) = 70.7918 pF/m, transition time (t_r) = 100 ps, and length (L) = 1 mm. Determine the interconnect model whether it is a transmission line or a distributed RC.

* * *

By putting the values of parasitic components, length, and transition time, Equation 7.1 is nonexistent. Hence the wire cannot model as a transmission line. Putting the values of R and L_s, it shows that the wire can be modeled as a distributed RC if $t_r > 3.04$ ps which is true in this case. Hence, the wire can be modeled as a distributed RC.

The details of on-chip interconnect in the DSM era are covered in literature. The scope of this chapter is narrowed to on-chip interconnect in network-on-chip (NoC) design and related issues. It has been reported in the work of Benini and Micheli (2006) that the impact of inductance in most on-chip interconnects is negligible and can be modeled as a distributed RC wire as shown in Figure 7.1.

The rest of this chapter is organized as follows: Section 7.2 describes the sources of different types of faults such as permanent faults, faults due to aging effect, and transient faults in DSM technology. Section 7.3 discusses the techniques to handle the permanent faults. Section 7.4 describes the

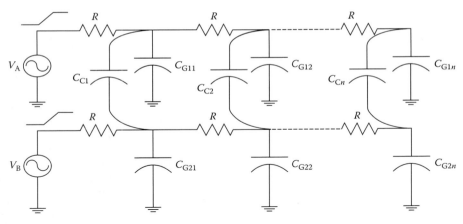

FIGURE 7.1
Parasitic components of a wire.

intra- and inter-router transient faults in detail. It also discusses different crosstalk avoidance techniques, soft error protection techniques, and error controlling techniques. Section 7.5 highlights a unified coding framework to address crosstalk avoidance, power minimization in bus, and error correction jointly. It also discusses different joint coding schemes. Section 7.6 presents the energy–reliability trade-off. Finally, Section 7.7 summarizes the whole chapter.

7.2 Sources of Faults in NoC Fabric

There are three types of failures that affect the system: permanent faults, faults due to aging effect, and transient faults. The permanent faults are mainly stuck-at faults (hard failures) and they remain in the system until repaired. Aging effect will cause transistor parameter degradation, typically as switching frequency degradation rather than a hard functional failure. Transient faults, however, generally occur in the run time. Soft errors, crosstalk noise, and timing faults are the major classifications of transient faults. Failures are usually described in terms of *mean time between failures* (MTBF) or *failures in time* (FIT). FIT is defined as one error per billion hours of device operation. MTBF is usually given in years of device operation. To put in perspective, the relation between MTBF and FIT is given as follows:

$$1\ \text{Year MTBF} = \frac{10^9}{24 \times 365.25} \approx 114{,}077\ \text{FIT}$$

7.2.1 Permanent Faults

Permanent faults, as the name suggests, originate from permanent damages in the circuit. These damages result in physical changes in the circuit whose behavior does not change with time. Broken wires, time-dependent dielectric breakdowns, and electromagnetic interference (EMI) are the examples of permanent failures on chip. Long on-chip interconnects, typical in customized, domain-specific irregular NoC topologies, are increasingly susceptible to EMI. Short switch-to-switch links in regular networks are more immune to such noise sources. Another cause of permanent fault is electromigration. Aluminum interconnects are highly affected from the electromigration effect, whereas modern copper interconnect-based ICs rarely fail due to electromigration effects. The permanent fault handling in NoC is described more detail in Section 7.3.

7.2.2 Faults due to Aging Effects

Device reliability issues such as negative-bias temperature instability (NBTI) and hot carrier injection (HCI) make circuit performance degrade as it ages and have more severe effects with shrinking device sizes and voltage margins. In Sections 7.2.2.1 and 7.2.2.2, each of them is described briefly.

7.2.2.1 Negative-Bias Temperature Instability

The instability of p-type metal oxide semiconductor (PMOS) transistor parameters (e.g., threshold voltage, transconductance, saturation current, etc.) under negative (inversion) bias and relatively high temperature has been well known since the 1970s and has become a significant reliability concern in present-day digital design.

When a PMOS transistor is biased in inversion ($V_{gs} = -V_{dd}$), the dissociation of Si—H bonds along the silicon–oxide interface causes the generation of interface traps. The rate of generation of these traps is accelerated by the temperature and the time of applied stress (PMOS is turn ON). These traps cause an increase in the threshold voltage (V_{th}) of the PMOS transistors. An increase in V_{th} causes the circuit delay to degrade, and when this degradation exceeds a certain limit, the circuit may fail to meet its timing specifications. This effect, known as negative-bias temperature instability, has become a reliability issue in high-performance digital IC design, especially in sub-130-nm technologies. Since a digital circuit consists of millions of nodes with various signal probabilities and activity factors, the degradation of timing paths is not uniform. Logic blocks in a circuit are currently designed by assuming a certain safety margin in the timing specifications to account for NBTI-induced performance degradation. Kumar et al. (2006) proposed an analytical model for measuring the impact of NBTI. Their simulation results on International Symposium on Circuits and Systems benchmarks under a 70-nm technology show that NBTI causes a delay degradation of about 8% in combinational logic-based circuits after 10 years.

7.2.2.2 Hot Carrier Injection

HCI describes the phenomena by which carriers gain sufficient energy due to increase in electric field and causes increase in velocity, which can leave the silicon and are injected into the gate oxide. This occurs as carriers move along the channel of metal oxide semiconductor field effect transistor and experience impact ionization near the drain end of the device. Electrons are trapped in the oxide and hence change the threshold voltage (increases for n-type metal oxide semiconductor [NMOS] and decreases for PMOS). The hot electron phenomenon can lead to a reliability problem, where the circuit might fail to meet the timing requirement after being in use for some time.

7.2.3 Transient Faults

As the technology is approaching toward DSM with shrinking feature sizes, scaling of supply voltages, increasing wire density, and faster clock rates, NoC suffers from following transient faults: (1) slowdown or speedup in delay due to crosstalk, (2) crosstalk noise, (3) single- or multi-event upset due to soft error, (4) delay due to process–voltage–temperature (PVT) variation, (5) synchronization failure, (6) delay due to power supply noise, (7) IR drop, and so on.

7.2.3.1 Capacitive Crosstalk

In modern DSM technologies with shrinking feature sizes and decreasing spacing between adjacent interconnects, the value of coupling capacitance becomes dominant, which causes capacitive crosstalk. There are two major deteriorating effects due to capacitive crosstalk—crosstalk noise and crosstalk delay. Cuviello et al. (1999) proposed a novel fault model, called maximum aggressor fault (MAF) model, which considers the effect of crosstalk between a set of aggressor lines and a victim line. For a link consisting of N wires, MAF model assumes the worst-case situation with one victim line and $(N-1)$ aggressor lines where all the aggressor lines are switching in a same direction. According to the MAF model, Figure 7.2 shows the possible errors on the victim wire in a three-wire model.

A crosstalk effect causing a positive (negative) glitch in the victim line (Y_2), which should ideally have a logic 0 (logic 1), due to a rising (falling) transition at the aggressor lines Y_1 and Y_3 as shown in Figure 7.2a. This glitch will consider as crosstalk noise when the peak value of the glitch is high enough to cross the switching threshold of the receiver. Moreover, the width of the glitch is also an important factor. Wider glitch may drive the load capacitance to a potential that can be interpreted as a different logic value. An analytical model of crosstalk positive glitch is shown in Figure 7.3. When a falling transition is applied to input A_{in}, the PMOS of the inverter driven by A_{in} can be modeled by its channel resistance, R_p, connecting A to V_{DD}; the corresponding NMOS device is off. The inverter driven by V_{in} can be modeled by

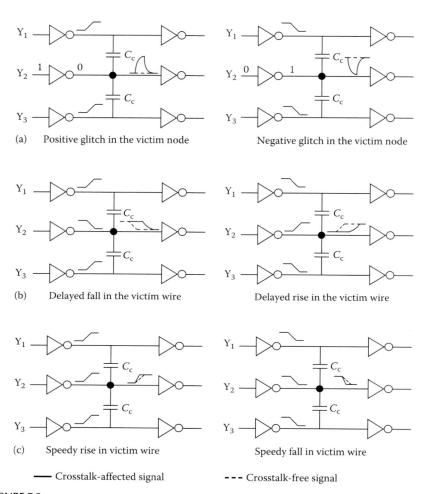

(a)   Positive glitch in the victim node   Negative glitch in the victim node

(b)   Delayed fall in the victim wire   Delayed rise in the victim wire

(c)   Speedy rise in victim wire   Speedy fall in victim wire

——— Crosstalk-affected signal   - - - Crosstalk-free signal

FIGURE 7.2
Effect of crosstalk according to MAF model: (a) crosstalk noise; (b) crosstalk delay; (c) crosstalk speedup.

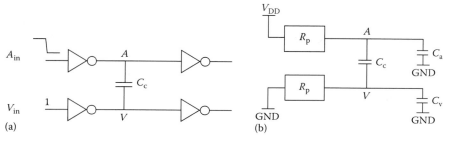

FIGURE 7.3
(a) Circuit model for crosstalk positive glitch analysis; (b) equivalent circuit.

the channel resistance of its NMOS device connecting V to GND. Figure 7.3a shows the circuit model for the situation just described. Figure 7.3b shows the equivalent circuit of Figure 7.3a.

From Figure 7.3b, the transfer function can be written as

$$\frac{V(s)}{A(s)} = \frac{s \times C_c}{(1/Rn) + s(C_c + C_v)}$$

Hence, the voltage induced at victim node V is a function of C_c and voltage of aggressor node A.

Crosstalk delay is categorized into crosstalk *slowdown* and crosstalk *speedup* as shown in Figure 7.2b and c, respectively. Crosstalk speedup occurs due to same transitions in the aggressor and victim wires, whereas crosstalk slowdown occurs due to opposite transitions in those wires. The amount of slowdown and speedup depends on two factors: (1) input transition and (2) skew between aggressor and victim wires. Chen et al. (1997) and Nazarian et al. (2005) described the effect of both the parameters on crosstalk speedup and slowdown.

At first, the effect of input transition is discussed, assuming that both signals switch simultaneously (skew = 0):

- When the slope of the input signal to the victim line is kept constant, faster the aggressor line changes, larger the speedup of the victim line.
- Similarly, faster aggressor causes larger worst-case slowdown.
- The maximum speedup and slowdown occur when the victim has the largest transition time, whereas the aggressor has the smallest transition time.
- If the transition time of both aggressor and victim wires is identical, slow transition has lesser effect than fast transition signals.

The effect of skew between the aggressor and victim wires on crosstalk speedup and slowdown is discussed next, assuming that both the signals have identical transition time:

- If the input skew is negative (aggressor switches first), the amount of speedup and slowdown increases as the skew increases from a negative value toward zero.
- The maximum crosstalk slowdown does not necessarily occur for zero input skew condition even for completely symmetric interconnects.
- If the input skew increases from zero to positive direction, the speedup and slowdown decrease. The speedup and slowdown will become zero after some fixed positive value.

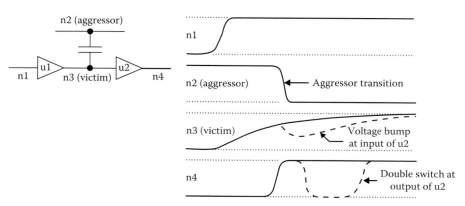

FIGURE 7.4
Double switching error.

Depending on the transition time (rise or fall) of the victim and aggressor nets, another ill effect of capacitive crosstalk is crosstalk double switching. Double-switching noise is the scenario that arises when a large bump occurs on a switching victim, which causes the output of the victim receiver to switch twice. The effect of a strong aggressor transition can be so large such that it can cause the victim net to cross the voltage threshold high enough to cause an incorrect capture of data at the receiver. These types of errors are called double-switching errors and are most often seen when very large bumps act on victim nets that are transitioning very slowly. This is shown in Figure 7.4.

In Figure 7.4, a rising transition on net n1 is propagated through buffers u1 and u2 to nets n3 and n4, respectively. Because of the low drive of buffer u1 and the capacitive load of net n3, the transition on n3 is relatively slow. In the presence of crosstalk (indicated by the dashed lines in the figure), an aggressor transition causes a voltage bump in the sensitive voltage region at the input of buffer u2. This causes the output of the buffer to switch twice.

There are two possible side effects, depending on whether the victim net goes to a clock pin or a data pin. When the victim net feeds a data pin, false data (glitch) can be clocked. If the victim net goes to the clock pin of a register, the register can be suffered either by false clocking on the inactive edge of a clock signal or by double-clocking on the active edge of a clock signal as shown in Figure 7.5.

The amount of crosstalk slowdown has been formulated by Sridhara and Shanbhag (2005). It is based on a different transition pattern in a three-wire model. The formula is given below:

$$T_l = \begin{cases} \tau_0[(1+\lambda)\Delta_1^2 - \lambda\Delta_1\Delta_2], & l = 1 \\ \tau_0[(1+2\lambda)\Delta_l^2 - \lambda\Delta_l(\Delta_{l-1}+\Delta_{l+1})], & 1 < l < n \\ \tau_0[(1+\lambda)\Delta_n^2 - \lambda\Delta_n\Delta_{n-1}], & l = n \end{cases}$$

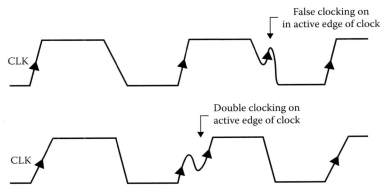

FIGURE 7.5
False clocking and double clocking due to double switching.

TABLE 7.1

Different Types of Crosstalk Delay

P	Relative Delay on Victim Wire	Transition Patterns in a Three-Wire Model
–	0	↑−−,−−↓,−−−,−−↑,↓−−,↑−↑,↑−↓,↓−↑,↓−↓
0	τ_0	↑↑↑,↓↓↓
1	$\tau_0(1+\lambda)$	↑↑−,↓↓−,−↑↑,−↓↓
2	$\tau_0(1+2\lambda)$	↑↑↓,↓↓↑,↑↓↓,↓↑↑,−↑−,−↓−
3	$\tau_0(1+3\lambda)$	↓↑−,↑↓−,−↑↑,−↓↑
4	$\tau_0(1+4\lambda)$	↑↓↑,↓↑↓

where:
  λ is the ratio of coupling capacitance to bulk capacitance
  τ_0 is the delay of a crosstalk-free wire
  Δ_l is the transition on wire l and its value is 1 for rising transition, −1 for
    falling transition, and 0 for no transition

In simplified form, if there is a transition in wire l, the propagation delay of lth wire is $T_l = (1 + p\lambda)\tau_0$, where the value of p lies between 0 and 4. Table 7.1 shows how the delay of victim wire (middle wire) varies with respect to signal transition in other two aggressor wires in a three-wire model. In the table, ↑, ↓, and − denote rising, falling, and no transition, respectively. There are a number of crosstalk avoidance techniques proposed in the literature that achieve different degrees of delay reduction, which will be discussed in Section 7.4.2.1.

7.2.3.2 Soft Errors

Soft errors are radiation-induced transient faults that are caused by thermal neutrons, high-energy neutrons generated from cosmic rays, and alpha particles generated by packaging materials. Both alpha particles

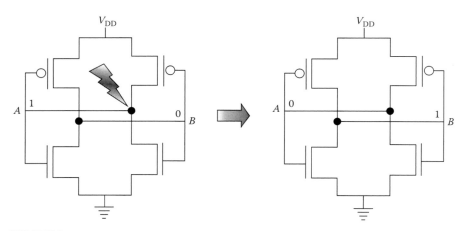

FIGURE 7.6
Soft error in a back-to-back inverter.

and neutrons generate electron–hole pairs along their path of traversal while hitting the transistor's diffusion. Neutrons are particularly trouble-some as they tend to generate more charge than alpha particles and can penetrate most man-made construction (a neutron can easily pass through five feet of concrete). This effect varies with both latitude and altitude. In London, the effect is 2 times worse than that on the equator. In Denver, with its high altitude, the effect is 3 times worse than that at sea-level San Francisco.

Traditionally, soft errors are considered as a major problem for dynamic RAM (DRAM). As the technology goes toward ultra-DSM level and the supply voltage also goes down, a significantly lower charge deposed by a particle strike suffices to flip the logic value of a node, thus creating a tran-sient pulse. The same phenomenon starts to affect the static RAMs (SRAMs). Unlike capacitor-based DRAMs, SRAMs are cross-coupled devices that have far less capacitance in each cell. The lower the capacitance, the greater the likelihood that an alpha particle or neutron will cause a single-event upset (SEU). Figure 7.6 shows how a soft error affects a back-to-back inverter-based memory. Initially, node A is at logic 1 and node B is at logic 0. Due to injection of the soft error, the logic value of node A flips which drives node B to invert the logic. This inverted bit will remain in the memory even after a soft error disappears.

Soft errors in latches and flip-flops are the major contributors of logic soft errors. Figure 7.7 focuses on the soft error in D-type latch. The same dis-cussion can also be extended to flip-flops. During positive cycle of a clock ($clk = 1$), the Q output of a D-type latch is strongly driven by the D input. Soft error during this period ($clk = 1$) can be treated as a glitch at the Q output but does not cause any SEU. In the negative cycle ($clk = 0$) when the Q output is

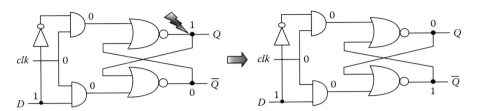

FIGURE 7.7
Soft error in D-type latch.

latched, soft error can invert the logic stored in the latch and causes a SEU as shown in Figure 7.7.

In combinational logic, a particle strike suffices to flip the logic value of a node, creating a single-event transient (SET) pulse. The transient pulse, after propagation through the logic, can be captured by a latch or flip-flop. With each technology generation, the transient pulses become wider with respect to logic transition time of the logic gates. In addition, as the clock frequency increases, the probability of latching a transient pulse increases as well. Due to these trends, the error rates in logic parts become as high as the error rates in memories. The details of logic soft error correction in NoC router are described in Section 7.4.1.1.

Depending on the soft error protection technique in a system, an effect of soft error can be classified into three types: (1) detected/corrected, (2) detected/uncorrected, and (3) undetected. The detected/corrected errors are typically associated with error correction logic. The detected/uncorrected errors often originate from parity protection since parity in most cases can detect but cannot correct the error. This type of error can result in a system reset or an application termination, depending on the severity of the error. The undetected errors can result in nonrecoverable errors causing system hangs or in *silent data corruption* (SDC) causing data integrity problems. Error detection and correction techniques can be used to cope with most of the logic soft errors, but at the same time, in a large system it is difficult to directly apply these techniques because of associated performance degradation and cost (power and area) overhead. Therefore, from a system point of view, it is very important to find out the regions that are highly susceptible to soft errors and apply the protection circuitry to meet the targeted MTBF (or FIT) at optimized performance and cost degradation.

Estimation of soft error rate (SER) in sequential circuit is very challenging since computation of the probability of erroneous system requires dynamic analysis of transients. SER estimation of each element in a system mainly depends on three factors: (1) *nominal FIT*, (2) *logic derating* (LD), and (3) *timing derating* (TD) (Nguyen and Yagil 2003). Full-chip FIT rate is the summation of individual elements' FIT rate on die. Particle strike must cause a glitch at the output of the gate (nominal FIT); this glitch has to propagate through the combinational logic to the flip-flop inputs (LD); and finally this erroneous glitch must be captured in a flip-flop, that is, the erroneous transient must have a sufficient overlap with the latching window of the flip-flop (TD). The SER of each element is given by the following relation:

$$\text{Individual elements' FIT} = \text{Nominal FIT} \times \text{TD} \times \text{LD}$$

In general, TD and LD are not independent, and hence accurate analysis requires the concept of TD–LD (Asadi and Tahoori 2006). A combinational logic (static CMOS based), by itself, cannot cause a SEU. The gate can suffer a glitch that might propagate downstream and be captured by a flop as shown in Example 7.2.

Example 7.2

Figure 7.8 shows the output of gate A (OR gate) is affected to soft error. The width of the soft error-induced glitch at this node is W. The propagation probability of the glitch at the output of gate D (AND gate) is the product of the probability at the output of gate B being 1 ($SP_B = 0.2$; the probability of the output of gate B being 1) and the propagation probability of the glitch (here, $1 \times 0.2 = 0.2$). Similarly, the propagation probability at the output of the gate E (OR gate) is calculated as $0.2 \times (1 - SP_C) = 0.2 \times 0.6 = 0.12$ (the probability of the output of gate C being 1). Hence, the LD factor is 0.12. The underlying assumption in this example is the value of all signals other than on-path signals (which propagated the erroneous glitch) is stable.

* * *

The TD is the probability that an erroneous value is captured in the flop. It can be calculated based on the setup (S) and hold (H) times of the flop, the glitch width (W_1), and the clock period (T). Glitch width at flop input (W_1) can be different from that at its origin (node A in Figure 7.8) due to various rise and fall transition delays for the gates along the path. The TD factor is defined as $TD = (S + H + W_1)/T$. The error propagation

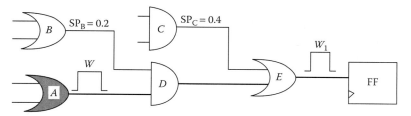

FIGURE 7.8
Propagation of glitch through logic gates to flip-flop (FF).

probability is calculated as the product of the LD and the TD. If there are multiple paths from the error site to the flip-flop, the overall error propagation probability is the product of individual paths' error propagation probability.

7.2.3.3 Some Other Sources of Transient Faults

7.2.3.3.1 PVT Variation

Process, supply voltage, and temperature variations are other important sources of timing fault for any VLSI circuit. The delay of a transistor in a slow process will be more than that in a fast process. With increasing temperature, both the transistor and interconnect delays will increase in 65-nm and above technologies. Of course, in 40-nm and lower geometries, temperature inversion can be observed. The delay of the transistor in 40-nm geometry will be more in cold condition than that in hot condition. But interconnect delay will keep on increasing with increasing temperature in those technologies. Thus, it is mandatory to do proper timing analysis for meeting the timing requirements in all possible sign-off corners.

7.2.3.3.2 Synchronization Failure

Any multicore design has several clock domains and the cores communicate with each other through globally asynchronous and locally synchronous (GALS) style. The communication can suffer due to synchronization errors, and hence synchronizers are required between the clock domains. Several types of synchronizers including dual-clock FIFO synchronizers, delay-line synchronizers, and simple pipeline synchronizers have been proposed by Dally and Poulton (1998). To avoid synchronization failure, synchronizers should be properly designed and well tested.

7.2.3.3.3 Power Supply Noise

Power supply noise is an important source of a transient error in any VLSI circuit. Noise in ground and power lines is known as ground bounce and power bounce, respectively. With fluctuating supply and ground voltages, the delay of the data path is affected, which may cause timing violation. Decoupling capacitance (on-chip or on-package) reduces power supply voltage noise.

7.2.3.3.4 IR Drop

Another major cause of signal integrity is the IR drop effect caused by wire resistance and current drawn from the power and ground grids. If the wire resistance is too large or the cell current is higher than predicted, an undesirable voltage drop may happen. The voltage drop causes the voltage supplied to the affected cells to be lower than required, which leads to larger gate and signal delays, which in turn can cause timing fault in the signal paths as well as clock skew. Voltage drop on power and ground grids can also affect the noise margins and compromises the signal integrity of the design. Therefore, special attention should be taken to resolve the IR drop effects during post-layout phase.

7.3 Permanent Fault Controlling Techniques

Permanent faults, at one level, are modeled as stuck-at faults, or as fail-stop faults. In the stuck-at fault, a node is stuck at either *logic 0* or *logic 1*. In the fail-stop model, a complete module (router or link) malfunctions and informs its neighbors about its out-of-order status (Dally and Towles 2004). Triple modular redundancy (TMR) is a well-known technique to handle stuck-at faults at links in which the faulty link, its two duplicate copies, and a voter are used to ensure protection against such errors.

In NoC, FIFOs are designed as either register based or SRAM based. For SRAM-based FIFOs, if any cell or row is permanently faulty, it can be repaired by a built-in self-repair (BISR) mechanism using redundant rows and columns (Wang et al. 2006). The detailed architecture of a BISR-based SRAM is beyond the scope of this book. Although physical faults are not as common and frequent as transient faults on-chip, in case a component fails, it is not always possible to repair or replace it on chip. In such a case, it is important to reroute the packets on alternate paths so that the communication infrastructure remains intact. Hence, to overcome the permanent faults, NoC must have to support adaptive (or dynamic) routing to avoid the faulty regions of the network and choose alternative paths dynamically. The main idea is to keep the chip in functioning state with graceful degradation of performance in the presence of faults.

Valinataj et al. (2009) proposed a deterministic, low-cost, deadlock-free, faulty-link-tolerant routing algorithm through dynamic reconfiguration to use new unique paths instead of the broken paths. For router permanent faults, Linder and Harden (1991) used a virtual channel (VC)-based fault-tolerant routing in two-dimensional (2D) mesh. Since the use of VC leads to more area and power consumption, several literatures have addressed fault-tolerant routing algorithms without using VCs. The packets are routed through alternate paths either by using a turn model (Chen and Chiu 1998; Fukushima et al. 2009; Glass and Ni 1996; Wu 2003) or by updating the routing table (Ali et al. 2007a; Fick et al. 2009), so that the communication infrastructure remains intact. Patooghy and Miremadi (2009) proposed a deadlock-free XYX routing for handling permanent faults in NoC. It makes a redundant copy of each packet at the source node and exploits two different routing algorithms, XY and YX, to route the original and redundant packets, respectively. Since two copies of each packet are received by the destination router, the erroneous packet is detected and replaced with the correct one.

7.4 Transient Fault Controlling Techniques

Primarily two types of transient faults can upset the on-chip network infrastructure: (1) intra-router errors occur within individual router components and (2) inter-router link errors occur during flit traversal.

7.4.1 Intra-Router Error Control

In NoC, VC-based routers are widely used. The details of a VC-based router architecture have been described in Chapter 3. It consists of a routing control unit, a switch arbiter (SA), a crossbar, an input buffer (FIFO), a VC allocator (VCA), a control logic, and handshaking signals. Park et al. (2006) focused on the possible errors in different modules of a NoC router and their controlling strategies as summarized below.

1. *Routing control unit error*: The routing control unit looks into the packet header to select the next router. The routing operation may be table based or algorithm based. A transient fault in the routing unit logic may cause a packet to be misdirected.

2. *SA error*: Depending on the packet header, the routing unit sends a request to the SA block. There may be a possibility that more than one requester can send their requests to access the same output channel. Thus, the SA block performs arbitration to select one of them and sends a grant signal to it. A transient fault in SA may give rise to the following conditions:

   a. A soft error in the control signals of the switch allocator can prevent flits from traversing the crossbar. This case is the least problematic, since the flits will keep on requesting access to the crossbar until they succeed.

   b. If a payload flit is mistakenly sent to a direction different from the header flit, it would cause flit/packet loss as it would deviate from the wormhole created by its header flit.

   c. An error can cause the arbiter to send a flit to multiple outputs (multicasting). If the flit is a payload flit, the same error will occur as in case (b). If the flit is a header flit, multiple VCs will mistakenly be reserved in all the receiving routers. The VCs will stay permanently reserved, thus reducing the effective buffer space in those routers.

3. *Crossbar error*: A transient fault in the crossbar may produce glitch in data transmission.

4. *FIFO error*: FIFO can be implemented as register banks or dedicated SRAM arrays. Due to the unidirectional nature of communication

in NoC, FIFOs are two-port memory—one write-only port and one read-only port. A transient fault can affect different parts of a FIFO, for example, the memory cells, the addressing mechanism, full and empty signal generation logic, and read/write enable signals. A single soft error within the FIFO will produce a SEU. A double (or more) error, however, would require more protection technique.

5. *VCA error*: The VCA, like the routing unit, operates only on header flits. All new packets request access to any one of the valid output VCs. The VCA arbitrates between all those packets requesting the same output VC. The VCA maintains the states of all successful allocations through a pairing between input VCs and allocated output VCs. Soft errors within the VCA may give rise to following scenarios:

   a. A soft error may assign one input VC to an invalid output VC. Such an assignment will block further traversal of the packet through the network.

   b. An unreserved output VC may be assigned to two different input VCs due to soft error. This will lead to packet mixing, and eventually packet/flit loss.

   c. Transient error may cause a reserved output VC to be assigned to a requesting input VC. This case is very similar to case (b) above.

6. *Error in control logic and handshaking signals*: Every router has several control logic inside it. These control logic blocks consist of combinational and sequential elements. The routers communicate with other neighboring routers by the handshaking signal lines to facilitate proper functionality and synchronization. Transient faults in the control logic and handshaking lines would disrupt the operation of the network.

As most of the transient faults inside the NoC router are caused by soft error, Section 7.4.1.1 describes the soft error correction techniques in latch, flip-flop, SRAM, and combinational logic.

7.4.1.1 Soft Error Correction

This section focuses on *logic soft error* protection—soft errors in latches, flip-flops, and combinational logic. Soft errors in SRAMs are protected using parity or error-correcting codes with interleaving. For a latch, as discussed in Section 7.2.3.2, the Q output is strongly driven by the D input during the active cycle of the clock and the latch is not susceptible to soft errors. In the inactive cycle when the output is latched to its previous state, soft error can cause a SEU. Mitra et al. (2006a) proposed a *built-in soft error resilience* (BISER) technique to protect the latches in the design using C-element. C-element is a special type of hardware where two PMOS and two NMOS transistors are connected in series. A two-input C-element and its truth table are shown in Figure 7.9. From the truth table, it is clear that if both the inputs of C-element

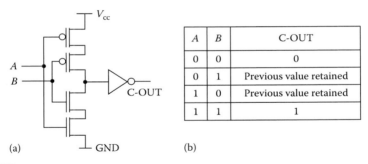

FIGURE 7.9
Structure of C-element (a) and the truth table (b).

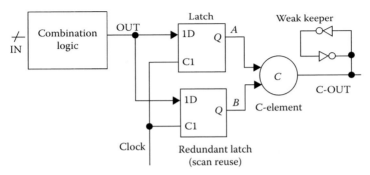

FIGURE 7.10
Soft error correction in latches.

are same, the output (C-OUT) will follow the inputs; otherwise, the output will latch at its previous value. Mitra et al. (2006a) presented the soft error correction in latches using C-element and a redundant latch as shown in Figure 7.10. If soft error affects any of the latches at a time, the output of the latches will differ and the C-OUT will retain to its previous value, and hence error will not propagate to the next stage. The correct logic value will be held at C-OUT by the keeper. Soft error in the keeper does not have a major effect because the C-element output will be strongly driven by the latch contents. The cost associated with the redundant latch is minimized by reusing on-chip resources such as scan for multiple functions at various stages of manufacturing and field use (Mitra et al. 2005).

As discussed in Section 7.2.3.2, soft error in combinational logic will cause a SET pulse. The transient pulse, after propagation through the logic, can be captured by a latch or a flip-flop. Hence, it is desirable to address soft error protection in flip-flops, latches, and combinational logic jointly. Otherwise, separate protection techniques for flip-flops, latches, and combinational logic introduce additional penalties and design complexity. Soft errors in combinational logic can be corrected using two techniques: (1) error correction using duplication and

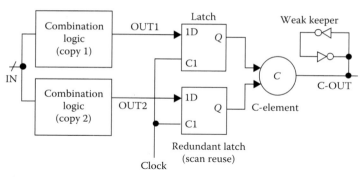

FIGURE 7.11
Soft error correction in combinational logic and latches using duplication.

(2) error correction using time-shifted outputs (Mitra et al. 2006b). Figure 7.11 shows the combined soft error correction in combinational logic and latches using duplication, assuming that soft error can affect any one of the two copies. Duplication technique improves the system-level soft error rate by 10 times over an unprotected design with negligible performance penalty (Zhang et al. 2006). However, the power and area costs of duplicating combinational logic can be significant. Usage of existing on-chip scan-out resources (for functional testing and design for debug) further reduces the area and power penalty.

Soft error correction using time-shifted output technique, however, does not require combinational logic duplication. The scheme is shown in Figure 7.12. Instead of duplicating the combinational logic, the output (OUT3) and its delayed version (delayed by τ), called OUT4, is used. If this path lies in the critical path of the design, the clock period must be increased by τ units to meet the timing requirement. The latch outputs are connected to a C-element as discussed above. Note that τ is a design parameter that can be tuned based on the reliability requirement.

The following scenario shows how time-shifted output technique can correct soft errors in combinational logic. Suppose that OUT3 and OUT4 settle

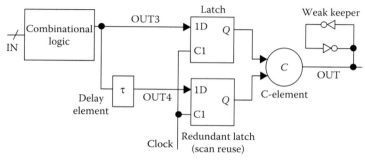

FIGURE 7.12
Soft error correction in combinational logic and latches using time-shifted output.

down to the correct value (because a path shorter than the timing critical path has been exercised here) and the C-element output is charged to a correct value, a hazard appears at OUT3 (and at OUT4 after time τ) due to a soft error inside the combinational logic. Suppose that the hazard settles down to the correct value within τ time units. In this case, correct logic value will be preserved on C-OUT since the hazards at OUT3 and OUT4 will not overlap. Figure 7.13 describes this scenario.

Figure 7.14 shows a flip-flop design for implementing soft error correction using time-shifted outputs. C-elements and keepers are added at the outputs

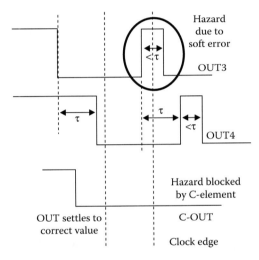

FIGURE 7.13
Soft error correction using time-shifted output.

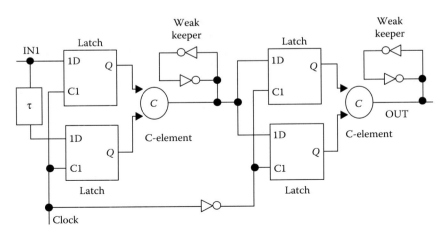

FIGURE 7.14
Soft error correction in combinational logic and flip-flop using time-shifted output.

of both master and slave stages to correct soft errors in combinational logic, as well as soft errors in master and slave latches.

7.4.2 Inter-Router Link Error Control

As discussed earlier in Section 7.2.3.1, capacitive crosstalk is one of the major sources of a transient fault in NoC links. It introduces crosstalk noise (glitch), crosstalk delay (both speedup and slowdown) that can violate the setup and hold requirements of any design, and double switching error. Several techniques have been proposed in literature to mitigate the ill effect of capacitive crosstalk and will be covered in Section 7.4.2.1.

7.4.2.1 Capacitive Crosstalk Avoidance Techniques

The capacitive crosstalk avoidance techniques can be categorized as follows:

1. Usage of shielding and duplicating wire
2. Increase of inter-wire spacing
3. Increase (decrease) of the driver strength of the victim (aggressor) nets
4. Usage of on-chip serialized link
5. Crosstalk avoidance code (CAC)

Each of the above techniques is described as follows:

1. *Usage of shielding and duplicating wire* is one of the most common approaches to mitigate capacitive crosstalk. In the shielding technique, a ground or supply wire is inserted after every signal wire. In the duplicating method, each signal wire is duplicated. Hence, shielding and duplicating techniques will cause double the bus width. Another shielding technique is *half shielding*, in which a ground wire is inserted after every two signal wires. Shielding and duplicating can be combined together to form a *duplicate-and-shield* technique. Though simple, this method has the disadvantage of requiring a significant number of extra wires. Applying the above methods in a three-wire model, there is no opposite transitions ($\uparrow \downarrow \uparrow$ or $\downarrow \uparrow \downarrow$) in all three wires, and hence can achieve different degrees of delay reduction. The glitch width and height are also lesser than the worst-case pattern. Table 7.2 depicts different shielding and duplicating methods described above for an 8-bit data word and also shows the length of the code word (n) along with the delay in the victim line, assuming that the impact of nonadjacent aggressors has negligible impact on the victim wire.

TABLE 7.2

Different Types of Shielding and Duplicating Techniques

Worst-Case Pattern in k-bit Data Word	Crosstalk Avoidance Techniques	Pattern Applied (Code Word = n)	Delay on Victim Wire
↑↓↑↓↑↓↑↓ ($k = 8$)	Half-shielding	↑↓G↑↓G↑↓G↑↓ ($n = 11$)	$\tau_0 (1 + 3\lambda)$
	Shielding	↑G↓G↑G↓G↑G↓G↑G↓ ($n = 15$)	$\tau_0 (1 + 2\lambda)$
Delay on victim wire = $\tau_0 (1 + 4\lambda)$	Duplicating	↑↑↓↓↑↑↓↓↑↑↓↓↑↑↓↓ ($n = 16$)	$\tau_0 (1 + 2\lambda)$
	Duplicating and shielding	↑↑G↓↓G↑↑G↓↓G↑↑G↓↓G↑↑G↓↓ ($n = 23$)	$\tau_0 (1 + \lambda)$

G, grounded shielding wire.

2. *Increase inter-wire spacing* is another common approach to reduce the value of coupling capacitance and hence capacitive crosstalk. Unlike shielding or duplicating, no extra wire has to be routed, and hence it is easier to implement. Increasing the inter-wire spacing will increase the wire pitch (pitch = width + space), and hence the wiring area. The crosstalk delay of a three-wire model, as discussed in Section 7.2.3.1, is $\tau_0 (1 + p\lambda)$, where λ is the ratio of coupling capacitance to bulk capacitance, and the value of p in worst case is 4. Shielding or duplicating will reduce the value of p from 4 to 3, 2, or 1 as shown in Table 7.2 without changing the value of λ. However, increasing inter-wire spacing will reduce the coupling capacitance and hence the value of λ, without reducing p.

3. *Driver strength* has a significant impact on signal transition time. When there is no transition in victim wire and sharp transition in aggressor wire, crosstalk noise will appear in the victim net. Reducing the strength of aggressor's driver will slow down its transition, which helps to reduce the crosstalk noise, but at the same time, the aggressor may violate its timing budget. Hence, driver strength should be properly adjusted. The similar explanation is also true for double switching error reduction. Increasing the driver strength of victim wire or reducing the driver strength of aggressor wire will solve the problem of double switching error. To address the crosstalk slowdown, increasing the driver strength of victim or using lower threshold transistor for victim's driver will reduce delay. Similarly, weaker driver or driver with high-threshold transistor for a victim wire will increase the data path delay and hence can address the crosstalk speedup.

4. On-chip serialization (OCS) technique (Lee et al. 2005) has already been discussed elaborately in Chapter 6. Using OCS, the number of interconnects reduces drastically. For example, 4:1 serializer converts 32-bit

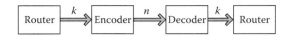

FIGURE 7.15
Bus encoding scheme for NoC interconnect.

data to 8-bit. Hence, the inter-wire spacing can be increased such that after serialization also the serialized link can occupy the same wiring area as earlier (unserialized link), thus reducing coupling capacitance.

5. *CAC* is another approach to mitigate the effect of capacitive crosstalk. It provides an elegant alternative to the above-mentioned approaches and a common framework to address signal integrity due to crosstalk. There are a number of CACs proposed in the literature that achieve different degrees of delay reduction. The fundamental idea of the CAC technique is to encode the data in such a way that restricts the worst-case transition (↑ ↓ ↑ or ↓ ↑ ↓) in the adjacent wires. Different CAC techniques have been studied for handling the above issue of interconnection links in the NoC paradigm considering the trade-off between the power efficiency, the wiring complexity, and the silicon area overhead. Coding maps k data bits to n code bits resulting in an (n, k) code, where $(n \geq k)$. The overall scheme is shown in Figure 7.15. All CACs are nonlinear. A binary code is linear if and only if the modulo-2 sum of the two code words is also a code word. The details of different CAC techniques are discussed in Sections 7.4.2.1.1 through 7.4.2.1.4.

7.4.2.1.1 *Forbidden Overlap Coding*

If the worst-case transitions (↑ ↓ ↑ or ↓ ↑ ↓) are avoided, the maximum coupling can be reduced to $p = 3$. This condition can be satisfied if and only if a code word having the bit pattern 010 does not make a transition to a code word having the pattern 101 at the same bit positions (Sridhara and Shanbhag 2007). The codes that satisfy the above condition are referred to as forbidden overlap codes (FOCs). The simplest method of satisfying the forbidden overlap condition is half-shielding, in which a ground wire is inserted after every two signal wires. Though simple, this method has the disadvantage of requiring a significant number of extra wires. Another solution is to encode the data links such that the code words satisfy the forbidden overlap condition. However, encoding all the bits at once is not feasible for wide links due to the prohibitive size and complexity of the codec (encoder and decoder) hardware. In practice, partial coding is adopted, in which the links are divided into subchannels that are encoded using CACs. The subchannels are then combined in such a way as to avoid crosstalk occurrence at their boundaries. Considering a 4-bit subchannel, the coding scheme is expressed in Table 7.3 (Pande et al. 2006a).

TABLE 7.3

Truth Table of FOC_{4-5}

Data Bits				Code Bits				
D3	D2	D1	D0	C4	C3	C2	C1	C0
0	0	0	0	0	0	0	0	0
0	0	0	1	0	0	1	0	0
0	0	1	0	0	0	0	0	1
0	0	1	1	0	0	1	0	1
0	1	0	0	0	0	0	1	1
0	1	0	1	0	0	1	1	1
0	1	1	0	1	0	0	1	1
0	1	1	1	1	0	1	1	1
1	0	0	0	1	0	0	0	0
1	0	0	1	1	0	1	0	0
1	0	1	0	1	0	0	0	1
1	0	1	1	1	0	1	0	1
1	1	0	0	1	1	0	0	0
1	1	0	1	1	1	1	0	0
1	1	1	0	1	1	0	0	1
1	1	1	1	1	1	1	0	1

For coding *32 bits* with a *4-bit* subchannel-based FOC (FOC_{4-5}), eight such blocks are needed. As a result of this, a 32-bit uncoded link will be converted to a 40-bit coded link. By contrast, half-shielding requires a 47-bit link. In the above FOC, two subchannels can be placed next to each other without any shielding as shown in Figure 7.16. This scheme does not violate the FO condition.

7.4.2.1.2 Forbidden Transition Coding

The maximum capacitive coupling, and hence the maximum delay, can be reduced even further by extending the list of nonpermissible transitions. By ensuring that the transitions between two successive codes do not cause adjacent wires to switch in opposite directions (i.e., if a code word has a *01* bit pattern, the subsequent code word cannot have a *10* bit pattern at the same bit position, and vice versa), the coupling factor can be reduced to $p = 2$.

This condition is referred to as forbidden transition condition, and the CACs satisfying it are known as forbidden transition codes (FTCs) (Sridhara and Shanbhag 2007). Inserting a shielding wire after each signal line can employ the simplest FTC. For wider links, a hierarchical encoding is more suitable, where the inter-switch links are divided into subchannels that are encoded individually. Considering a *3-bit* subchannel, the coding scheme is expressed in Table 7.4 (Pande et al. 2006a). In this case also, the subchannels are combined in such a way that there is no forbidden transition at the

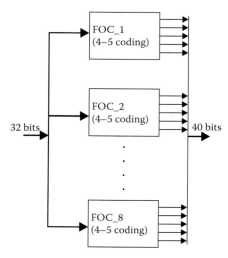

FIGURE 7.16
Combining adjacent subchannels in FOC encoding.

TABLE 7.4

Truth Table of FTC_{3-4}

Data Bits			Code Bits			
D2	D1	D0	C3	C2	C1	C0
0	0	0	0	0	0	0
0	0	1	0	1	0	0
0	1	0	0	0	0	1
0	1	1	0	1	0	1
1	0	0	0	1	1	1
1	0	1	1	1	0	0
1	1	0	1	1	0	1
1	1	1	1	1	1	1

boundaries between them. Consequently, a 32-bit uncoded link will be converted to a 53-bit coded link as shown in Figure 7.17. By contrast, shielding requires a 63-bit link.

7.4.2.1.3 Forbidden Pattern Coding

The same reduction of the coupling factor as for FTCs ($p = 2$) can be achieved by avoiding 010 and 101 bit patterns for each of the code words. This condition is referred to as forbidden pattern condition, and the corresponding CAC is known as a forbidden pattern code (FPC) (Sridhara and Shanbhag 2007). The simplest FPC code is realized by duplication, where each data bit is transmitted using two adjacent wires. Considering a 4-bit subchannel, the coding

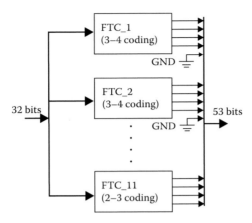

FIGURE 7.17
Combining adjacent subchannels in FTC encoding.

TABLE 7.5

Truth Table of FPC_{4-5}

Data Bits				Code Bits				
D3	D2	D1	D0	C4	C3	C2	C1	C0
0	0	0	0	0	0	0	0	0
0	0	0	1	0	0	0	0	1
0	0	1	0	0	0	1	1	0
0	0	1	1	0	0	0	1	1
0	1	0	0	0	1	1	0	0
0	1	0	1	0	0	1	1	1
0	1	1	0	0	1	1	1	0
0	1	1	1	0	1	1	1	1
1	0	0	0	1	0	0	0	0
1	0	0	1	1	0	0	0	1
1	0	1	0	1	1	0	0	0
1	0	1	1	1	0	0	1	1
1	1	0	0	1	1	1	0	0
1	1	0	1	1	1	0	0	1
1	1	1	0	1	1	1	1	0
1	1	1	1	1	1	1	1	1

scheme is expressed in Table 7.5 (Pande et al. 2006a). While combining the subchannels, it is ensured that there is no forbidden pattern at the boundaries. As a result of this, similar to the above two CACs, FPC also adds redundant bits to the uncoded link and a 32-bit uncoded link is converted to a 54-bit coded link as shown in Figure 7.18. By contrast, duplication requires a 64-bit link.

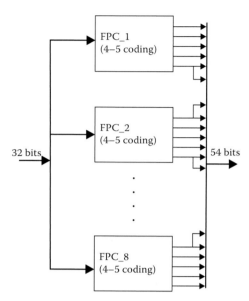

FIGURE 7.18
Combining adjacent sub-channels in FPC encoding.

7.4.2.1.4 One Lambda Coding

A minimum value of $p = 1$ can be achieved using one lambda coding (OLC). This can be achieved if the transitions in victim wire and its two adjacent wires show (1) a forbidden transition condition, (b) a forbidden pattern condition, and (c) an additional constraint that two adjacent bit boundaries in the codebook cannot both be 01 type or 10 type (forbidden adjacent boundary pattern condition) (Sridhara and Shanbhag 2007). The simplest OLC is duplication and shielding, where every bit is duplicated and shield wires are inserted between adjacent pairs of duplicated bits. However, OLC results in excessively large number of output wires; hence, no analysis has been performed on it in literature.

In NoC, message passing communication is followed by wormhole switching approach where packets are decomposed into *flits* (flow control units) over which flow control is performed. Transient faults, except crosstalk-induced faults, in NoC links can be handled in two ways: (1) error detection–retransmission and (2) error correction. Both the schemes are discussed in Sections 7.4.2.2 and 7.4.2.3, respectively.

7.4.2.2 Error Detection and Retransmission

Link errors have been studied extensively by the researchers, since they have so far been considered the dominant source of errors in on-chip network fabrics. They have been tackled within the context of two central themes:

detection and retransmission. Error detection and retransmission can be either end to end or switch to switch (Murali et al. 2005).

In the end-to-end error detection scheme, a parity bit or cyclic redundancy check (CRC) bit is added to the packet. The parity or CRC encoder is added to the sender network interface (NI) and the decoder is added to the receiver NI. The error detection mechanism is performed only at the receiver NI. The receiver NI sends a nonacknowledgment (NACK) or acknowledgment (ACK) signal (labeled as *credit* signal in Figure 7.19a) back to the sender, depending on whether the data contained an error or not. The sender will retransmit the packet if an error is detected at the receiver NI. But if the error corrupts the header flit, which contains the information of destination address, the packet will perhaps not reach the destination. Therefore, the header flit is protected by parity or CRC codes, which the switch checks at each hop traversal. If a switch detects an error on a packet's header flit, it drops the packet. To account for errors on the ACK or NACK signals, a time-out mechanism is incorporated for retransmission at the sender. The overall scheme is shown in Figure 7.19a.

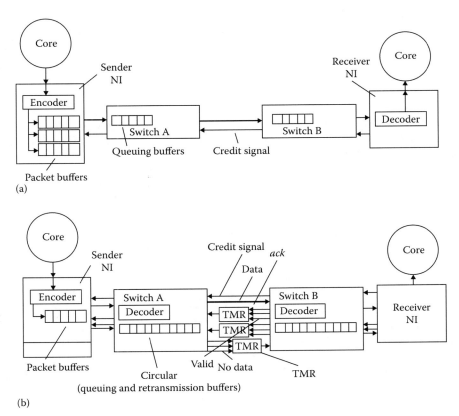

FIGURE 7.19
Retransmission scheme: (a) End to end; (b) switch to switch.

However, in the switch-to-switch error detection scheme, error detection is performed at each switch input port and data retransmission occurs between adjacent switches. There can be two types of switch-to-switch flow control schemes: parity or CRC at flit level (*ssf*) and at packet level (*ssp*). In the *ssp* error detection scheme, the transmitting switch adds parity or CRC bits to the packet's tail flit, whereas in *ssf*, the transmitting switch adds parity or CRC bits to each flit. To handle the fault in ACK/NACK signal (*credit* signal), a triple modular redundancy (TMR) technique is used as shown in Figure 7.19b. Murali et al. (2005) showed that the average packet latency is higher in end-to-end rather than in switch-to-switch flow control. In switch-to-switch flow control, *ssp* has more latency than *ssf*. In NoC, *ssf* flow control technique is widely used.

The error detection and retransmission methodology is discussed in Sections 7.4.2.2.1 and 7.4.2.2.2.

7.4.2.2.1 Error Detection Methodology

For error detection, appending a parity bit (odd or even) to the data is the most common example. But using a parity bit, all burst errors cannot be detected. However, cyclic redundancy codes (CRCs) take advantage of the considerable burst error detection capability provided by cyclic codes. Common CRC polynomials can detect the following types of errors: (1) all single-bit errors, (2) all double-bit errors, (3) all odd number of errors, and (4) any burst error for which the burst length is less than or equal to the polynomial length.

Linear feedback shift register (LFSR) with serial data feed has been used since the 1960s to implement the CRC algorithm in hardware. For parallel data transfer, parallelism has been introduced in CRC-generating hardware (Albertengo and Sisto 1990; Campobello et al. 2003; Joshi et al. 2000; Pei and Zukowski 1992; Shieh et al. 2001; Sprachmann et al. 2001). To present the parallel CRC architecture, the LFSR is considered as a synchronous finite-state machine. State S holds the checksum bits. The message is fed to the input I and the combinatorial network calculates the next state S_{next} from the current state and the new input. The state machine calculates a new checksum every clock cycle. For example, the LFSR-based hardware of the generator polynomial $g(x) = x^3 + x^2 + 1$ is shown in Figure 7.20a and its parallel implementation is shown in Figure 7.20b. A parallel architecture for the LFSR is able to include more than one bit of the message in a single clock cycle. For example, a 4-bit input data word ($i_3 i_2 i_1 i_0$) with the same generator polynomial is shown in Figure 7.21.

Therefore, for a *32-bit* input, 32 XOR networks will be cascaded to form a combinational network. The problem of this circuit is that it has large critical path delay. Thus, the operating clock frequency will degrade. To improve the performance of this circuit, a byte-wise CRC technique is used (Perez 1983). In the byte-wise CRC technique, 32-bit data are broken into four bytes, and for each byte, CRC operation is performed. Thus, four ACK signals will be generated from the receiver. The ANDing of all the ACK signals is the final ACK and this signal will come back to the sender.

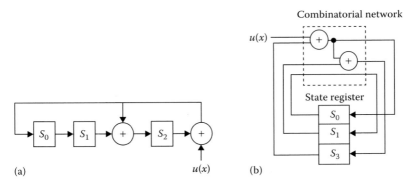

FIGURE 7.20
Hardware implementation of $g(x) = x^3 + x^2 + 1$: (a) LFSR based; (b) finite-state machine based.

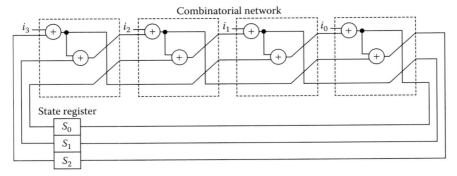

FIGURE 7.21
Parallel state machine of $g(x) = x^3 + x^2 + 1$ with 4-bit input.

7.4.2.2.2 Retransmission Methodology

As the transmitter transmits each flit, it retains a copy in a retransmit flit buffer until a correct receipt is acknowledged. As each flit is received, the receiver checks the flit for errors. If the flit is received correctly, an acknowledgment is sent to the transmitter. Upon receiving this acknowledgment, the transmitter discards its copy of the flit. If a flit is received in error, the sender will retransmit the faulty flit. There may be a possibility that the flits are not reaching to the destination in order. For example, first payload flit might be received before the retransmitted header flit of its packet. To avoid such complication, it is easier to roll back transmission to the faulty flit and retransmit all flits from the retransmit flit buffer starting at that point. This policy is called *go-back-N* retransmission. In hop-by-hop (HBH) error control (Park et al. 2006), go-back-N retransmission mechanism is used.

Another retransmission strategy is the *selective repeat*. When it is used, any corrupted flit received at the receiver is discarded, but unaffected flits are buffered. When the sender times out, only the unacknowledged flits are retransmitted. If such a flit arrives correctly, the receiver can deliver to the network layer, in sequence, all the flits it has buffered. This retransmission scheme needs a larger buffer space. In NoC, go-back-N retransmission is widely used (Ali et al. 2007b; Bertozzi and Benini 2004; Pullini et al. 2005).

7.4.2.3 Error Correction

The error detection and retransmission scheme mentioned above suffers from higher area requirement as it requires retransmission buffers. Retransmission will give rise to multiple communications over the same link, and hence ultimately it will not be energy efficient. Moreover, the performance of the system is degraded due to retransmission. In the DSM era, area and power are the major issues from the design perspective. A forward error correction (FEC) technique, however, does not have this type of problem. *Single error correction* (SEC), for example, *Hamming code* (Lin and Costello 1983), is widely used as FEC in any VLSI design. The TMR method for error correction in NoC links was proposed by Huang et al. (2008). Error correction is possible if the Hamming distance between any two code words in the codebook is greater than one. In general, if the minimum Hamming distance between any two code words is H, then all the $(H - 1)$ errors appearing on the bus can be detected and $(H - 1)/2$ errors can be corrected. Error-correcting code (ECC) is a linear code (Sridhara and Shanbhag 2005).

With increasing complexity, decreasing power supply voltage, and higher switching speeds, single-error-correcting capabilities will not sufficiently increase the error resilience of the system implemented in the current and forthcoming ultra-DSM (UDSM) technologies. Hence, multiple error-correcting (MEC) codes are necessary to address this issue. MEC schemes add more redundancy than the SEC ones. The major challenge in applying the existing MEC schemes in high-speed NoCs are the delay of the codec (encoder and decoder) module. Bose–Chaudhuri–Hocquenghem (BCH) code, Reed–Solomon (RS) code, Viterbi code, and Turbo code are widely used for MEC in off-chip communication. The codec delay of these codes is quite large, and the summation of wire and codec delay will fail to meet the targeted clock cycle budget in high-speed NoCs. Hybrid techniques (Murali et al. 2005) provide both error correction and retransmission and allow for more robust protection of data. Hybrid solutions compensate for the limitations of ECCs. For example, SEC and double error detection (DED) codes can correct at most one error, but can detect double-bit errors. Therefore, upon detection of a double-bit error, the SEC/DED unit may invoke a retransmission mechanism.

In the current and forthcoming UDSM technologies, the burst error is likely to better capture the nature of many on-chip errors (Benini and Micheli 2006). In the work of Zimmer and Jantsch (2003), a conventional fault model for on-chip buses (Hedge and Shanbhag 2000) has been replaced by a new fault model to support the burst error. It also proposes a burst error correction technique for NoC links. The overall scheme is shown in Figure 7.22.

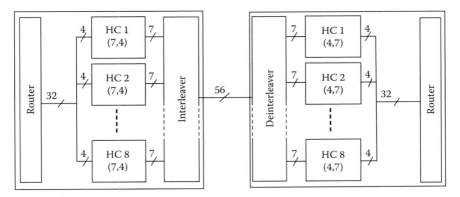

FIGURE 7.22
Burst error correction in link. HC, Hamming code.

In Figure 7.22, a 32-bit output data from a transmitting router is split into eight groups. Each group is encoded separately by using Hamming(7,4) encoder. Thus, eight bits (one from each group) can be corrected by this scheme. To support burst error correction, the output of the Hamming encoder is fed to an interleaver having an interleaving degree of 8. The deinterleaver module rearranges its input to get back the *56-bit* encoded output. This *56-bit* data is again split into eight groups. Each group is decoded separately by using the Hamming decoder. If the burst error affects any eight contiguous wires in the communicating channel, this scheme corrects it properly.

7.5 Unified Coding Framework

With shrinking feature sizes, scaling of supply voltages, increasing interconnect density, and faster clock rates, the NoC suffers from three major problems: (1) power consumption due to transition, (2) capacitive crosstalk, and (3) reliability. Sridhara and Shanbhag (2005) proposed a unified framework from which codes are derived that jointly optimizes power, delay, and reliability. The overall scheme is shown in Figure 7.23.

Low-power code (LPC), CAC, and ECC can be combined into a system if the following conditions are satisfied.

1. CAC needs to be the outermost code as, in general, it involves nonlinear and disruptive mapping from data to code word.
2. LPC can follow CAC as long as it does not destroy the peak coupling transition constraint of CAC. The additional information bits generated by LPC need to be encoded through a linear CAC to ensure that they do not suffer from crosstalk delay.

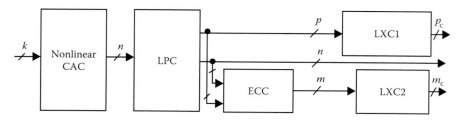

FIGURE 7.23
A unified framework.

3. ECC needs to be systematic to ensure that the reduction in transition activity and the peak coupling transition constraint are maintained.

4. The additional parity bits generated by ECC need to be encoded through a linear CAC to ensure that they do not suffer from crosstalk delay.

A framework satisfying the above conditions is shown in Figure 7.23. LXC1 and LXC2 are linear CACs based on either shielding or duplication. Nonlinear CACs cannot be used because error correction has to be done prior to any other decoding at the receiver. In Figure 7.23, a k-bit input is coded using CAC to get an n-bit code word. The n-bit code word is further encoded to reduce the average transitions through LPC, resulting in p additional low-power information bits. ECC generates m parity bits for the $(n + p)$ code bits. The m parity bits and p low-power bits are further encoded for crosstalk avoidance to obtain m_c and p_c bits, respectively, that are sent over the bus along with n code bits. The total number of wires required to encode a k-bit bus is thus $(n + p_c + m_c)$. A variety of codes based on the unified framework that allow for a trade-off between delay, power, area, and reliability are presented in Sections 7.5.1 through 7.5.4.

7.5.1 Joint CAC and LPC Scheme (CAC + LPC)

Combining LPC and CAC codes (Sridhara and Shanbhag 2005) is a hard problem as both are nonlinear codes and, even when such a combination is possible, the resulting code is complex. For example, it is not possible to combine a bus-invert (BI) coding with FTC or FOC as inverting an FTC or FOC code word destroys its crosstalk avoidance property. However, FTC reduces the average coupling power dissipation as it avoids the high-power-consuming opposing transitions on adjacent wires. Thus, FTC can independently be used for crosstalk avoidance and low power. While comparing with shielding and duplicating techniques, CAC will introduce an extra codec overhead in terms of both area and energy consumption. To address this issue, Pande et al. (2006a) incorporated the above-mentioned CAC techniques in a 64-IP

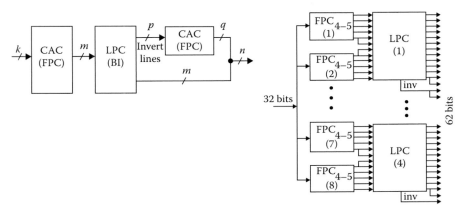

FIGURE 7.24
Joint CAC and LPC scheme.

mesh-based NoC at 130-nm technology. A thorough cost evaluation has been carried out with varying λ, where λ is the ratio of coupling capacitance to bulk capacitance. Any coding scheme is energy efficient if $(E_{uncoded} - E_{coded})/E_{codec} > 1$, where $E_{uncoded}$, E_{coded}, and E_{codec} denote the energy consumptions of uncoded link, coded link, and codec module, respectively. They observed that for FOC to be energy efficient at 130-nm technology, the value of λ has to be greater than 4. At $\lambda = 4$, FOC is not energy efficient, whereas both FPC and FTC are energy efficient. Energy savings in FTC scheme is the highest and FPC occupies the second position among the three. However, FPC can be combined with BI-based LPC schemes. This is because inverting an FPC code word maintains the FP condition. In a joint code with FPC and BI-based LPC, the invert bits are encoded using duplication code to avoid crosstalk delay in the invert bits, as shown in Figure 7.24. The joint code is a concatenation of the two component codes and no further optimization is possible. The codec overhead of the joint code is the sum of codec overheads of the component codes, and hence, the joint code is complex.

7.5.2 Joint LPC and ECC Scheme (LPC + ECC)

The joint LPC and ECC scheme is shown in Figure 7.25a. Here, BI and Hamming SEC codes are combined. The scheme is called BI Hamming (BIH) code (Sridhara and Shanbhag 2005). These codes are suitable for long DSM buses where reduction of power consumption is important but voltage scaling is not possible due to the presence of DSM noise. A property of XOR operation is that if an odd (even) number of inputs of an XOR gate are inverted, then the output is inverted (unchanged). Therefore, after computing the invert bit from the BI code, the output odd parity bit is conditionally inverted using the invert bit. Thus, parity generation and invert bit

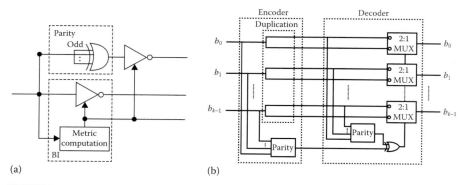

FIGURE 7.25
(a) Joint LPC and ECC: BIH; (b) joint CAC and ECC: DAP.

computation can occur in parallel, reducing the total delay to the maximum of the two and the delay of an inverter.

7.5.3 Joint CAC and ECC Scheme (CAC + ECC)

To make the system tolerant against transient errors other than crosstalk, in addition to CAC, there is a need to incorporate the FEC codes into the NoC data stream. There are a few joint CAC and SEC codes, among which the duplicate-add-parity (DAP) code (Sridhara and Shanbhag 2005), FTC + Hamming code (Sridhara and Shanbhag 2005), boundary shift code (BSC) (Patel and Markov 2003), and modified dual-rail (MDR) code (Rossi et al. 2005) reduce the maximum coupling to $p = 2$. In FTC + Hamming code, FTC(4,3) and hamming codes with shielding are used. However, the coding overhead of this joint coding scheme is large as the joint code is a concatenation of the two individual codes. A significantly lower overhead joint coding scheme is DAP where the incoming bits are duplicated to reduce crosstalk delay (FPC). The duplication scheme has a Hamming distance of 2 as any two distinct code words differ in at least two bits. By appending the parity bit, the Hamming distance can be increased to 3, and thus, a single error can be corrected. Figure 7.25b shows the encoder and decoder implementation of the overall scheme.

The MDR code is very similar to the DAP (Pande et al. 2006b). In the dual-rail code, considering a link of k information bits, $m = k + 1$ check bits are added, leading to a code word length of $n = k + m = 2k + 1$. The $k + 1$ check bits are defined with the following equations:

$$c_i = d_i, \text{ for } i = 0 \text{ to } (k-1)$$

$$c_k = d_0 \oplus d_1 \oplus d_2 \oplus \cdots \oplus d_{k-1}$$

In MDR, two copies of parity bits c_k are placed adjacent to the other code word bits to reduce crosstalk.

The BSC is very similar to DAP in that it uses duplication and one parity bit to achieve crosstalk avoidance and SEC. However, the fundamental difference is that at each clock cycle, the parity bit is placed on the opposite side of the encoded flit. Table 7.6 shows examples of different CAC + ECC code words. The parity bits are indicated in bold. Pande et al. (2006b) observed that the energy saving in DAP and MDR is almost the same for mesh and folded torus topologies at 130-nm technology. The BSC scheme has less energy saving compared to DAP and MDR.

Another joint CAC and ECC code, triplication error correction coding, was proposed by Huang et al. (2008). The coding scheme shown in Figure 7.26 is a SEC code by triplication of each bit. For the triplication error correction coding, the Hamming distance of each bit is equal to 3. Therefore, each bit can be corrected by itself if there are no more than one error bits in the three triplicate bits. The error bit can be corrected by a majority gate. The function of the majority gate is shown in Figure 7.26.

TABLE 7.6

Coded Flit Structure for a Different CAC + ECC Scheme

Clock Cycle	Flit	BSC	DAP	MDR
1	0010	**1** 00001100	**1** 00001100	**11** 00001100
2	0010	00001100 **1**	**1** 00001100	**11** 00001100
3	1100	**0** 11110000	**0** 11110000	**00** 11110000
4	0100	00110000 **1**	**1** 00110000	**11** 00110000
5	0011	**0** 00001111	**0** 00001111	**00** 00001111

Note: The parity bits are indicated in bold.

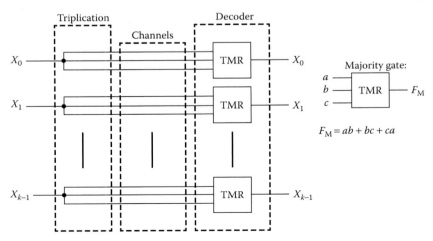

FIGURE 7.26
Triplication error correction coding.

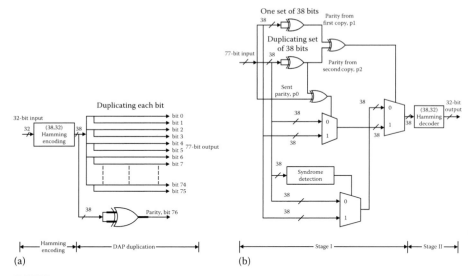

FIGURE 7.27
CADEC: (a) encoder; (b) decoder.

To address MEC codes along with crosstalk avoidance capabilities (CAC/ MEC), Ganguly et al. (2007) proposed crosstalk avoidance and double error correction (CADEC) codes. The encoding scheme is a combination of Hamming encoding followed by DAP or BSC. The 32-bit word is first encoded by a standard (38,32) shortened single-error-correcting Hamming code and then encoded by either DAP or BSC scheme, and also adds a parity bit calculated from one Hamming copy. The CADEC encoder is shown in Figure 7.27a. A standard Hamming code has a Hamming distance of 3 between adjacent code words. On duplication, the Hamming distance becomes 6 and after adding the extra parity bit, this distance becomes 7. Thus, it has the ability of triple error correction. But the authors restricted themselves up to double error correction due to higher complexity involved in triple error correction. The CADEC decoding scheme consists of the following steps:

1. The parity bits of the individual Hamming copies are calculated (p1 and p2) and compared.

2. If these two parities obtained in step 1 differ, then the copy whose parity matches with the transmitted parity (p0) is selected as the output copy of the first stage.

3. If the two parity bits are equal, then any one copy is sent forward for syndrome detection; if the syndrome obtained for this copy is zero, then it is selected as the output of the first stage. Otherwise, the alternate copy is selected.

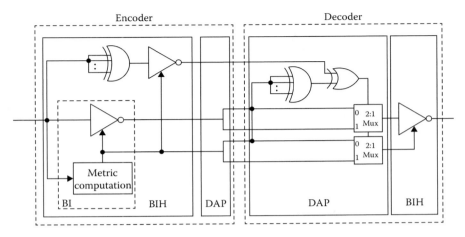

FIGURE 7.28
Joint CAC, LPC, and ECC.

4. The output of the first stage is sent for (38,32) Hamming decoder, finally producing the decoded output of the CADEC scheme.

The circuit-level implementation of the decoding scheme for CADEC is shown in Figure 7.27b. It can be observed that the channel width after CADEC encoder increases to almost 2.5 times of original uncoded data, and hence, it may introduce congestion in signal routing.

7.5.4 Joint CAC, LPC, and ECC Scheme (CAC + LPC + ECC)

Figure 7.28 shows a scheme for joint CAC, LPC, and ECC code to achieve low-power, crosstalk avoidance, and error correction properties. Here joint LPC and ECC (BIH) are combined with joint CAC and ECC (DAP). The joint code is referred as a DAP bus-invert (DAPBI) code. The invert bit is duplicated to ensure error correction and crosstalk avoidance for the bit.

7.6 Energy and Reliability Trade-Off in Coding Technique

It may be noted that an accurate characterization of error phenomena due to DSM noise (such as power grid fluctuations, EMI, crosstalk, or particle hits) is difficult as it requires the knowledge of various noise sources and their dependence on physical and electrical parameters. Reducing supply voltage will increase the bit error probability. Every time a transfer occurs across a wire, it can make an error with a certain probability ε. The following assumptions can be made to simplify the modeling problem:

1. A Gaussian distributed noise voltage V_N with variance σ_N^2 and zero mean is added to the signal waveform to represent the cumulative effect of all noise sources.
2. The variance σ_N^2 of the noise voltage V_N is independent of V_{DD}.
3. Errors occurring on different link lines are supposed to be independent.

The probability of bit error (ε) is given by (Hedge and Shanbhag 2000)

$$\varepsilon = Q\frac{V_{sw}}{2\sigma_N}$$

where V_{sw} is the voltage swing and $Q(x)$ is the Gaussian pulse:

$$Q(x) = \frac{1}{\sqrt{2\pi}}\int_\infty^\infty e^{-(y^2/2)}dy$$

This model accounts for the decrease of noise margins and hence an increase in the bit error rate ε (BER) as shown in Figure 7.29.

Figure 7.29 indicates that as V_{DD} reduces, the two curves approach each other, thereby increasing the overlap area, and hence they increase the probability of bit error (ε). By incorporating the error correction technique, the supply voltage can be reduced to save power without compromising the reliability of the system.

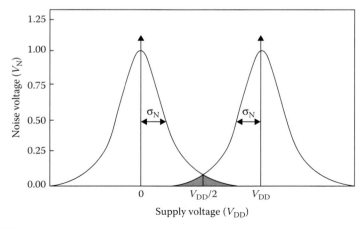

FIGURE 7.29
Dependency between bit error probability and supply voltage.

As error correction techniques enhance the reliability of on-chip interconnects to some degree, these will allow the designers to go for power consumption and reliability trade-off. For a given σ_N^2, the bit error probability increases by decreasing the voltage swing of signals. The error correction techniques allow decreasing the voltage swing of signal and guaranteeing the reliability at the same time if and only if Equation 7.3 satisfies where $P_{uncoded}(\varepsilon)$ is the probability of word errors in the uncoded case with full swing voltage and $P_{ecc}(\bar{\varepsilon})$ is the residual word error probability with ECC at lower swing voltage (V_{DD}):

$$P_{uncoded}(\varepsilon) \geq P_{ecc}(\bar{\varepsilon}) \qquad (7.3)$$

In order to obtain the lowest supply voltage for a specific error correction technique under the same level reliability of uncoded data, the supply voltage can be written as

$$\overline{V_{DD}} = V_{DD} \times \frac{Q^{-1}(\bar{\varepsilon})}{Q^{-1}(\varepsilon)} \text{ such that } P_{ecc}(\bar{\varepsilon}) = P_{uncoded}(\varepsilon) \qquad (7.4)$$

In the above equation, V_{DD} is the nominal supply voltage in the absence of any coding technique such that $P_{ecc}(\bar{\varepsilon}) = P_{uncoded}(\varepsilon)$. Therefore, to compute the $\overline{V_{DD}}$ or a specific coding scheme, the residual word error probability needs to be computed. For example, for a k-bit link, the residual word error probabilities for a small bit error rate (ε) of uncoded, hamming, DAP, CADEC, and Self-corrected green coding schemes are $k\varepsilon$, $k^2\varepsilon^2$, $3k(k+1)\varepsilon^2/2$, $k^2(k-4)\varepsilon^3$, and $(3k\varepsilon^2 - 2k\varepsilon^3)$, respectively.

Ganguly et al. (2007) showed that as CADEC has higher error correction capability compared to DAP, it allows maximum voltage swing reduction.

Although an individual bit error probability in the random bit error rate model is independent of each other, in a burst error scenario this consideration is no longer valid and demands for a burst word error rate model. The burst word error probability of the above error correction scheme has been formulated below.

Assuming an individual bit error rate to be ϵ, the probability that a burst error has affected i consecutive lines in an n-bit link ($i \leq n$) is $\epsilon^i (1-\epsilon)^{n-i}$. Thus, the total burst error probability of any of the i consecutive lines getting affected can be computed by varying the set of lines under consideration. Thus, for a specific value of i ($i < n$), the word error probability is $n \times \epsilon^i \times (1-\epsilon)^{(n-i)}$, whereas, for $i = n$, there is only one combination. Figure 7.22 shows a burst error correction scheme with an interleaving degree of 8, which can correct errors in at most eight adjacent wires whose link width n is 56. If the burst length is more than 8, the above scheme will fail. Therefore, the burst word error probability can be written as

$$P_{burst\ word\ error} = 56 \times \left[\sum_{i=9}^{55} \epsilon^i (1-\epsilon)^{56-i} \right] + \epsilon^{56} \qquad (7.5)$$

For smaller values of ϵ, the above expression converges to $P_{\text{burst word error}} \approx 56\epsilon^9$. At this point, the metric of interest is the energy reduction due to coding techniques compared to the uncoded case. Different coding schemes have different number of bits after encoding. Hence, a fair comparison in terms of energy-saving demands to consider the redundant bits. The energy consumption due to uncoded link (E_{uncoded}) forms a reference to evaluate different coding techniques.

$$\text{Energy reduction } (\%) \ = \left[\frac{E_{\text{uncoded}} - (E_{\text{coded}} + E_{\text{codec}})}{E_{\text{uncoded}}} \right] \times 100\% \qquad (7.6)$$

Reducing the supply voltage of the router without knowing the workload will slow down the router, which in turn affects the overall performance of the network. Thus, for a fair comparison of network energy consumption, nominal supply voltage has to be applied to the router and codec module, whereas the driver of the interconnection link has to be driven by the lowered voltage using static voltage scaling to save power consumption in the links. Moreover, when the signal crosses from a low-voltage domain to a high-voltage domain, a level shifter has to be inserted at the receiving router. Due to reduction of supply voltage in the driver and repeater, the link delay will increase. Moreover, the codec will also introduce an extra delay. Thus, it has to be ensured that after applying this technique, timing requirement of that path is still meeting.

To address the energy consumption of the joint codes compared to the uncoded link, Ganguly et al. (2007) evaluated the energy consumption of uncoded, DAP, BSC, MDR, and CADEC techniques on 64-IP mesh-based and folded torus-based NoCs at 130-nm technology keeping the word error probability of 10^{-20}. The voltage swing reduction is computed for each type of coding scheme and the reduce voltage is applied to the drivers and repeaters of the links. It has been observed that energy consumption is lesser in all the coding techniques compared to the uncoded link. CADEC achieves more energy saving compared to other two joint codes. The energy savings of DAP, BSC, and MDR are almost similar.

7.7 Summary

Signal integrity and reliability issues in NoC need to be addressed efficiently to solve the problems of transmission errors and power consumption. A considerable amount of work has been performed by various research groups across the globe to solve these problems. This chapter has presented a comprehensive review of the same. It has introduced the sources of different types of faults that affect the DSM technology and their controlling techniques in

NoC platform. Among the signal integrity issues, it has presented a detailed study of capacitive crosstalk effects on the NoC link and avoidance techniques. This chapter also focused on the effects of soft error and correction techniques in latches, flip-flops, combinational logic, and SRAM. It has discussed about the error detection and correction approaches due to transient faults. To address crosstalk avoidance, power minimization in the links, and reliability in communication jointly, it highlights a unified coding framework. A number of joint coding schemes have been discussed to address the signal integrity issues. As error correction techniques enhance the reliability of on-chip interconnects to some degree, these will allow the designers to go for power consumption and reliability trade-off. The error correction techniques allow decreasing the voltage swing of signal, and at the same time guaranteeing the reliability of the encoded data under the same reliability level of uncoded data. The energy reliability trade-off of different coding techniques has also been discussed. Thus, this chapter surveys the full gamut of research carried out to address signal integrity and reliability issues in NoC design of DSM technology.

Chapter 8 will focus on the challenges of NoC testing in achieving a sufficient fault coverage under a set of fault models relevant to NoC characteristics, under constraints such as test time, test power dissipation, less area over head, and test resources.

References

Albertengo, G. and Sisto, R. 1990. Parallel CRC generation. *IEEE Micro*, vol. 10, no. 5, pp. 63–71.

Ali, M., Welzl, M., and Hessler, S. 2007a. A fault tolerant mechanism for handling permanent and transient failures in a network on chip. *Proceedings of IEEE International Conference on Information Technology*, April 2–4, Las Vegas, NV, IEEE, pp. 1027–1032.

Ali, M., Welzl, M., Hessler, S., and Hellebrand, S. 2007b. An efficient fault tolerant mechanism to deal with permanent and transient failures in a network on chip. *International Journal of High Performance Systems Architecture*, vol. 1, no. 2, pp. 113–123.

Asadi, H. and Tahoori, M. 2006. Soft error derating computation in sequential circuits. *Proceedings of IEEE/ACM International Conference on Computer Aided Design*, November 5–9, San Jose, CA, IEEE, pp. 497–501.

Benini, L. and Micheli, G. D. 2006. *Networks on Chips: Technology and Tools*. Morgan Kaufmann Publishers, San Francisco, CA.

Bertozzi, D. and Benini, L. 2004. Xpipes: A network-on-chip architecture for gigascale system-on-chip. *IEEE Circuits and Systems Magazine*, vol. 4, no. 2, pp. 18–21.

Campobello, G., Patane, G., and Russo, M. 2003. Parallel CRC realization. *IEEE Transactions on Computers*, vol. 52, no. 10, pp. 1312–1319.

Chen, K. H. and Chiu, G. M. 1998. Fault-tolerant routing algorithm for meshes without using virtual channels. *Journal of Information Science and Engineering*, vol. 14, pp. 765–783.

Chen, W., Gupta, S. K., and Breuer, M. A. 1997. Analytic models for crosstalk delay and pulse analysis under non-ideal inputs. *Proceedings of International Test Conference*, November 1–6, Washington, DC, IEEE, pp. 809–818.

Cuviello, M., Dey, S., Bai, X., and Zhao, Y. 1999. Fault modeling and simulation for crosstalk in system-on-chip interconnects. *Proceedings of IEEE/ACM International Conference on Computer Aided Design*, November 7–11, San Jose, CA, IEEE, pp. 297–303.

Dally, W. J. and Poulton, J. H. 1998. *Digital Systems Engineering*. Cambridge University Press, Cambridge, UK.

Dally, W. J. and Towles, B. 2004. *Principles and Practices of Interconnection Networks*. Morgan Kaufmann Publishers, San Francisco, CA.

Fick, D., Orio, A. D., Chen, G., Bertacco, V., Sylvester, D., and Blaauw, D. 2009. A highly resilient routing algorithm for fault-tolerant NoCs. *Proceedings of Design, Automation and Test in Europe Conference*, April 20–24, Nice, France, IEEE, pp. 21–26.

Fukushima, Y., Fukushi, M., and Horiguchi, S. 2009. Fault-tolerant routing algorithm for network on chip without virtual channels. *Proceedings of IEEE International Symposium on Defect and Fault Tolerance in VLSI Systems*, October 7–9, Chicago, IL, IEEE, pp. 313–321.

Ganguly, A., Pande, P. P., Belzer, B. and Grecu, C. 2007. Addressing signal integrity in networks on chip interconnects through crosstalk-aware double error correction coding. *Proceedings of IEEE Computer Society Annual Symposium on VLSI*, March 9–11, Porto Alegre, Brazil, IEEE, pp. 317–324.

Glass, C. J. and Ni, L. 1996. Fault-tolerant wormhole routing in meshes without virtual channels. *IEEE Transaction on Parallel and Distributed Systems*, vol. 7, no. 6, pp. 620–635.

Hedge, R. and Shanbhag, N. R. 2000. Toward achieving energy efficiency in presence of deep submicron noise. *IEEE Transactions on Very Large Scale Integration (VLSI) Systems*, vol. 8, no. 4, pp. 379–391.

Huang, P. T., Fang, W. L., Wang, Y. L. and Hwang, W. 2008. Low power and reliable interconnection with self-corrected green coding scheme for network-on-chip. *Proceedings of ACM/IEEE International Symposium on Networks-on-Chip*, April 7–10, Newcastle upon Tyne, IEEE, pp. 77–83.

Ismail, Y. I., Friedman, E. G., and Neves, J. L. 1998. Figures of merit to characterize the importance of on-chip inductance. *Proceedings of Design Automation Conference*, June 15–19, San Francisco, CA, IEEE, pp. 560–565.

Joshi, S. M., Dubey, P. K., and Kaplan, M. A. 2000. A new parallel algorithm for CRC generation. *IEEE International Conference on Communication*, June 18–22, New Orleans, LA, IEEE, pp. 1764–1768.

Kumar, S. V., Kim, C. H., and Sapatnekar, S. S. 2006. An analytical model for negative bias temperature instability. *Proceedings of IEEE/ACM International Conference on Computer Aided Design*, pp. 493–496.

Lee, S. J., Lee, K., and Yoo, H. J. 2005. Analysis and implementation of practical, cost-effective networks on chips. *IEEE Design & Test Computers*, vol. 22, no. 5, pp. 422–433.

Lin, S. and Costello, D. J. 1983. *Error Control Coding: Fundamentals and Applications*. Prentice Hall, New Jersey.

Linder, D. H. and Harden, J. C. 1991. An adaptive and fault-tolerant wormhole routing strategies for k-ary n-cubes. *IEEE Transactions on Computer*, vol. 40, no. 1, pp. 2–12.

Mitra, S., Zhang, M., Mak, T. M., Seifert, N., Zia, V., and Kim, K. S. 2005. Logic soft errors: A major barrier to robust platform design. *Proceedings of International Test Conference*, November 8, Austin, TX, IEEE, pp. 687–696.

Mitra, S., Zhang, M., Seifert, N., Mak, T. M., and Kim, K. S. 2006a. Soft error resilient system design through error correction. *IFIP International Conference on VLSI*, October 16–18, Nice, France, IEEE, pp. 332–337.

Mitra, S., Zhang, M., Waqas, S., Seifert, N., Gill, B., and Kim, K. S. 2006b. Combinational logic soft error correction. *Proceedings of IEEE International Test Conference*, pp. 29.2.1–29.2.9, October, Santa Clara, CA, IEEE.

Murali, S., Theocharides, T., Vijaykrishnan, N., Irwin, M. J., Benini, L., and Micheli, G. D. 2005. Analysis of error recovery schemes for Networks on Chips. *Proceedings of IEEE Design and Test of Computers*, pp. 434–442, September–October, IEEE.

Nazarian, S., Pedram, M., and Tuncer, E. 2005. An empirical study of crosstalk in VDSM technologies. *Proceedings of ACM Great Lakes symposium on VLSI*, April 17–19, New York, NY, ACM, pp. 317–322.

Nguyen, H. T. and Yagil, Y. 2003. A systematic approach to SER estimation and solutions. *Proceedings of IEEE International Reliability Physics Symposium*, pp. 60–70.

Pande, P. P., Ganguly, A., Feero, B., Belzer, B., and Grecu, C. 2006a. Design of low power & reliable networks on chip through joint crosstalk avoidance and forward error correction coding. *Proceedings of IEEE International Symposium on Defect and Fault-Tolerance in VLSI Systems*, October 2006, Arlington, VA, IEEE, pp. 466–476.

Pande, P. P., Zhu, H., Ganguly, A., and Grecu, C. 2006b. Energy reduction through crosstalk avoidance coding in NoC paradigm. *Proceedings of EUROMICRO Conference on Digital System Design*, August 30–September 1, Dubrovnik, IEEE, pp. 689–695.

Park, D., Nicopoulos, C. A., Kim, J., Vijaykrishnan, N., and Das, C. R. 2006. Exploring fault-tolerant network-on-chip architectures. *Proceedings of International Conference on Dependable Systems and Networks*, June 25–28, Philadelphia, PA, IEEE, pp. 93–102.

Patel, K. N. and Markov, I. L. 2003. Error-correction and crosstalk avoidance in DSM busses. *IEEE Transactions on Very Large Scale Integration (VLSI) Systems*, Special Issue for System Level Interconnect Prediction, vol. 12, no. 10, pp. 1–5.

Patooghy, A. and Miremadi, S. G. 2009. XYX: A power & performance efficient fault-tolerant routing algorithm for network on chip. *Proceedings of Parallel, Distributed, and Network-based Processing*, February 18–20, Weimer, IEEE, pp. 245–251.

Pei, T. B. and Zukowski, C. 1992. High-speed parallel CRC circuits in VLSI. *IEEE Transactions on Communications*, vol. 40, no. 4, pp. 653–657.

Perez, A. 1983. Byte-wise CRC Calculation. *IEEE Micro*, vol. 3, no. 3, pp. 40–50.

Pullini, A., Angiolini, F., Bertozzi, D., and Benini, L. 2005. Fault tolerance overhead in network-on-chip flow control schemes. *Proceedings of Integrated Circuits and Systems Design*, September 4–7, Florianópolis, Brazil, IEEE, pp. 224–229.

Rossi, D., Metra, C., Nieuwland, A. K., and Katoch, A. 2005. New ECC for crosstalk impact minimization. *IEEE Design and Test of Computers*, vol. 22, no. 4, pp. 340–348.

Shieh, M. D., Sheu, M. H., Chen, C. H., and Lo, H. F. 2001. A systematic approach for parallel CRC computations. *Journal of Information Science and Engineering*, vol. 17, pp. 445–461.

Sprachmann, M. 2001. Automatic generation of parallel CRC circuits. *IEEE Design and Test of Computers*, vol. 18, no. 3, pp. 108–114.

Sridhara, S. R. and Shanbhag, N. R. 2005. Coding for system-on-chip networks: A unified framework. *IEEE Transactions on Very Large Scale Integration (VLSI) Systems*, vol. 13, no. 6, pp. 655–667.

Sridhara, S. R. and Shanbhag, N. R. 2007. Coding for reliable on-chip buses: A class of fundamental bounds and practical codes. *IEEE Transactions on CAD of Integrated Circuits and Systems*, vol. 26, no. 5, pp. 977–982.

Valinataj, M., Mohammadi, S., Safari, S., and Plosila, J. 2009. A link failure aware routing algorithm for networks-on-chip in nano technologies. *Proceedings of IEEE Conference on Nanotechnology*, July 26–30, Genoa, Italy, IEEE, pp. 687–690.

Wang, L. T., Wu, C. W., and Wen, X. 2006. *VLSI Test Principles and Architectures*. Morgan Kaufmann Publisher. San Francisco, CA.

Wu, J. 2003. A fault-tolerant and deadlock-free routing protocol in 2D meshes based on odd-even turn model. *IEEE Transactions on Computers*, vol. 52, no. 9, pp. 1154–1169.

Zhang, M., Mitra, S., Mak, T. M., Seifert, N., Wang, N. J., Shi, Q., Kim, K. S., Shanbhag, N. R., and Patel, S. J. 2006. Sequential element design with built-in soft error resilience. *IEEE Transactions on Very Large Scale Integration (VLSI) Systems*, vol. 14, no. 12, pp. 1368–1378.

Zimmer, H. and Jantsch, A. 2003. A fault model notation and error-control scheme for switch-to-switch buses in a network-on-chip. *Proceedings of the International Conference on Hardware/Software Codesign and System Synthesis*, October 1–3, Newport Beach, CA, IEEE, pp. 188–193.

8

Testing of Network-on-Chip Architectures

8.1 Introduction

Testing network-on-chip (NoC) architectures is very challenging. In fact, the problem is difficult for any system-on-chip (SoC) design paradigm. In SoC, the *intellectual property* (*IP*) cores are integrated into a system design. A SoC integrator does not possess the detailed knowledge about the implementation of individual cores in the system. The netlist-level description, required for running a test generation tool, is generally not provided by the core vendors due to IP rights. The layout-level description of the cores provided by the vendors cannot act as input for the test generation process. Hence, the system integrator has to depend upon the test sets provided by the core vendors. However, core vendors, in the absence of knowledge about the final integrated SoC platform, cannot generate a very compact test set for their core. The core vendors do provide a test set that can ensure the correctness of only the core, not the whole integrated system. The test set provided by the core vendor needs to be applied to the input of the core and the responses are to be observed. This poses a challenge as the input–output lines of the cores, deeply embedded inside the chip, are not directly accessible from the system pins. In SoC platforms, the problem is often resolved by having dedicated *test access mechanism* (*TAM*) for the chip. The TAM is accessible from the input/output pins of the chip. The individual cores can be accessed via this access mechanism. Several different access mechanisms have been proposed; however, the most elegant solution is to have one or more dedicated bus(es) for the test access. In test mode, the test signals carried by the TAM lines are applied to the core, instead of functional inputs. Similarly, the responses from the core, instead of being applied on other cores, are transferred to the chip output through the TAM for observation. This solves the access problem. However, in the absence of system-level test patterns for the entire chip, the core-level testing has to be done exhaustively. This requires application of all the test patterns for individual cores. This leads to a huge test time for a moderate to complex SoC. Hence, test time reduction is another important challenge. The reduction in test time is often achieved via test parallelism—having multiple parallel test sessions.

The overall TAM structure is divided into a number of buses to carry out the test in parallel.

However, in a NoC environment, usage of extra TAM is not advisable. The chip itself contains a full network for transport of test packets through it to the individual cores. Hence, it is imperative to use the existing communication backbone for carrying the test patterns from the test equipment connected to some pins of the chip to the individual cores and to transfer the responses from the cores to the test equipment for analysis. The two main issues in NoC testing are as follows:

1. *Efficient testing of NoC communication fabric*: In a NoC, a significant portion of the chip is occupied by the routers and links constituting the NoC fabric. In the absence of direct signal communication between the cores, the success of message-based communication between them is fully dependent on the correct functioning of the network resources. The fabric, in turn, consists of two different types of components—the routers and the links. The routers, apart from having the logic circuitry, also possess the first-in first-out (FIFO) buffers. Hence, testing the communication fabric is a difficult task as it necessitates different fault models for the components, such as memory fault models for the FIFO, logic fault models for the router circuitry, and link fault models for the links.

2. *Testing of functional cores*: In this case, the test patterns are provided by the core vendors. The challenge is to apply these patterns to the cores. Once the network infrastructure is tested to be okay, it is utilized to transport the test patterns and responses through the network.

8.2 Testing Communication Fabric

Correctness of communication fabric is the first thing to be ensured to guarantee the correct operation of the NoC. Test strategies for the fabric must perform the following two operations (Pande et al. 2005): testing of switch blocks and testing of interswitch wire segments. Grecu et al. (2007) suggested a comprehensive framework to test both these components using suitable fault models and testing mechanisms. The scheme has two interesting features. First, the two tests are integrated together so that the portion of the NoC determined to be correct can be utilized to test successive portions and establish their correctness. This recursive formulation helps in arriving at an elegant solution to the test transportation problem. Second, the NoC infrastructure is exploited successfully to introduce parallelism in the test session. This reduces the overall test time of the NoC significantly.

8.2.1 Testing NoC Links

Cuivello et al. (1999) suggested a novel fault model for links in the deep submicron (DSM) technology. The model, known as *maximum aggressor fault* (*MAF*) model, corresponds to the crosstalk effects between a set of parallel lines in DSM SoCs. This model has been discussed in detail in Chapter 7 (Figure 7.2). In this model, a signal transition in one single line (known as victim line) is affected through crosstalk by transitions in the neighboring lines (known as aggressors). For a link with N wires, in the worst case, a victim line can get affected by transitions in the $(N - 1)$ aggressors. As it can be observed from Figure 7.2, each line can be affected in six different ways. Thus, testing a single line needs six two-pattern tests. For example, if we assume a three-wire model (in which the middle wire is the victim and the other two are aggressors), to test a *delayed rise* case, we need to apply the pattern 101 followed by 010. For a link with large number of wires (which is common for NoC) and sufficiently wide neighborhood, this requires a prohibitive amount of test effort to ensure good test coverage (Bushnell and Agrawal 2005).

The test sequence for link testing using the MAF model exhibits some important properties that can be utilized in compact and efficient design of test packets:

1. In each test vector, the logic value in the victim line is the complement of that in the aggressor lines. All aggressor lines are assigned the same logic value. This is required to have the maximum aggression effect on the victim.

2. After applying an exhaustive set of test sequences for a victim, the sequence for testing the next line can be obtained easily by shifting or rotating the test patterns in the previous sequence by exactly one bit.

3. Transition from one test vector to another can often be concatenated such that the total number of test vectors needed to test the MAF faults gets reduced. For example, the "Fast-to-fall" needs the test vector 111 followed by 000, whereas the "Fast-to-rise" needs the test vector 000 followed by 111. Thus, the three-vector sequence 111–000–111 can test both the faults, reducing the required number of test vectors from four to three. As shown in Table 8.1, instead of requiring 12 vectors per wire, application of only 8 vectors per wire suffices to check all six MAF faults.

The MAF tests can be carried out in eight distinct states, s1–s8, as shown in Figure 8.1. The finite-state machine shown in the figure generates eight different patterns in a cyclical fashion. Thus, in one cycle, one line can be tested as a victim, whereas the others act as aggressors. The selection of the victim wire can be achieved through the victim-line counter field that controls the

TABLE 8.1

Test Sequence for Wire *i*

	Lines		
i − 1	*i*	*i* + 1	**State in Figure 8.1**
1	1	1	s1
0	0	0	s2
1	1	1	s3
0	1	0	s4
1	0	1	s5
0	1	0	s6
0	0	0	s7
1	0	1	s8

Note: Link faults: s1→s2: Fast-to-fall, s2→s3: Fast-to-rise, s3→s4: Negative glitch, s4→s5: Slow-to-fall, s5→s6: Slow-to-rise, s7→s8: Positive glitch.

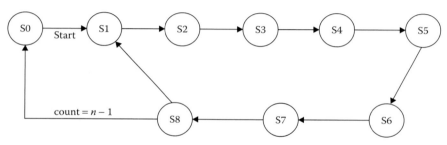

FIGURE 8.1
State machine for MAF pattern generation.

test hardware. In the first cycle, the first wire of the link is selected as the victim, whereas the others act as aggressors. The eight test patterns are applied in the eight states s1 through s8. This tests all MAF faults on line 1. Next, in the second cycle of eight states, the second wire is the victim. The test patterns are generated in such a fashion that they are 1-bit-shifted version of those generated in the first cycle. The procedure is repeated until all the lines of the link have been tested.

8.2.2 Testing NoC Switches

A NoC switch consists of the following components:

- Combinational logic for the arbitration, routing, error control, and so on
- FIFO buffers, implemented as register banks or dedicated static random access memory arrays

Test patterns can be generated for the combinational part using any *automated test pattern generation (ATPG)* tool. Normally, scan-based testing strategies are adopted for testing these blocks. For the FIFO part, a FIFO falls under the category of restricted two-port memories. Since the data flow through the NoC links is unidirectional in nature, each FIFO possesses one write-only port and one read-only port. The functionality of the FIFO can then be considered to be divided into three ways: the memory cell array, the addressing mechanism, and the FIFO-specific functions (such as empty/full conditions). The memory array fault models can include stuck-at faults, transition faults, bridging faults, and so on (Bushnell and Agrawal 2005).

Assuming that a FIFO is b-bit wide and has n locations, individual test patterns are of b bits each. For example, to detect a bridging fault between bits b_i and b_j $(i \neq j)$, four specific test patterns are needed: 0101..., 1010..., 0000..., and 1111.... To test for dual-port coupling faults, the following sequence is used: $w\{\Uparrow_1^{n-1} (wr)\}r$, for each of the four test patterns. The first write operation (w) sets the *read* and *write* pointers to FIFO cells 0 and 1, respectively. The next $(n - 1)$ simultaneous read and write operations (wr) sensitize coupling between adjacent cells. The last read operation clears the FIFO and prepares it for testing with the next pattern.

8.2.3 Test Data Transport

The test data for testing the NoC switches and links are to be transported through the NoC itself. Hence, it is essential that the communication infrastructure available in the NoC be used for this purpose as well. As a result, it is necessary to test the switches and links of the infrastructure in phases. The switches and links found to be okay up to a certain stage of testing are to be reused to transport test patterns to test the next set of switches and links. The transport is therefore dependent on the routing scheme and algorithm followed by the routers in the NoC. It is expected that the same routing policy, used in normal functioning of the NoC, be utilized in the test mode as well. The test mode cannot demand for a new routing policy, not required otherwise in the functional mode of operation of the NoC. The addressing scheme supported by the basic router can be of the following types:

1. *Unicast mode*: This is the commonly available mode of packet transfer in NoC. Packets arriving at an input port of a switch are decoded and forwarded to one of its output ports, based upon the routing algorithm and the destination address noted in the packet header.

2. *Multicast mode*: In this mode, the packets have multiple destinations. Packets with multicast address are decoded and forwarded to the switch outputs depicted by the multicast decoder. Multicast packets can reach their destinations in a more efficient and faster manner than the scheme based on unicast transmission, in which several unicast messages are sent to different destinations. This is illustrated in Figure 8.2.

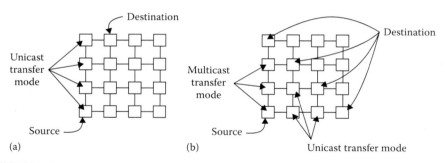

FIGURE 8.2
(a) Unicast data transfer; (b) multicast data transfer.

If the underlying NoC routers support multicast mode of transmission, it can be utilized for testing as well. Otherwise, it is suggested that a *multicast wrapper unit* (MWU) be designed. It is transparent to the packets in normal operation mode. In the test mode, the *routing logic block* (RLB) of the switch is completely bypassed. This is achieved by a set of multiplexers and demultiplexers that aid in bypassing the RLB. As test scheduling is done offline, for each NoC switch, the subset of input and output ports participating in multicast test data transfer is known beforehand. The MWU design for the switch can be restricted accordingly. For example, in Figure 8.2, only a few of the switches need multicast facility. However, a proper scheduling of testing the components (i.e., switches and links) is necessary. In order to search for this optimal schedule, it is required to consider the following test times for individual components.

- $T_{l,L}$: The latency of interswitch link L (generally equal to one clock cycle if the link is reasonably short)
- $T_{t,L}$: The time to test the link L
- $T_{l,S}$: The switch latency (the number of clock cycles required for a flit to traverse a NoC switch from input to output)
- $T_{t,S}$: The time required to perform the actual testing of the switch [Time needed to test the RLB (T_{RLB}) + Time to test FIFO (T_{FIFO})]

For example, in a unicast mode of transmission, the total test time $T_{2,3}^u$ to test both the switches S_2 and S_3, as shown in Figure 8.3, is given by

$$T_{2,3}^u = 2(T_{l,L} + T_{l,S}) + 2T_{t,S}$$

For a multicast environment, the test time $T_{2,3}^m$ to test the switches S_2 and S_3 is given by

$$T_{2,3}^m = (T_{l,L} + T_{l,S}) + T_{t,S}$$

The time to test a switch $T_{t,S}$, or a link $T_{t,L}$, is determined by the fault model and the design of these components. However, the test transport time can be controlled

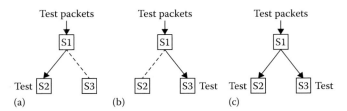

FIGURE 8.3
(a,b) Unicast transport; (c) multicast transport.

by proper scheduling of tests. In the following, the problem of test transport time minimization is taken up for both unicast and multicast environments.

8.2.4 Test Transport Time Minimization—A Graph Theoretic Formulation

The minimization of test delivery times to the NoC elements can be formulated as a graph theoretic problem in which the NoC infrastructure is represented as a graph. A NoC can be visualized as a graph $G = (S,L)$ in which each vertex $s_i \in S$ is a NoC switch and each edge $l_i \in L$ is an interswitch link. Each switch is associated with a pair of values $(T_{l,S}, T_{t,S})$ corresponding to the switch latency and switch test time, respectively. Each link is similarly labeled with a pair $(T_{l,L}, T_{t,L})$ corresponding to the link latency and link test time. Now, for each NoC component, the shortest path from an arbitrary node to the element, traversing only the previously tested fault-free components, is determined. By repeating the process for all possible nodes in the network and choosing the solution that requires the shortest test time, the minimum test transport time problem can be solved.

To frame the search operation, a symbolic toggle t is defined for each of the edges and vertices of the graph. The variable t can take up only two values: **N** or **T**. When $t = $ **N**, the cost of the associated edge/vertex is its latency term. For $t = $ **T**, the cost is the test time term. A modified version of Dijkstra's shortest path algorithm can be used to determine the shortest paths.

Initially, the toggle t for all edges and vertices are set to **T**. The algorithm starts with an arbitrarily chosen start node of the graph. In the execution of the search algorithm, every time an element is encountered with toggle $t = $ **T**, the cost function is updated with the test time term of the component, and t is switched to **N**. However, if an element with $t = $ **N** is encountered, its latency term is used to update the cost function, and t remains unchanged. Compared to the classic Dijkstra's algorithm, the following differences are incorporated into the search procedure.

1. The vertices also possess weights.
2. Weights of vertices and edges change dynamically during the graph traversal.

3. After a directed edge is traversed, the edge in the opposite direction is traversed as soon as possible.
4. All toggles t must be set to **N** by the end of the algorithm. This ensures that all the edges and switches are tested by the end of the algorithm.

Depending upon whether unicast or multicast transport is employed, the test cost function has to be computed differently. In the following, the two strategies are presented for minimum cost test scheduling—one for unicast and the other one for multicast.

8.2.4.1 Unicast Test Scheduling

The problem of unicast test scheduling can be stated as follows:

Given the graph $G(S,L)$ and the pairs $(T_{l,S}, T_{t,S})$ and $(T_{l,L}, T_{t,L})$ and assuming that only one vertex or edge with toggle $t = \mathbf{T}$ can be visited at a time, determine a graph traversal sequence that covers all vertices and edges and has a minimum associated test cost function $F_{TC,u}$.

The restriction that only one toggle can be switched at a time ensures unicast transport. The unicast cost function $F_{TC,u}$ can be defined recursively as follows: If $F_{TC,u}^{old}$ be the cost before the current NoC element is tested and $F_{TC,u}^{new}$ be the cost after including the test cost of the current NoC element, then

$$F_{TC,u}^{new} = F_{TC,u}^{old} + \sum_{L \,\in\, \text{Links in the path}} T_{l,L} + \sum_{S \,\in\, \text{Swiches in the path}} T_{l,S} + \begin{cases} T_{t,L} & \text{if current element is a link} \\ T_{t,S} & \text{if current element is a switch} \end{cases}$$

In the following, the unicast test scheduling algorithm is presented. As in Dijkstra's algorithm, $d[u]$ denotes the distance of the current test source to the switch u under test. It represents the test injection time corresponding to u. At the end of the algorithm, the switch s_{min} corresponding to the minimum value of $F_{TC,u}$ is selected as the final source.

Algorithm *Unicast*

Input: $G(S,L)$ and the pairs $(T_{l,S}, T_{t,S})$ and $(T_{l,L}, T_{t,S})$.
Output: s_{min} corresponding to the minimum test time $F_{TC,u}$.
Begin
    For each $s \in S$ do *Unicast_min(S, L, s, weight)*
                      /* weight contains the test pairs $(T_{l,S}, T_{t,S})$ and $(T_{l,L}, T_{t,S})$ */
    Find s_{min} corresponding to minimum $F_{TC,u}$
End
Procedure *Unicast_min(S, L, start, w)*
Begin

For each vertex $s \in S$ do
Begin
        $s.toggle = \mathbf{T}$;
        $s.weight = T_{t,S}$;
        $d[s] = \infty$;
        $s.parent = UNDEFINED$;
End
$d[start] = 0$;
For each edge $e = (u,v) \in L$ do
Begin
        $e.toggle = \mathbf{T}$;
        $e.weight = T_{t,L}$;
End
$F_{TC,u} = 0$;
Put all vertices of S in minpriority queue Q based on d values;
While Q is not empty do
Begin
        $u = Extract\_Min(Q)$;
        For each edge $e = (u,v)$ outgoing from u do
        Begin
                If $(d[u] + e.weight < d[v])$ then
                Begin
                        $d[v] = d[u] + e.weight$;/* This affects the queue Q */
                        Update $F_{TC,u}$;
                        $v.toggle = \mathbf{N}$;
                        $e.toggle = \mathbf{N}$;
                        $u.weight = T_{l,S}$;
                        $e.weight = T_{l,L}$;
                        $v.parent = u$;
                End;
        End;
End;
Return $F_{TC,u}$;
End.

Figure 8.4 shows a four node NoC along with the links. Starting with node S_1, unicast testing of components proceeds in the following sequence: S_1, I_1, I_2, S_2, I'_1, S_3, I'_2, I_5, I'_5, I_3, S_4, I'_3, I_4, and finally I'_4. Accordingly, the test time requirement increments in the following sequence: $T_{t,S}$, $T_{t,S} + T_{l,S} + T_{t,L}$, $T_{t,S} + 2T_{l,S} + 2T_{t,L}$, $2T_{t,S} + 3T_{l,S} + 2T_{t,L} + T_{l,L}$, $2T_{t,S} + 5T_{l,S} + 3T_{t,L} + 2T_{l,L}$, $3T_{t,S} + 6T_{l,S} + 3T_{t,L} + 3T_{l,L}$, $3T_{t,S} + 8T_{l,S} + 4T_{t,L} + 4T_{l,L}$, $3T_{t,S} + 10T_{l,S} + 5T_{t,L} + 5T_{l,L}$, $3T_{t,S} + 12T_{l,S} + 6T_{t,L} + 6T_{l,L}$, $3T_{t,S} + 14T_{l,S} + 7T_{t,L} + 7T_{l,L}$, $4T_{t,S} + 16T_{l,S} + 6T_{t,L} + 9T_{l,L}$, $4T_{t,S} + 19T_{l,S} + 8T_{t,L} + 11T_{l,L}$, $4T_{t,S} + 21T_{l,S} + 9T_{t,L} + 12T_{l,L}$, and $4T_{t,S} + 24T_{l,S} + 10T_{t,L} + 14T_{l,L}$. The same is

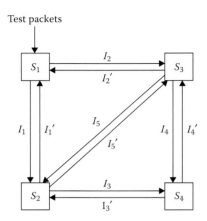

FIGURE 8.4

A four-switch network with unidirectional links.

repeated for the cases starting with S_2, S_3, and S_4. The minimum of all these cases is taken as the final solution.

8.2.4.2 Multicast Test Scheduling

The multicast test scheduling problem can be stated as follows:

Given the graph $G(S,L)$ and the pairs $(T_{l,S}, T_{t,S})$ and $(T_{l,L}, T_{t,S})$ and assuming that all vertices or edges with toggle $t = \mathbf{T}$ and are adjacent to edges/vertices whose toggle equals $\mathbf{N}$ can be visited at a time, determine a graph traversal sequence that covers all vertices and edges and has a minimum associated test cost function $F_{TC,m}$.

The flexibility that more toggles can be switched at a time provides the multicast transport mechanism, assuming that the NoC supports multicast. The multicast test cost can be defined recursively as follows:

$$F_{TC,m}^{\text{new}} = F_{TC,m}^{\text{old}} + \underset{\text{all adjacent elements}}{\text{Max}} \left(\sum_{L \in \text{Links in the path}} T_{l,L} + \sum_{S \in \text{Swiches in the path}} T_{l,S} \right)$$

$$+ \underset{\text{all adjacent elements}}{\text{Max}} \left(\begin{cases} T_{t,L} \text{ if current element is a link} \\ T_{t,S} \text{ if current element is a switch} \end{cases} \right)$$

The multicast transport algorithm is similar to the unicast one, differing essentially in the cost function and toggle update. The details can be found in the work of Grecu et al. (2007). For the graph in Figure 8.4, starting at node S_1, the components are tested in groups as follows: $\{S_1\}$, $\{I_1, I_2\}$, $\{S_2, S_3\}$, $\{I'_1, I'_2, I_3, I_4, I_5, I'_5\}$, and $\{S_4\}$. The corresponding multicast test costs after each group of component testing are $T_{t,S}$, $T_{t,S} + T_{l,S} + T_{t,L}$, $2T_{t,S} + 2T_{l,S} + T_{t,L} + T_{l,L}$, $2T_{t,S} + 4T_{l,S} + 2T_{t,L} + 2T_{l,L}$, and $3T_{t,S} + 6T_{l,S} + 2T_{t,L} + 4T_{l,L}$.

8.3 Testing Cores

Once the NoC fabric has been tested and found okay, the cores of the NoC are tested. The network resources (switches and links) are used to transport the test patterns. The cores embedded deep inside the chip are difficult to access, whereas the cores near the periphery of the chip may be direct accessibility from the chip input/output pins. As a result, the input/output channels of the *automated test equipment* (*ATE*) can be connected to the cores in the periphery. These cores are conveniently called I/O ports for testing. For a core deep inside the NoC, its test patterns can be transported to a core acting as input port. From this core, the patterns are transferred over the network to the core under test. Similarly, the responses from the core under test are transferred to the output core for subsequent delivery to the ATE for analysis. Figure 8.5 shows a NoC with two input cores and two output cores. A pair of input/output cores can be connected to an ATE channel for testing purpose. Since in Figure 8.5 there are two I/O ports, two cores can be tested in parallel, provided that the resources (switches and links) needed are not conflicting.

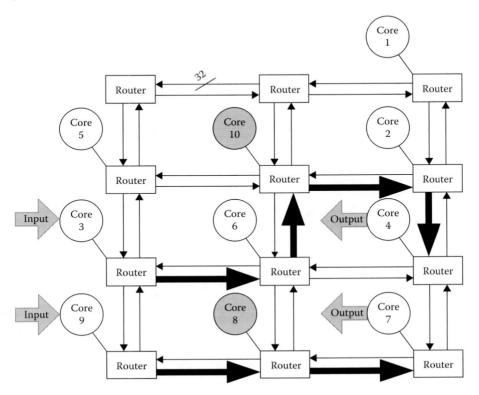

FIGURE 8.5
NoC with input/output cores for testing and two routing paths.

A proper scheduling of cores for testing is necessary to perform their testing in an efficient manner.

Apart from this scheduling problem, it is also necessary to have a proper interface between the cores and the network links. The cores may have different numbers of input–output lines, compared to the link width. This is taken care of by putting the cores inside test wrappers discussed in Section 8.3.1.

8.3.1 Core Wrapper Design

In a SoC test environment, individual cores need to be encapsulated within a wrapper. A wrapper is an integral part of IEEE 1500 working group proposal. It is a layer of design-for-testability (DfT) logic that connects the TAM to a core under test. The 1500 wrapper has four main functions: *Normal* operation, *Intest* (testing of core itself), *Extest* (core external test, such as testing of interconnects originating from or ending at the core), and *Bypass*. The typical structure of 1500 wrapper (Marinissen et al. 2002) is shown in Figure 8.6.

The core designers, at the time of designing the core, also incorporate some DfT features in it in the form of scan chains. The number of scan chains created is often dependent on the core designer. Apart from these scan inputs, a test pattern contains bits for the primary inputs as well. Hence, application of a test pattern necessarily means filling up the scan flip-flops and primary inputs of the core. The total time needed for this depends upon the NoC channel width (flit size), the maximum length of scan chain, the total number of scan chains, and the number of primary inputs and outputs. A wrapper for a core combines the core internal scan chains, primary inputs, and primary outputs into wrapper scan chains. Ideally, all the wrapper chains should be of equal length. Moreover, the total number of wrapper chains should be equal to the flit size, such that each flit can fill up exactly one bit of a scan chain. If the length of the longest wrapper chain is n, after n flits, one full test pattern will get loaded into the scan chains. The pattern is then applied to the core. Responses are collected into the wrapper scan chains and are shifted out as flits to the test sink. Figure 8.7 shows two wrapper chains. Both the figures create two wrapper chains. The first one is unbalanced, in which one chain is of length 6, whereas the other one is of length 18. The second design is more balanced—both the chains are of length 12.

The number of flits for a test pattern is dependent upon the maximum length of scan chains. Partitioning of core scan chains, primary inputs, and outputs into wrapper chains constitute a major role in determining this maximum wrapper scan chain length. Given a flit size, determining the scan partitioning to optimize the overall test time is *NP-hard* (Marinissen et al. 2000). The wrapper design problem has been addressed in the work of Iyengar and Chakrabarty (2002). The overall problem can be stated as follows:

Given a core with n functional inputs, m functional outputs, and sc internal scan chain of lengths $l_1, l_2, \ldots, l_{sc}$, respectively, and NoC channel width k,

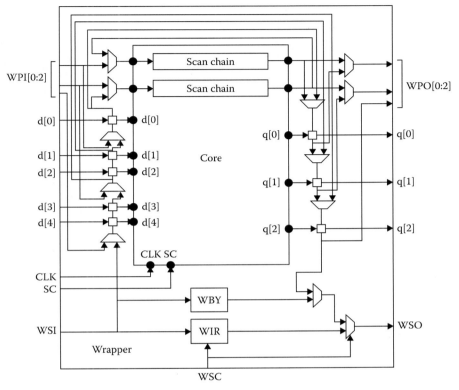

FIGURE 8.6
1500 Wrapper. CLK, functional clock; d, functional inputs; q, Functional outputs; SC, Scan Enable Control; WIR, Wrapper Instruction Register; WBY, Wrapper Wide Bypass; WPI, Wrapper Parallel Input; WPO, Wrapper Parallel Output; WSI, Wrapper Serial Input; WSC, Wrapper Scan Control; WSO, Wrapper Serial Output.

assign $n + m + sc$ wrapper scan chain elements to $k' \leq k$, such that $\max\{s_i, s_o\}$ is minimized, where s_i and s_o are the lengths of the longest wrapper scan-in and scan-out chains, respectively. Also, try to minimize k'.

The wrapper design algorithm proposed in the work of Iyengar and Chakrabarty (2002) uses an approximation algorithm based on the best fit decreasing (BFD) heuristics (Garey and Johnson 1979). The algorithm has three main parts: (1) partition the internal scan chains among a minimal number of wrapper scan chains to minimize the longest wrapper scan chain length, (2) assign the functional inputs to the wrapper chains created in part (1), and (3) assign the functional outputs to the wrapper scan chains created so far. To solve the problem, first the internal scan chains are sorted in descending order of their length. Each internal scan chain is then attached to a wrapper scan chain, whose length after this assignment is closest to, but not exceeding, the current maximum wrapper scan chain length. In other words,

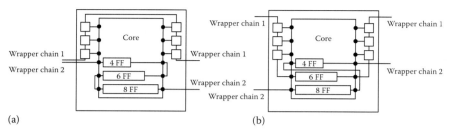

FIGURE 8.7
Two wrappers: (a) balanced; (b) unbalanced. FF, flip-flop.

at any point of time, the next scan chain gets attached to wrapper chain in which it fits the best. If no such wrapper chain is available, the internal scan chain is attached to the current shortest wrapper scan chain. The process is repeated for the functional input and outputs. The procedure *Design_wrapper* (Iyengar and Chakrabarty 2002) is noted as follows:

Procedure Design_wrapper

Input: Number of functional inputs (n)
      Number of functional outputs (m)
      Scan chain lengths $l_1, l_2, ..., l_{sc}$
Output: Wrapper scan chains
Begin
Step 1: Sort the internal scan chains in descending order of length
Step 2: For each internal scan chain l picked up in order do
      Step 2.1: Find wrapper scan chain S_{max} with current maximum length
      Step 2.2: Find wrapper scan chain S_{min} with current minimum length
      Step 2.3: Assign l to the wrapper scan chain S, such that,
            (length(S_{max}) – length(S) + length(l)) is minimum
      Step 2.4: If there is no such S, assign l to S_{min}
Step 3: Repeat Step 2 for the functional inputs
Step 4: Repeat Step 2 for functional outputs
End

For example, suppose Core A has 8 functional inputs $a[0:7]$, 11 functional outputs $z[0:10]$, 9 internal scan chains of lengths 12, 12, 8, 8, 8, 6, 6, 6, 6 flip-flops, and a scan enable control sc. Assume the flit size to be 4. Hence, in a single clock, 4 bits of data can reach the core from the NoC channel. The scan elements in the core are partitioned among four wrapper scan chains using the algorithm, as shown in Table 8.2. This partition yields a longest scan-in chain of length 20 and a longest scan-out chain of length 21, both of which are optimal values for a 4-bit NoC channel width. The worst-case complexity of the *Design_wrapper* algorithm is $O(sc \log sc + sc \cdot k)$.

TABLE 8.2

Example of *Design_wrapper* Algorithm

	Wrapper Scan Chains			
	1	2	3	4
Internal scan chains	12 + 6	12 + 6	8 + 6 + 6	8 + 8
Wrapper input cells	2	2	0	4
Wrapper output cells	3	3	0	5
Scan-in length	20	20	20	20
Scan-out length	21	21	20	21

Next, the time needed to test a core is computed. For this, it is assumed that the core has p test patterns, the maximum length of scan chain (corresponding to the number of flits per packet) is l, and the numbers of hops from source to core and core to sink are $h_{src \to core}$ and $h_{core \to sink}$, respectively. It is further assumed that the cores are scheduled for testing using a nonpreemptive scheduling strategy. The scheduler assigns each and every core a fixed path during the scheduled test time. All resources of this path are reserved throughout the test time of the core. Test data are moved along the reserved path in a pipelined fashion. Since each flit is unpacked in one cycle, the time required for a test vector to be sent to the core plus the test response to be received at the sink is be given by

$$h_{src \to core} + (l-1) + 1 + h_{core \to sink} + (l-1)$$

That is, first the l flits corresponding to a test pattern are sent in a pipelined fashion from the source to the core. In the next cycle, the pattern is applied to the core. Next, the response bits are transferred to the test sink in a pipelined fashion. During testing, test stimulus and responses move in a pipelined fashion with an overlap in the shifting-in of the current test pattern and shifting-out of the previous response. For a set of p patterns, the test time needed consists of the following:

1. Time to shift in the first pattern, $h_{src \to core} + (l-1)$
2. Time to apply the first pattern, 1
3. Time for overlapped transmission of the test bits of the current pattern and the response bits of the previous pattern plus the test application time for the current pattern, for next $(p-1)$ packets, which is given by

$$[1 + \text{Max}\{h_{src \to core} + (l-1), h_{core \to sink} + (l-1)\}] \times (p-1)$$

4. Time required to shift out the response of pattern p, $h_{core \to sink} + (l-1)$.

Thus, the total time required for testing is given by

$$h_{src \to core} + (l-1) + 1 + [1 + Max\{h_{src \to core} + (l-1), h_{core \to sink} + (l-1)\}] \times (p-1)$$
$$+ h_{core \to sink} + (l-1) = [1 + Max\{h_{src \to core}, h_{core \to sink}\} + (l-1)]$$
$$\times p + [Min\{h_{src \to core}, h_{core \to sink}\} + (l-1)]$$

Test scheduling problem can be formulated as follows:

> Given a NoC with n cores having their test parameters (number of test patterns, scan chain length, etc.) and k number of I/O pairs, determine an assignment of cores to I/O pairs and the time schedule to minimize the overall test time of all the cores present in the NoC.

There are a few variants of the problem reported in the literature. A multi-frequency test scheduling assumes that in test mode, individual cores can be made to operate at different frequencies, thus requiring proportional test times. Some formulations put a limit on the total peak power that can be sustained by the chip, whereas some other formulations are concerned about minimizing the peak temperature of the NoC or making the temperature uniform in the chip. The solution strategies proposed can broadly be divided into the following categories:

1. Exact solution via mathematical tools, such as integer linear programming (ILP)
2. Heuristic algorithms
3. Evolutionary algorithms, such as particle swarm optimization (PSO)

8.3.2 ILP Formulation

ILP formulation of an optimization problem can provide its exact solution at the cost of execution time of the solver. For the problem of core testing in NoC, Salamy and Harmanani (2011) reported an ILP formulation. The overall problem addressed is as follows:

> Given a system of N_c cores, N_p input/output ports, and a set of clock rates F_c (at which individual cores may operate during testing), map the cores to the input/output ports and obtain a test schedule so that the overall test time is minimized.

To start with, a few definitions are noted. T_{iuc} = Test time of core i on input/output pair u under clock frequency c, $1 \leq i \leq N_c$,

$$1 \leq p \leq N_p \text{ and } c \in F_c$$

S_i = Start time of core i
I_{ix} = Input/output pair of core i, $1 \leq x \leq N_p$

The constraints formulated are as follows:

1. Each core in the system must be assigned to one and only one input/output pair in the system.

$$I_{ix} = \begin{cases} 1, & \text{if core } i \text{ is assigned to pair } x \\ 0, & \text{otherwise} \end{cases} \qquad (8.1)$$

$$\sum_{x=1}^{N_p} I_{ix} = 1, \forall i \qquad (8.2)$$

2. Each core must be assigned a clock rate.

$$F_{ic} = \begin{cases} 1, & \text{if clock } c \text{ is assigned to core } i \\ 0, & \text{otherwise} \end{cases} \qquad (8.3)$$

$$\sum_{c} F_{ic} = 1, \forall i \qquad (8.4)$$

The test time of core i on input/output port x is given by

$$\sum_{c} T_{ixc} \cdot F_{ic}.$$

3. Two cores cannot be tested in parallel if they share some common resources. Two cores i and j assigned to input/output pairs x and y can be tested in parallel if and only if either of the following two conditions is valid:

$$S_{ix} \geq S_{jy} + \sum_{c} T_{jyc} \cdot F_{jc}$$

or

$$S_{jy} \geq S_{ix} + \sum_{c'} T_{ixc'} \cdot F_{ic'}$$

Let the binary variable Z_{ixjy} represent the conflict of cores i and j assigned to input/output ports x and y, respectively. Test between two cores conflict if they need some common NoC resource (e.g., router, link). $Z_{ixjy} = 1$ if the tests conflict; it is 0 otherwise. To prevent the conflicting cores getting scheduled simultaneously, either of the following two conditions needs to be valid:

$$Z_{ixjy}\left(S_{ix} - S_{jy} - \sum_{c} T_{jyc} \cdot F_{jc}\right) \geq 0 \qquad (8.5)$$

or

$$Z_{ixjy}\left(S_{jy} - S_{ix} - \sum_{c'} T_{ixc'} \cdot F_{ic'}\right) \geq 0 \qquad (8.6)$$

where:
$$1 \leq i,j \leq N_c$$
$$1 \leq x,y \leq N_p$$
$$c,c' \in F_c$$

The constraints (8.5) and (8.6) are not linear. To linearize them, two new binary variables K_{ixjy1} and K_{ixjy2} are introduced with the constraint that $K_{ixjy1} + K_{ixjy2} = 1$. This leads to a new constraint combining the two constraints (8.5) and (8.6).

$$Z_{ixjy} \cdot K_{ixjy1}\left(S_{ix} - S_{jy} - \sum_c T_{jyc} \cdot F_{jc}\right) + Z_{ixjy} \cdot K_{ixjy2}\left(S_{jy} - S_{ix} - \sum_c T_{ixc'} \cdot F_{ic'}\right) \geq 0 \quad (8.7)$$

The above constraint is still nonlinear as it contains multiplication of two variables K and S. To transform it into a linear one, the product $K_{ixjy1} \cdot S_{ix}$ is replaced by a new variable M_{ixjy1} that takes up a value of 1 if $K_{ixjy1} = 1$ and $S_{ix} = 1$. This can be enforced by adding three additional constraints noted as follows:

$$M_{ixjy1} \leq K_{ixjy1}$$

$$M_{ixjy1} \leq S_{ix}$$

$$M_{ixjy1} \geq K_{ixjy1} + S_{ix} - 1$$

Similar transformations are to be introduced to replace the products $K_{ixjy1} \cdot S_{jy}$ by M_{ixjy2}, $K_{ixjy2} \cdot S_{jy}$ by M_{ixjy3}, and $K_{ixjy2} \cdot S_{ix}$ by M_{ixjy4}.

The objective function is to minimize the overall test time of the NoC. The overall test time is equal to the maximum of finish times for all cores. Thus, the objective function can be written as

$$\text{Minimize } C = \text{Maximum}\left(S_{ix} + \sum_c T_{ixc} \cdot F_{ic}\right), \text{ for all } i, x, c$$

Again, the objective function is not linear. To linearize, it is required to minimize C along with a set of constraints noted as follows:

$$C \geq \sum_{x=1}^{N_p} I_{ix}\left(S_{ix} + \sum_c T_{ixc} \cdot F_{ic}\right), \text{ for all } 1 \leq i \leq N_c \qquad (8.8)$$

The above constraint is still not linear. To linearize this, it is necessary to replace the product $I_{ix} \cdot S_{ix}$ by an additional binary variable R_{ix} with the following constraints:

$$R_{ix} \le I_{ix}$$

$$R_{ix} \le S_{ix}$$

$$R_{ix} \ge I_{ix} + S_{ix} - 1$$

This completes the ILP formulation for the core test scheduling problem. All constraints and objective function are now linear in nature. However, the solution takes a considerably large amount of CPU time prohibiting its usage only to small NoCs having a few cores.

8.3.3 Heuristic Algorithms

Ahn and Kang (2006) has proposed a NoC test scheduling strategy using multiple test clocks. Cota et al. (2004) is one of the first works to suggest the usage of on-chip networks to transport test data for cores. Cota and Liu (2006) proposed a set of heuristic algorithms for different versions of the core test scheduling problem. The first one is a technique that uses a dedicated routing path for the test packets to move through the NoC. All tests are applied with full pipeline in a nonpreemptive fashion. The heuristic starts by creating an ordered list of cores and I/O pairs. The cores are sorted in decreasing order of test time. I/O pairs are permuted and every permutation is tried out. Different permutations of I/O pairs represent different priorities of their allocation to a core. That is, if core C_i is the next one to be scheduled, the I/O pairs I_1 and I_2 are free, and I_1 appears earlier than I_2 in the current permutation, C_i will be tested via interface I_1. For each permutation, attempt is made to assign the next core to the first available I/O pair. If no I/O pair is free, current time is updated to the next most recent time tag, at which some already scheduled core finishes its testing. At that time, the resources allocated to the core will become free, and thus may make the testing of new core possible. If an I/O pair is available, a routing path is created and the algorithm checks if it conflicts with any other path for the cores currently being tested. In case of a conflict, the next core in the sequence is considered. If all cores are tried out and none of them could be scheduled, the current time will be updated. All the remaining cores are again tried out for scheduling. The process continues till all cores are scheduled.

If the system has N_c number of cores and M number of I/O pairs, the complexity of the algorithm is $O(M!N_c)$. To explore larger search space, it is suggested that some other core orders be tried out. The proposed algorithm tries out a user-defined number of core permutations. The overall algorithm is detailed as follows:

Algorithm NoC_Schedule

1.	Start with sorted cores in decreasing order of test time;
2.	Permute all possible order of I/O pairs;
3.	For specified number of permutations of I/O pairs do
4.	While there are unscheduled cores do
5.	For each unscheduled core do
6.	Find a free I/O pair;
7.	If no free I/O pair then
8.	Update current time; Repeat from 4;
9.	Else
10.	Check the corresponding routing path;
11.	If path is blocked
12.	If all cores have been attempted
13.	Update current time; Repeat from 4;
14.	Else
15.	Try next core in the list;
16.	Else
17.	Assign core to the path; Update time labels;
18.	Repeat from 3 for a user-defined number of core permutations;

The nonpreemptive test scheduling algorithm discussed so far lacks flexibility, in the sense that the minimum manageable unit in test scheduling is the full test application time of a core. For example, the power consumption of a core during test is generally much higher than that during normal mode of operation. This happens as the successive functional inputs are generally correlated, while in order to maximize fault coverage, the successive test patterns in a test sequence are highly uncorrelated. This excessive power dissipation and lack of heat transfer can create hot spots within the chip. Applying the entire test suite continually can increase the temperature significantly. Hence, it may be necessary to split the test into multiple sessions and put idle times in between for cooling. In a nonpreemptive test, the test resources are held by the core currently being tested, this causes wastage of test time and resource utilization. A preemptive test can overcome this situation by performing tests in a preemptive fashion.

Many a times, for testing complex cores, multiple test sets are used. For example, a core may be tested by both *built-in self-test* (*BIST*) and external test sessions. The tests may also need to be partially ordered. The BIST being on-chip may be applied at a much higher frequency than the external testing. The BIST is applied first, as it can detect the random-detectable faults (the faults that can be detected by random patterns) easily. For the remaining random pattern-resistant faults, the test patterns generated by dedicated algorithms are applied through the external tester. Testing of memory cores may be carried out earlier than the logic cores. Once tested, the memory cores can be used to test the logic cores. Larger cores occupying more amount of chip area are likely to possess more defects than smaller cores. Hence, it may be more desirable to test the larger cores first. Thus, the core order part needs to be reconsidered while formulating the NoC test scheduling algorithm.

Another important constraint is that of peak power consumption during test. The chip will have a predefined safety level of power dissipation. The scheduling algorithm must ensure that this power limit is not violated at any time during test. The total power consumed during testing of a core has two main components: the power consumed by the core and the power consumed by the network in transporting the test packets for the core. The power consumed by the core depends on various factors, such as the core design, the test vectors, and the order of test vectors, which are mostly determined by the core vendor providing the test patterns. For the purpose of scheduling, the power consumed by individual cores can be assumed to be available. The power consumed by the wrapper can also be taken along with the core, since the wrapper is active only when the core is being tested. However, the power consumed by the network to transport each test packet (P_{packet}) can be expressed as follows:

$$P_{packet} = nb_{routers} \times P_{router} + nb_{channels} \times P_{channel}$$

where:
  $nb_{routers}$ is the number of routers in the path established in the network for the packet
  $nb_{channels}$ is the number of channels in the path
  P_{router} is the power consumed by a single router per cycle
  $P_{channel}$ is the power consumed by a single channel per cycle

The router power consumption depends on the supply voltage (V_{dd}), the load capacitance (C_L), the frequency of operation (f), the number of flip-flops (nb_{ff}), the number of logic gates (nb_{gt}), and their corresponding expected switching activities (σ_{ff}) and (σ_{gt}), respectively.

$$P_{router} = C_L \times V_{dd}^2 \times f \times [(\sigma_{ff} + 1) \times nb_{ff} + \sigma_{gt} \times nb_{gt}]$$

The channel power consumption is given by the following expression. Here, the load capacitance of the channel is given by the product of number of wires in the channel (ch_w), the length of the channel (ch_l), and the width of the wire ($wire_w$). σ_w is the switching factor of the wire.

$$P_{channel} = V_{dd}^2 \times f \times \sigma_w \times (ch_l \times wire_w \times ch_w)$$

Since power consumption is calculated per cycle, packet length is not that significant.

Based on the above constraints and power consumption metric, Cota and Liu (2006) proposed an improved test scheduling algorithm for cores in NoC. The cores may be BISTed and have multiple test sessions with individual sessions being preemptive or nonpreemptive. To start with, a few definitions are presented that are used in the algorithm:

1. A set $C = \{i, 1 \leq i \leq N_c\}$ of cores in the NoC.
2. A set $T = \{T_{j_i}, 1 \leq j \leq N_{t_i}, 1 \leq i \leq N_c\}$ of test sessions for the cores in C. Each core i can have N_{t_i} test sessions, each of which can be BISTed, external, preemptive, or nonpreemptive.
3. For each test session k, a six-tuple $I = \{(wc, cl, p, pwr, preemp, payload)_k, 1 \leq k \leq |T|\}$ is defined, where wc is the number of wrapper scan chains, cl is the maximum length of scan chains, p is the number of test patterns, pwr is the power consumption during test, $preemp$ indicates whether the test is preemptive or not, and $payload$ is the size of the test packet.
4. A set $Prec = \{(p_1,..., p_n)_k, 1 \leq n, k \leq |T|\}$ of precedence constraints for each test session k. For any test to be scheduled, all test sessions in its precedence list must have been finished.
5. A graph $G = (V,E)$ corresponding to the topology of the NoC. Each vertex corresponds to a router to which a number of cores may be connected. An edge corresponds to a communication channel between the two routers for the vertices.
6. A list of I/O ports corresponding to some vertices in G, indicating that these cores can be used as I/O ports during testing.

The scheduling process first creates an ordered list of test sessions and a list of I/O pairs. For each core, a list of possible access paths is created, sorted by the number of routers on each path. While allocating a path to a core for test, the availability of shorter paths is checked earlier than longer ones. Once a test packet p is picked up for scheduling, one of the following situations can occur.

1. p *belongs to a nonpreemptive test*: The first available I/O pair that can be used by this packet is chosen. If no such I/O pair is available, the next ready packet is picked up for scheduling. The delivery time of packet p is set to the time when the first I/O pair in the list becomes available. If an I/O pair is available, the entire test session to which p belongs to gets scheduled (since the test is nonpreemptive). Power consumption of the packet is calculated. It is checked that no power violation occurs in the duration for which the test session runs. The network channels identified by the I/O pair are marked unavailable during the transmission time of the packet. The corresponding response packet is then automatically scheduled.
2. p *belongs to a preemptive test*: The shortest available path for this packet is selected and the duration for its transmission is calculated. If no path is available for p, the next ready packet is picked up for possible

scheduling and the delivery time of p is set to the time at which the first path from the list of possible paths for the packet becomes available. If resources are available, the power consumption is calculated. It is checked that no power violation occurs during the transmission of the packet. In such a case, the network channels for transmission of this packet are marked unavailable for the duration of transmission. The response packet is set to be ready at appropriate time after the transmission of the test packet is over and the pattern is applied to the core. The next test packet of the core is set to be ready at a time, ensuring that the new vector will not arrive before the previous vector is processed.

3. p *refers to an autonomous BIST session*: A single flit containing the BIST enable signal and other BIST information (such as LFSR [Linear Feedback Shift Register] seed values) must be sent to the core. Two cases are possible: In the first case, each BISTed core possesses its own BIST controller. The transmission of packet p is similar to that of a preemptive test. The chosen path is occupied for transmission of a single flit. The corresponding response packet is set to be ready at a time equal to the time to transfer this flit and the total number of cycles for which the BIST is set to run. In the second case, a number of BISTed cores may share a single BIST engine. The BIST sessions are now to be scheduled as nonpreemptive tests with precedence constraints. Once a BIST session is started for a core, it cannot be preempted and the BIST engine is devoted to test some other core.

The overall algorithm for the combined preemptive, nonpreemptive, and BISTed cores is presented in the following. It tries out a number (N_1) of permutations of BIST test packets, a number (N_2) of permutations of external test packets, and all permutations of I/O pairs to explore the search space.

Algorithm NoC_Schedule_Preemp_Non-Preemp

1.	UBP = Create ordered list of unscheduled BIST test packets.
2.	UEP = Create ordered list of unscheduled external test packets.
3.	IOP = Create list of I/O pairs.
4.	For each core i in C do
5.	Create ordered list of all possible access paths;
6.	For N_1 permutations of UBP list do
7.	For N_2 permutations of UEP list do
8.	For every permutation of IOP list do
9.	While there are unscheduled packets in UBP ∪ UEP do
10.	L_t = Selected packets ready for schedule in UBP ∪ UEP satisfying precedence constraints and delivery times;
11.	Select test packet p for core i in L_t;

12.	If p is non-preemptive then
13.	Find a free I/O pair;
14.	If no such I/O pair is available then
15.	Update delivery time of p; Repeat from 11;
16.	else if p is preemptive or p is BIST then
17.	Find a free access path to/from core i;
18.	If no such path is available then
19.	Update delivery time for p; Repeat from 11;
20.	Calculate duration of packet transmission;
21.	Calculate power consumption for packet transmission;
22.	If power constraint is met then
23.	Assign packet to the chosen path;
24.	Update schedule and time tags;
25.	If p is non-preemptive then
26.	Assign response packet to chosen I/O pair;
27.	Update schedule and time tags;
28.	else if p is preemptive or p is BIST then
29.	Define delivery time of next packet of core i;
30.	Update L_i;
31.	else
32.	Update packet delivery time; Repeat from 11;
33.	If all packets have been attempted then
34.	Update current time; Repeat from 9;
35.	End.

8.3.4 PSO-Based Strategy

Apart from the ILP and heuristic approaches, discussed so far for testing cores, the meta-search techniques can also be employed to determine a test schedule. Farah and Harmanani (2010) has suggested a simulated annealing based strategy to generate a schedule for testing the cores in the NoC. In the following, a discrete PSO (DPSO)-based formulation will be discussed for the NoC core test scheduling problem to minimize the test application time. A similar DPSO formulation was discussed in Section 5.6. Hence, only the particle structure and evolution mechanism is discussed in Section 8.3.4.1.

8.3.4.1 Particle Structure and Fitness

A particle corresponds to possible test scheduling order of the cores. It has two components—the core order part (*core part*) and the I/O assignment

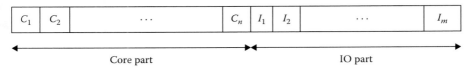

FIGURE 8.8
Particle structure.

part (*IO part*). An example of a particle structure is shown in Figure 8.8. The *core part* is a permutation of core numbers C_1 through C_n. It corresponds to the order in which the scheduling procedure (noted next) attempts to assign time slots to the cores. The next part is an array of size equal to the number of cores. If there are m I/O pairs, an individual entry in I/O pair is an integer in the range of 1 to m. To evaluate the fitness of a particle, start with the first core in the core part. It is scheduled from time zero. Appropriate resources of the NoC (links) are reserved for the purpose. At a certain point, suppose that up to core i in the core order part has been scheduled. The scheduling time of the core $i + 1$ is determined by consulting the I/O pair part. If the corresponding I/O pair is k, the next available time slot for k is determined, so that the resources are available to schedule the test of core $i + 1$. When all cores are scheduled, the highest time for any of the I/O pairs constitutes the overall test time for the NoC. The overall testing time forms the fitness function.

8.3.4.2 Evolution of Generations

The particles evolve through generations to create new particles that are expected to result in overall test scheduling time closer to the optimum. In the first generation, the initial population is created randomly and the fitness of individual particles is evaluated. The local best (*lbest*^{k}) of each particle is set to be the same as the initial particle. The global best (*gbest*$_i$) of a generation is the particle giving the minimum test time. Successive generations are evolved through a series of operations called *swap* operation. The local best of each particle and the global best of the generation are modified if the corresponding values in the current generation are less than in the previous generation.

> *Swap operator*: For a particle P, core as well as I/O pair part are indexed by 0 to $N - 1$ (N being the number cores). Let the swap operator $SO_{j,k}$ ($0 \leq j, k \leq N - 1$) swap the jth and kth positions of particle P to create a new particle P_{New}. For example, consider the particle $P = \{0, 4, 3, 6, 1, 5, 7\}$, where a number represents the core number. The swap operator $SO_{5,6}$ swaps the cores at positions 5 and 6, creating a new particle $P_{New} = \{0, 4, 3, 6, 5, 1, 7\}$. Similar swaps can also be applied to the I/O pair part.

Swap sequence: A swap sequence SS is made up of one or more swap operators. Swap operation in the swap sequence is applied to a particle for creating new particle. For example, let the swap sequence $SS = \{SO_{5,6}, SO_{2,4}\}$ be applied upon the particle $P = \{0, 4, 3, 6, 2, 1, 5, 7\}$. It creates new particle $P_{New} = \{0, 4, 3, 6, 2, 5, 1, 7\}$. To align a particle P^k with its local best, the swap sequence is identified. Let this be SS_k^{lbest}. Then another swap sequence is identified to align the particle with the global best. Let this be SS_i^{gbest}. Now the swap sequence SS_k^{lbest} is applied on particle P_k with probability α. Let the modified particle be P_k^{lbest}. Then the swap sequence SS_i^{gbest} is applied on P_k^{lbest} with probability β. This creates new particle P_{New}. Its fitness is evaluated and the local best is updated for kth particle, if it is the better than the previous local best for the particle. If the best fitness in a generation is better than the global best of the previous generation, the global best is also updated.

8.4 Summary

Testing NoC is one of the most important and intricate problems to be solved in NoC-based SoC design. Since the entire chip is being integrated, it is not possible to assume the correctness of individual components (as in a board-based design). Both the communication infrastructure and the cores need to be tested. The infrastructure test includes testing the routers and links of the NoC. The problem involves having different fault models and test mechanisms for these components. The problem is NP-hard. However, testing of cores require proper scheduling of the test sessions. Different variants of the problem have been solved—preemptive versus nonpreemptive, BIST versus external test, power aware (Liu et al. 2005a), thermal aware (Liu et al. 2005b, 2006), and so on. A number of test scheduling mechanisms have been reported, such as ILP, heuristics, and metasearch techniques. Overall, a number of dedicated approaches may need to be combined to get reasonably good reduction in test time and cost.

References

Ahn, J. H. and Kang, S. 2006. Test scheduling of NoC based SoCs using multiple test clocks. *ETRI Journal*, vol. 28, no. 4, pp. 475–485.
Bushnell, M. L. and Agrawal, V. D. 2005. *Essentials of Electronic Testing for Digital, Memory and Mixed-Signal VLSI Circuits.* Springer-Verlag, Boston, MA.

Cota, E., Carro, L., and Lubaszewski, M. 2004. Reusing an on-chip network for the test of core-based systems. *ACM Transactions on Design Automation of Electronic Systems*, vol. 9, no. 4, pp. 471–499.

Cota, E. and Liu, C. 2006. Constraint driven test scheduling for NoC based systems. *IEEE Transactions on Computer-Aided Design of Integrated Circuits and Systems*, vol. 25, no. 11, pp. 2465–2478.

Cuivello, M., Dey, S., Bai, X., and Zhao, Y. 1999. Fault modeling and simulation for crosstalk in system-on-chip inerconnects. *IEEE/ACM International Conference on Computer Aided Design*, IEEE, pp. 297–303.

Farah, R. and Harmanani, H. 2010. A method for efficient NoC test scheduling using deterministic routing. *IEEE International SOC Conference*, IEEE, pp. 363–366.

Garey, M. R. and Johnson, D. S. 1979. *Computers and Intractability: A Guide to the Theory of NP-Completeness*. W. H. Freeman and Co., San Francisco, CA.

Grecu, C., Ivanov, A., Saleh, R., and Pande, P. 2007. Testing network-on-chip communication fabrics. *IEEE Transactions on Computer-Aided Design of Integrated Circuits and Systems*, vol. 26, no. 12, pp. 2201–2214.

Iyengar, V. and Chakrabarty, K., 2002. Test wrapper and test access mechanism co-optimization for system-on-chip. *Journal of Electronic Testing: Theory and Applications*, vol. 18, pp. 213–230.

Liu, C., Iyengar, V., and Pradhan, D. K. 2006. Thermal-aware testing of network-on-chip using multiple frequency clocking. *24th IEEE VLSI Test Symposium*. IEEE.

Liu, C., Iyengar, V., Shi, J., and Cota, E. 2005a. Power aware test scheduling in network-on-chip using variable rate on-chip clocking. *IEEE VLSI Test Symposium*. IEEE.

Liu, C., Veeraraghavan, K., and Iyengar, V. 2005b. Thermal-aware test scheduling and hot spot temperature minimization for core based systems. *20th IEEE International Symposium on Defect and Fault Tolerance in VLSI Systems*. IEEE.

Marinissen, E. J., Goel, S. K., and Lousberg, M. 2000. Wrapper design for embedded core test. *International Test Conference*, IEEE, pp. 911–920.

Marinissen, E. J., Kapur, R., Lousberg, M., McLaurin, T., Ricchetti, M., and Zorian, Y. 2002. On IEEE P1500's standard for embedded core test. *Journal of Electronic Testing: Theory and Applications*, vol. 18, pp. 365–383.

Pande, P., Grecu, C., Ivanov, A., Saleh, R., and Micheli, G De. 2005. Design, synthesis, and test of networks on chip. *IEEE Design and Test of Computers*, vol. 22, no. 5, pp. 404–413.

Salamy, H. and Harmanani, H. M. 2011. An optimal formulation for test scheduling network-on-chip using multiple clock rates. *IEEE Canadian Conference on Electrical and Computer Engineering*. IEEE.

9

Application-Specific
Network-on-Chip Synthesis

9.1 Introduction

The network-on-chip architectures, discussed so far in the book, are developed around regular topologies, with the inherent assumption that the cores are all of equal size. The main advantage of such a regular NoC architecture is topology reuse and reduced design time. The assumptions of equal core size and communication bandwidth requirement hold for homogeneous cores. However, application-specific system-on-chip (SoC) architectures, in general, contain heterogeneous cores and memory elements with widely varying sizes. Hence, even though the system-level NoC architecture is regular, after floorplanning the final topology becomes irregular. Maintaining a regular structure at this level of layout necessitates large area overhead. The links also become longer, resulting in increase in delay and power consumption in them. This necessitates looking for alternative NoC topologies, specific for the application. Such NoCs are known as *application-specific NoC (ASNoC)*. ASNoC provides the facility to incorporate custom NoC architectures, optimized for the target problem domain. It is not necessary to conform to any regular topology. As a result, ASNoCs often provide architectures superior to the regular ones, in terms of power and area consumption under identical performance requirements. The routers can also be parameterized (such as the number of ports, the physical link length and width, and the number of virtual channels), and thus can be reused in the design.

In ASNoC synthesis, the application is specified as a set of tasks with different communication requirements between them. The tasks are mapped onto a computation *architecture*. The computation architecture consists of a set of processing and memory cores. The tasks are distributed among the processing cores. As a result, the cores now need to communicate between themselves.

9.2 ASNoC Synthesis Problem

The overall ASNoC synthesis problem can be stated as follows:
  Given the following:

- A directed communication graph $G(V,E)$, where each $v_i \in V$ denotes an intellectual property (IP) core of the design, and each directed edge $e_k = (v_i, v_j) \in E$ denotes a communication trace from v_i to v_j. For every $v_i \in V$, the height and width of the core are denoted by H_i and W_i, respectively

- For every $e_k \in E$, $\omega(e_k)$ denotes the bandwidth requirement and $\sigma(e_k)$ denotes the latency constraint in hops for the edge

- A router architecture with η number of I/O ports per router, peak bandwidth Ω per port, two quantities Ψ_i and Ψ_o denoting the power consumed per megabytes per second of traffic flowing through the router in input and output directions, respectively

- A physical link power model denoted by Ψ_l per megabytes per second per millimeter

- Two constants H and W denoting the height and width constraints on the overall system-level floorplan dimension

- Two ratios γ_{min} and γ_{max} denoting the lower and upper bounds on the aspect ratio of the layout, respectively

Let R denote the set of routers in the synthesized architecture, E_r is the set of links between two routers, and E_v be the set of local links connecting cores to the routers. The objective of the NoC synthesis problem is to generate a system-level floorplan and a network topology $T(R, V, E_r, E_v)$ such that

- For every $e_k \in E$, there exists a route p in T that satisfies $\omega(e_k)$ and $\sigma(e_k)$.
- The bandwidth constraints on the ports and routers are satisfied.
- The bounding box of the floorplan satisfies H and W.
- The aspect ratio of the floorplan lies between γ_{min} and γ_{max}.
- The total system-level communication cost/power for communication is minimized.

The floorplanning subproblem is a variant of quadratic assignment problem (Garey and Johnson 1979), while the interconnection network generation problem is an instance of Steiner forest problem (Ravi et al. 2001). Both these problems are NP-hard. As a result, many exact and heuristic methods have been developed to solve the problem. The overall problem may be solved as an integrated problem in which the floorplan and router network are designed simultaneously. The other option is to start with a given floorplan of the IP cores and then insert routers at appropriate positions within the floorplan so that the interconnect network gets synthesized.

9.3 Literature Survey

A survey of ASNoC design techniques was presented by Ascia et al. (2004). It enumerates the advantages of custom topologies over standard ones for ASNoC. While some of the design techniques attempt to modify a regular topology to take into consideration the variation in core sizes and communication pattern, others try to synthesize the most suited topology. A performance-aware design methodology of inserting long-range interconnects on top of a regular mesh-based network for synthesis of application-specific architectures was presented by Ogras and Marculescu (2005a). Challenges in application specific NoC design have been surveyed in Benini (2006). In the following, the major ASNoC synthesis works will be reviewed.

First, we will look into the strategies that do not consider the floorplan of the NoC. A holistic approach for application-specific NoC synthesis has been presented in Leary and Chatha (2010). It results in solution with minimum dynamic power, at most twice the number of routers and leakage, compared to the optimal solution. Average communication latency and jitter are also low. A multiobjective NoC synthesis approach has been presented in Li and Harmanani (2010). A heuristic was proposed by Pinto et al. (2003) for a constraint-driven communication synthesis of on-chip networks. A quadratic programming-based approach along with a clustering algorithm was proposed for this purpose. However, no floorplan information is utilized during the topology generation process. A method to reduce hardware cost of NoC via link aggregation was presented by Korotkyi et al. (2012). It creates ASNoC with nonuniform distribution of physical links in logical connections, depending on the amount of network traffic that is passed through them. A network partitioning technique based on Fiduccia–Mattheyses (FM) partitioning algorithm was proposed in the work of Morgan et al. (2009) to reduce the area cost. A linear programming based technique was proposed in the work of M'zah and Hammami (2011) to generate an area-optimized NoC architecture constrained to propagation delay time and bandwidth requirements. Different topology generation techniques were analyzed and compared in the work of Morgan et al. (2008) with respect to area and average delay. This work analyzed network partitioning techniques and long-range generation techniques, thus determining the best scheme to be used for NoC topology generation. A tool, Q8WARE, was presented by Sami and Mohammad (2006), which is a small-scale version of Application Specific Integrated Circuit assembly line. It deals with synthesis of ASNoC with hardware reusability constraints. A genetic algorithm (GA)-based technique was presented by Srinivasan and Chatha (2005) for synthesis of custom NoC architectures that support guaranteed throughput traffic. The energy consumption of NoC is optimized by minimizing the cumulative traffic flowing through the ports of all routers. The total area consumption is minimized by reducing the total number of routers used. A methodology for generating energy-efficient application-specific architectures was presented by Filippopoulos et al. (2010). It uses application partitioning to reduce processing element dependencies. Topology exploration and buffer sizing techniques are utilized to generate custom topology with reduced power consumption. A NoC topology generation and analysis method was presented by Dumitriu and Khan (2009) that addresses throughput requirements by considering the latency present in the system. An irregular application-specific topology generation algorithm was presented by Ar et al. (2009) to reduce power consumption. It clusters the given application, based on the communication characteristics, and then constructs the topology by connecting clusters to each other, one by one. A partitioning approach based on trees has been presented in Binijie et al. (2011) for NoC synthesis. Genetic algorithm-based techniques have been proposed in Choudhary et al. (2010, 2011) for application specific NoC design.

Next, we look into the floorplan-aware methodologies. Alabei (2010) has proposed a custom NoC synthesis procedure that uses B*-tree for floorplan representation. A multiobjective mathematical model for NoC synthesis has been suggested in Abderazek et al. (2007). A custom NoC instantiation tool *Xpipes Compiler* was presented by Jalabert et al. (2004), which is based on the inputs as specified by the designer. A floorplan-aware tool was presented by Reza et al. (2009) for NoC design and synthesis, integrated with a graphical user interface for interacting among abstract traffic flow specification, topology synthesis, and floorplanning. A GA-based technique was presented by Leary et al. (2009) for ASNoC design with an objective of minimizing the power consumption. This work assumes the routers to be at the corners of the cores. A branch-and-bound algorithm was proposed by Ogras and Marculescu (2005b) for customized communication architecture synthesis. It uses the core coordinates from the floorplan. A GA-based topology synthesis method was proposed by Lai et al. (2010) to minimize power consumption by taking floorplan information. This work also assumes the routers to be at the corners of the cores. A multiobjective approach to topology design based on Tabu search technique was presented by Tino and Khan (2011) to meet power and performance requirements of an application. A synthesis-oriented design flow, xpipes lite, was presented by Stergiou et al. (2005) for the generation of synthesizable simultaneous models for ASNoCs. In the work of Ahoen et al. (2004), a physical floorplan is used during topology design to reduce power consumption on wires. However, the area and power consumption of the switches are not taken into consideration.

In the work of Srinivasan et al. (2005), a slicing tree-based floorplanner is used during the topology design process. This work assumes that the switches to be located at a corner of the cores and the network components are not considered in the floorplanning process. NoC topology generation algorithms were presented by Srinivasan et al. (2006) based on slicing structures where switch locations are restricted to the corners of the cores. A two-step topology generation procedure was proposed by Murali et al. (2006) using a min-cut partitioner to cluster highly communicating cores on the same switch and a path allocation algorithm to connect clusters together to minimize power consumption. An iterative refinement strategy to generate an optimized NoC topology that supports both packet-switched network and point-to-point communications was presented by Chan et al. (2008). This assumes the network interfaces for the processing cores to be located on the corners, whereas the router nodes are in the center. A partition-driven floorplanning algorithm that uses a heuristic to insert switches and an algorithm for inserting NIs, limited to mosaic type of floorplans was proposed by Bei et al. (2010). Two heuristic algorithms were proposed by Shan and Lin (2008) to examine different set partitions. Partitioning is carried out based on the communication flow and a physical network topology has to be generated for each partition. A three-stage synthesis approach was presented by Zhong et al. (2011) that integrates communication requirements, physical information among cores, and partitioning into the floorplanning phase to explore the optimal switch number for clustering of cores with minimized link and switch power consumption. A complete synthesis flow was illustrated by Bertozzi et al. (2005) for customized NoC architectures. It partitions the flow into three major steps: topology mapping, selection, and generation. Tools, such as SUNMAP and Xpipes Compiler, are provided for their automatic execution. Thermal- and nonthermal-aware ASNoC synthesis frameworks that combine multiple algorithms and heuristics to efficiently explore the solution space were presented by Kwon et al. (2011). This work describes both thermal- and nonthermal-aware approaches for router placement. Both the techniques assume that switches can only be placed at the interconnection of cores in the floorplan. A topology generation method was presented by Khan and Tino (2012) by employing analytical models and simulation tools to design low-power, high-performance custom NoCs. Hu et al. (2005) has presented a work on energy-efficient NoC synthesis through topology exploration and wire optimization. The above works generate floorplan as part of the synthesis process.

In the following section, we will look into a few strategies to solve the ASNoC synthesis problem. The first one addresses system-level floorplanning to minimize NoC power consumption subject to layout constraints. This will be followed by discussion on custom topology and route generation. We will also discuss on a scheme to intelligently put routers in a given NoC floorplan to optimize communication cost and energy consumed.

9.4 System-Level Floorplanning

This section discusses a mixed integer linear programming (MILP)-based approach to solve the NoC-centric floorplanning problem (Srinivasan et al. 2006). Since the interconnection architecture is not known at this stage, the interconnect power can be approximated in terms of communication via point-to-point links between communicating cores. Another important factor is to satisfy the latency constraints for communication between cores. It may be difficult to satisfy the latency constraints for cores placed far away. Apart from power and latency, we can also minimize the overall layout area. Hence, the minimization goal is a linear combination of power–latency function and the area of the layout, as shown in the following equation:

$$\alpha \times \left[\sum_{\forall e(u,v) \in E} \text{dist}(u,v) \times \Psi_l \times \frac{\omega(e)}{\sigma^2(e)} \right] + \beta \times \left[X_{\max} + Y_{\max} \right] \qquad (9.1)$$

where:
   $\text{dist}(u,v)$ is the distance between cores u and v
   α and β are constants
   $X_{\max}$ and $Y_{\max}$ represent the boundaries in X and Y directions respectively
   Rest of the variables are as defined in Section 9.2

The objective function puts more emphasis on latency constraint compared to the bandwidth. The values of α and β determine the relative weight given to power minimization compared to area minimization.

9.4.1 Variables

9.4.1.1 Independent Variables

For each core $v_i \in V$, let $(X_{i,min}, Y_{i,min})$ denote the lower left coordinate of the placed core.

9.4.1.2 Dependent Variables

- For each core $v_i \in V$, let $(X_{i,max}, Y_{i,max})$ denote the upper right coordinate of the placed core. Hence,

$$X_{i,\max} = X_{i,\min} + W_i \, ; \, Y_{i,\max} = Y_{i,\min} + H_i$$

- For each pair of cores $v_i, v_j \in V$, let $DX_{i,j}$ and $DY_{i,j}$ represent the differences between the X and Y coordinates of the top right corner of the placed cores. Thus,

$$DX_{i,j} = X_{i,\max} - X_{j,\max}; \ DY_{i,j} = Y_{i,\max} - Y_{j,\max}$$

- For each pair of cores $v_i, v_j \in V$, let $X_{i,j}$ and $X'_{i,j}$ be the binary variables given by

$$X_{i,j} = \begin{cases} 1, \text{if } X_{i,\min} \geq X_{j,\max} \\ 0, \text{otherwise} \end{cases}$$

$$X'_{i,j} = \begin{cases} 1, \text{if } X_{j,\max} > X_{i,\min} \\ 0, \text{otherwise} \end{cases}$$

- $X_{i,j}$ and $X'_{i,j}$ can be obtained as follows, taking MAXVAL as a large integer:

$$X_{i,\min} - X_{j,\max} - X_{i,j} \cdot \text{MAXVAL} < 0$$

$$X_{j,\max} - X_{i,\min} - X'_{i,j} \cdot \text{MAXVAL} \leq 0$$

$$X_{i,j} + X'_{i,j} = 1$$

- Let $Y_{i,j}$ and $Y'_{i,j}$ denote similar quantities along the Y coordinates.

9.4.2 Objective Function

The objective function for floorplanning is to minimize the following:

$$\alpha \times \left[\sum_{\forall e(u,v) \in E} \left(|DX_{i,j}| + |DY_{i,j}| \right) \times \Psi_l \times \frac{\omega(e)}{\sigma^2(e)} \right] + \beta \times [X_{\max} + Y_{\max}]$$

$|DX_{i,j}|$ is modeled by introducing two new variables $DX^+_{i,j}$ and $DX^-_{i,j}$.

$$DX^+_{i,j} - DX^-_{i,j} = DX_{i,j}$$

and

$$DX^+_{i,j} + DX^-_{i,j} = |DX_{i,j}|$$

9.4.3 Constraints

- No two cores v_i and v_j should overlap when they are placed on the layout. This can be stated through the following constraints. At least one of them must hold true.

$$X_{i,\min} \geq X_{j,\max}$$

$$X_{j,\min} \geq X_{i,\max}$$

$$Y_{i,\min} \geq Y_{j,\max}$$

$$Y_{j,\min} \geq Y_{i,\max}$$

Therefore,

$$DX_{i,j} + DX_{j,i} + DY_{i,j} + DY_{j,i} \geq 1$$

- The layout should satisfy the aspect ratio constraints. Thus,

$$Y_{\max} \geq \gamma_{\min} \times X_{\max}$$

$$Y_{\max} \leq \gamma_{\max} \times X_{\max}$$

- The layout should not violate X and Y boundaries. Thus, for each node $v_i \in V$,

$$X_{i,\max} \leq X_{\max}$$

and

$$Y_{i,\max} \leq Y_{\max}$$

9.4.4 Constraints for Mesh Topology

The cores in a mesh topology are aligned along a grid. The height and width of a grid are determined by the largest core in that particular row or column, respectively. Hence, we need additional constraints to restrict the start positions of cores at some valid grid points only. Figure 9.1a shows a valid mesh floorplan, whereas 9.1b is an invalid one.

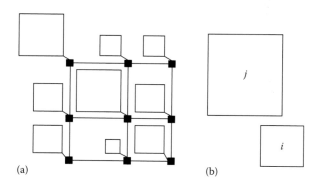

(a)   (b)

FIGURE 9.1
(a) Valid mesh-based floorplan; (b) illegal layout of mesh-based topology. (Redrawn from Srinivasan, K., et al., *IEEE Transactions on Very Large Scale Integration (VLSI) Systems*, 14(4), 407–420, 2006.)

For every pair of cores v_i and $v_j \in V$, the binary variables $GX_{i,j}$ and $GY_{i,j}$ are defined as follows:

$$GX_{i,j} = \begin{cases} 1, \text{if } X_{i,\max} > X_{j,\max} \\ 0, \text{otherwise} \end{cases}$$

$$GY_{i,j} = \begin{cases} 1, \text{if } Y_{i,\max} > Y_{j,\max} \\ 0, \text{otherwise} \end{cases}$$

The following two inequalities need to be satisfied for the cores to get aligned to grid positions:

$$GX_{i,j} + X'_{i,j} \le 1$$

$$GY_{i,j} + Y'_{i,j} \le 1$$

9.5 Custom Interconnection Topology and Route Generation

The floorplan information generated so far does not include the routers in it. The next stage of the formulation will select router locations in the floorplan with an objective of minimizing the power consumption. Since there can be a large number of potential locations for placement of routers, it is desirable to reduce the number of candidate locations via some means, so that the MILP formulation for router location identification can run faster. To start with, bounding boxes are created for each core placed as a rectangle in the floorplan. A bounding box is a rectangular enclosure of the core rectangle (called a node in subsequent discussion) such that the bounding boxes of two neighboring nodes are touching each other. Figure 9.2 shows such a situation. The bounding box of node 4 extends to the top boundary of node 3 and so on. Once the bounding boxes are designed, a *channel intersection graph* (Sait and Youssef 1994) will be constructed. The graph has vertices corresponding to two perpendicular boundaries of the floorplan. The edges of the graph correspond to the bounding boxes of the floorplan. Routers can be placed at each vertex of the channel intersection graph. Figure 9.2b shows the possible placement of routers, represented as filled circles in the diagram. Many of the routers are redundant, in the sense that they are placed very close to each other. Hence, in the final layout, it is very much unlikely that both the closely placed routers will be utilized to connect cores. The routers may be removed when:

1. Routers are placed along the perimeter of the layout.
2. Routers are placed less than a specified distance apart.

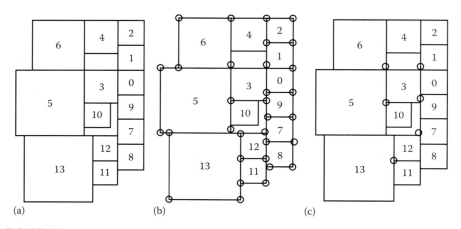

(a)   (b)   (c)

FIGURE 9.2
Router allocation for custom topology. (Redrawn from Srinivasan, K., et al., *IEEE Transactions on Very Large Scale Integration (VLSI) Systems*, 14(4), 407–420, 2006.)

Figure 9.2c shows the situation when routers along the perimeter and those placed less than a specified distance apart have been removed.

We will next look into an ILP formulation for the minimization of the communication power consumption of the NoC. This power is given by the sum of the power consumed by the routers and the physical links. Power consumed by a router is given by the product of the bandwidth of data flowing through its ports and the characterization function specifying the power consumption per unit bandwidth. Power consumption of a physical link is equal to the product of the bandwidth of data flowing through the link, the length of the link, and the characterization function specifying power consumption per unit bandwidth per unit length.

9.5.1 Variables

9.5.1.1 Independent Variables

- *Number of routers*: Let $r_i \in R$, $0 \le i < R_{max}$, be a router. Each router is assumed to be similar, having η number of ports and peak bandwidth Ω per port, all ports being bidirectional.

- *Ports of a router*: Let $p_{i,j}$, $0 \le i < \eta$, represent the jth port of the ith router.

- *Node-to-port mapping variables*: Let $NR_{k,i,j}$ be a {0,1} variable that is 1, if node v_k is mapped to port $p_{i,j}$ of router r_i; otherwise it is 0.

- *Port-to-port mapping variables*: Let $RR_{i,j,k,l}$ be a {0,1} variable that is 1, if port $p_{i,j}$ of router r_i is linked to port $p_{k,l}$ of router r_k; otherwise it is 0.

- *Variable for flow of traffic out of a port*: For each edge $(v_i, v_j) \in E$, let $O_{i,j,k,l}$ be a $\{0,1\}$ variable that is 1, if traffic from node v_i to node v_j flows out of port $p_{k,l}$; otherwise it is 0.
- *Variable for flow of traffic into a port*: For each edge $(v_i, v_j) \in E$, let $I_{i,j,k,l}$ be a $\{0,1\}$ variable that is 1, if traffic from node v_i to node v_j flows into port $p_{k,l}$; otherwise it is 0.

9.5.1.2 Derived Variables

- *Variable for total traffic flowing out of a port*: Let $BO_{k,l}$ represent the total traffic flowing out of port $p_{k,l}$. It can be written as

$$BO_{k,l} = \sum_{\forall e_m = \{v_i, v_j\} \in E} \omega(e_m) \times O_{i,j,k,l}$$

- *Variable for total traffic flowing into a port*: Let $BI_{k,l}$ represent the total traffic flowing into port $p_{k,l}$. It can be written as

$$BI_{k,l} = \sum_{\forall e_m = \{v_i, v_j\} \in E} \omega(e_m) \times I_{i,j,k,l}$$

- *Variable for flow of traffic on a link*: Let $Z_{i,j,k,l,m,n}$ be a $\{0,1\}$ variable that is 1, if traffic (v_i, v_j) leaves port l of router k and that port is connected to port n of router m. It can be represented as

$$Z_{i,j,k,l,m,n} = O_{i,j,k,l} \times RR_{k,l,m,n}$$

- The equation can be linearized using the following two rules:

$$O_{i,j,k,l} + RR_{k,l,m,n} \geq 2 \times Z_{i,j,k,l,m,n}$$

$$O_{i,j,k,l} + RR_{k,l,m,n} \leq Z_{i,j,k,l,m,n} + 1$$

9.5.2 Objective Function

The objective is to minimize the power consumption of the NoC due to cumulative traffic flowing through all routers. The overall function is as follows:

Minimize $(P_R + P_L)$, P_R being the router power and P_L being the link power, is given by

$$P_R = \Psi_i \times \sum_{\forall r_i \in R} \sum_{\forall p_{i,j}} BI_{i,j} + \Psi_o \times \sum_{\forall r_i \in R} \sum_{\forall p_{i,j}} BO_{i,j}$$

$$P_L = \Psi_L \left(\begin{array}{l} \displaystyle\sum_{i,j,k,l,m,n} \omega(i,j) \times RD_{k,m} \times Z_{i,j,k,l,m,n} \\[2ex] + \displaystyle\sum_{i,j,k,l} ND_{i,k} \times \omega(i,j) \times NR_{i,k,l} \\[2ex] + \displaystyle\sum_{i,j,k,l} ND_{j,k} \times \omega(i,j) \times NR_{j,k,l} \end{array} \right)$$

where:

Ψ_i and Ψ_o are the weights that denote power consumed per megabytes per second of traffic flowing through a router port in input and output directions, respectively

Ψ_L is the link power per unit length per megabytes per second

$RD_{k,m}$ is the distance between routers r_k and r_m

$ND_{i,k}$ denotes the distance between core node i and router r_k

9.5.3 Constraints

- *Port capacity*: The bandwidth usage of an input/output port should not exceed its capacity.

$$\forall i \in R, \forall p_{i,j}, BI_{i,j} \le \Omega, BO_{i,j} \le \Omega$$

- *Port-to-port mapping*: A port can be mapped to a core node or to any one port of a different router.

$$\forall p_{i,j}, \sum_{\forall r_k \in R, k \ne i} \sum_{\forall p_{k,l}} RR_{k,l,i,j} + \sum_{\forall v_m \in V} NR_{m,i,j} \le 1$$

$$\forall p_{i,j}, \forall r_k \in R, k \ne i, RR_{k,l,i,j} = RR_{i,j,k,l}$$

- The first inequality captures the situation that a port may not be mapped to any other port or core node. The second equation models the symmetry of the variable *RR*.
- *Node-to-port mapping*: A core node should be mapped to exactly one port.

$$\forall v_i \in V, \sum_{\forall r_k \in R_i} \sum_{\forall p_{k,l}} NR_{i,k,l} = 1$$

- *Traffic routing*: For every $e_k = (v_i, v_j) \in E$, there exists a path $p = \{(v_i, r_i), (r_i, r_j), \ldots, (r_k, v_j)\}$ in *T*. The condition can be captured by the following set of constraints:

- If a core node is mapped to a port of a router, all traffic emanating from that node must enter into that port. The same is the case for a destination node. Hence, for each router r_k, for all $p_{k,l}$ and for all (v_i, v_j) in E, $I_{i,j,k,l} \geq NR_{i,k,l}$ and $O_{i,j,k,l} \geq NR_{j,k,l}$
- If a core node is mapped to a port of a router, no traffic from any other port can enter or leave that port.

$$\forall (v_i, v_j) \in E, \forall v_m \in V, m \neq i, m \neq j, \forall p_{k,l}$$

$$NR_{m,k,l} + I_{i,j,k,l} \leq 1 \text{ and } NR_{m,k,l} + O_{i,j,k,l} \leq 1$$

- If a traffic enters a port of a router, it should not enter from any other port of that router. The same is true for a traffic leaving a port of a router. The constraint ensures that the traffic does not get split across multiple ports. For each router r_k and for all $(v_i, v_j) \in E$,

$$\sum_{\forall p_{k,l}} I_{i,j,k,l} \leq 1 \text{ and } \sum_{\forall p_{k,l}} O_{i,j,k,l} \leq 1$$

- If a traffic enters a port of a router, it has to leave exactly one of the other ports of that router. Similarly, if a traffic leaves a port of a router, it must have entered from exactly one of the other ports of the router. The constraint conserves the flow of traffic. For each router r_k, for all $p_{k,l}$ and for all $(v_i, v_j) \in E$,

$$\sum_{\forall p_{k,m}, m \neq l} O_{i,j,k,m} \geq I_{i,j,k,l} \text{ and } \sum_{\forall p_{k,m}, m \neq l} I_{i,j,k,m} \geq O_{i,j,k,l}$$

- If two ports of different routers are connected, a traffic leaving from one port should enter the other and vice versa (assuming bidirectional links). For each pair of routers $\{r_k, r_m\}$, $k \neq m$, for all $p_{k,l}$, for all $p_{m,n}$, and for all $(v_i, v_j) \in E$,

$$RR_{k,l,m,n} + I_{i,j,k,l} - O_{i,j,m,n} \leq 1 \text{ and } RR_{k,l,m,n} - I_{i,j,k,l} + O_{i,j,m,n} \leq 1$$

- If two different ports are connected, a traffic can leave exactly one of the two ports. Similarly, a traffic can enter only one of the two ports. For each pair of routers $\{r_k, r_m\}$, $k \neq m$, for all $p_{k,l}$, for all $p_{m,n}$, and for all $(v_i, v_j) \in E$,

$$RR_{k,l,m,n} + I_{i,j,k,l} + I_{i,j,m,n} \leq 2 \text{ and } RR_{k,l,m,n} + O_{i,j,k,l} + O_{i,j,m,n} \leq 2$$

- If a traffic enters a port of a router, the port must be mapped to a core node or to a port of a different router. That is, if $I_{i,j,k,l}$ is 1 for some $(v_i, v_j) \in E$, some $NR_{i,k,l}$ should be 1, or some $RR_{m,n,k,l}$ should be 1, where

$p_{m,n}$ exists. Similarly, if a traffic leaves a port of a router, the port must be mapped to a node or a port of a different router. That is,

$$NR_{j,k,l} + \sum_{\forall r_m \in R} \sum_{\forall p_{m,n}} RR_{k,l,m,n} \geq I_{j,i,k,l} \text{ and } NR_{i,k,l} + \sum_{\forall r_m \in R} \sum_{\forall p_{m,n}} RR_{k,l,m,n} \geq O_{j,i,k,l}$$

- *Latency*: It can be stated as

$$\forall e_k = (v_i, v_j) \in E, \sum_{\forall r_k \in R} \sum_{\forall p_{k,l}} O_{i,j,k,l} \leq \sigma(e_k)$$

The MILP formulation discussed above can produce optimum solution, however, takes exponential time for large communication trace graphs. A clustering heuristic was proposed by Roy (1978) to reduce the time requirement by partitioning the trace graph into clusters of nodes. Cluster size is constrained by the maximum number of nodes in a cluster, as specified by the designer. For each edge $e \in E$, the clustering algorithm first assigns a distance metric, given by, $DF_e = \sigma_e^2/\omega_e$. The clustering procedure, as proposed by Srinivasan et al. (2006), attempts to put nodes with low latency and high bandwidth close to each other, that is, in the same cluster. Once the clusters are formed, for every edge that cuts across a cluster boundary, one dummy node will be inserted in each of the corresponding clusters. If two or more such cut edges share a node in a cluster, a single dummy node is inserted in the cluster for all of them. Figure 9.3 shows

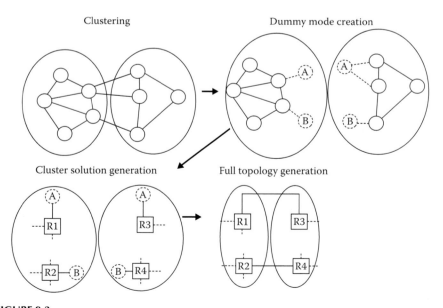

FIGURE 9.3
Clustering-based approach. (Redrawn from Srinivasan, K., et al., *IEEE Transactions on Very Large Scale Integration (VLSI) Systems*, 14(4), 407–420, 2006.)

such a situation. The latency constraint of the original communication trace is split into half across the edges attached to the dummy node pair for a cut edge. The bandwidth constraint is duplicated. The MILP formulation is now run for each of the clusters to generate the topologies for the subgraphs. The topologies are then combined by establishing physical links between the dummy nodes, thus generating the complete ASNoC.

9.6 ASNoC Synthesis with Flexible Router Placement

The ASNoC synthesis procedure discussed in Section 9.5 generates a floor-plan in which routers are placed only at the corners of the tiles containing a core and its associated routers. This may often lead to multihop connection due to the link length constraints. In the work of Soumya and Chattopadhyay (2013), the problem is addressed to generate a flexible placement of routers in a given floorplan of the NoC containing only the cores.

The advantage of having flexibility in choosing router locations compared to placing them at the corners can be understood by the example noted in Figure 9.4. It corresponds to the benchmark application. Picture in Picture, having eight cores C0–C7. Figure 9.4a shows the communication trace graph of the application, in which the edges are annotated with bandwidth requirements between the corresponding tasks in megabytes per

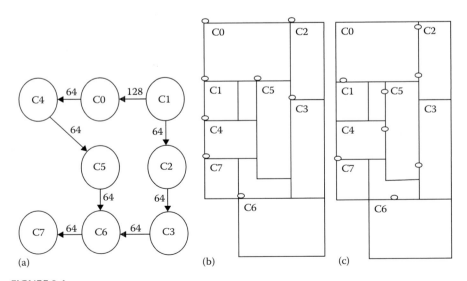

FIGURE 9.4
(a) Communication trace graph; (b) routers at corners; (c) flexible router placement. (Redrawn from Soumya, J. and Chattopadhyay, S., *Journal of Systems Architecture*, 59, 361–371, 2013.)

second. Figure 9.4b shows a floorplan for the cores. It also includes the routers (shown as circles) at the corners corresponding to the cores. Figure 9.4c shows the floorplan with router locations found by the approach proposed by Soumya and Chattopadhyay (2013). (It may be noted that the floorplan is an input to the router location problem solved in this section, rather than evolving the floorplan itself.) When the routers are located at the corners of each core, it may require longer interconnects between the routers, or multiple hops for a communication between cores. For example, let us consider the communication between C3 and C6. For Figure 9.4b, placing routers only at the corners, it requires either a longer interconnect or needs to travel in three hops (using routers attached with C3, C5, C4, and C6). However, in the router placement approach discussed in this section (Figure 9.4c), the routers are placed on the floorplan such that smaller interconnects are needed between the routers and the core-to-core communications also need less number of hops. The same communication from C3 to C6 requires only a single hop in this case. This becomes possible due to the intelligent placement of routers that also considers the link length constraint put into the synthesis process. In this case, it could obtain a 35.71% decrease in the overall communication cost.

It is assumed that the cores corresponding to the tasks of the application are already laid out on a two-dimensional grid, corresponding to the floorplan of the chip. Now, the routers will be inserted into this grid floorplan of the application. Individual routers are assumed to be of size equal to the smallest square in the grid. This essentially makes all routers to be of same size, though the constraint can be relaxed easily by considering routers with varying complexities, thus requiring different amount of area. The maximum allowed link length (L_{MAX}) is taken as another constraint. If a router is located at grid point (i,j) and the second one at (k,m), a link can exist between them only if the distance $\sqrt{(i-k)^2 + (j-m)^2}$ is less than L_{MAX}.

9.6.1 ILP for Flexible Router Placement

This section presents an ILP formulation for the problem of finding suitable router locations in a given NoC floorplan.

9.6.1.1 Variables

- R is the set of probable router locations on the grid floorplan.
- For each pair of probable router locations r_i and r_j, let $rdist_{r_i,r_j}$ be the distance between them.
- For each core c_i and a probable router location r_s, let $cdist_{c_i,r_s}$ be the distance from the center of the floorplan of c_i to the router r_s.
- L_{MAX} is the maximum link length permitted between two routers or between a core and the router associated with it.

- BW_{ij} is the bandwidth requirement between cores c_i and c_j.
- sr_s is a binary variable, which is 1 if location r_s is selected to hold a router and 0 otherwise.
- $m_{c_i}^{r_s}$, a binary variable, which is 1 if core c_i is connected to router at r_s and 0 otherwise.
- $cl_{c_i}^{r_s}$ is a binary variable, which is 1 if link exists between core c_i and router at location r_s and 0 otherwise.
- $rl_{r_i r_j}$ is a binary variable, which is 1 if link exists between locations r_i and r_j and 0 otherwise.
- $P_{c_i c_j}^{r_s r_t}$ is a binary variable, which is 1 if path exists between routers r_s and r_t, to which cores c_i and c_j have been attached, and 0 otherwise.
- $n_i^{r_s r_t}$ is a binary variable, which is 1 if router r_i is a part of the path from router r_s to r_t and 0 otherwise.
- $l_{r_i r_j}^{r_s r_t}$ is a binary variable, which is 1 if the link between routers r_i and r_j is part of the path from r_s to r_t and 0 otherwise.
- $D_{r_s r_t}$ is an integer variable identifying the distance between router locations r_s and r_t, in terms of the number of hops. It can take up values from 0 to the number of routers in the network.

9.6.1.2 Objective Function

The objective is to minimize the communication cost by selecting suitable router locations. The objective function can be formulated as follows: If cores c_i and c_j are mapped to router locations r_s and r_t and a path exists between them in the network, $P_{c_i c_j}^{r_s r_t}$ is equal to 1. This multiplied by $D_{r_s r_t}$ gives the number of hops of the communication from c_i to c_j. The number of hops multiplied by bandwidth gives the communication cost, which has to be minimized over all the edges in the core graph. Thus, the overall objective function is

$$\text{Minimize} \sum_{e_{ij} \in E} BW_{ij} \left(\sum_{r_s, r_t \in R} D_{r_s r_t} \times P_{c_i c_j}^{r_s r_t} \right)$$

9.6.1.3 Constraints

The following is the set of constraints framed to solve the router location selection problem:

- *Mapping constraints*
  - Each core has to be mapped onto only one router.

$$\forall c_i \in C, \sum_{r_s \in R} m_{c_i}^{r_s} = 1$$

- Each router, selected for synthesis, should have one core mapped onto it.

$$\forall r_s \in R, \sum_{c_i \in C} m_{c_i}^{r_s} = 1$$

- A core can be mapped to a router only when that router location is selected for synthesis.

$$\forall r_s \in R, sr_s - m_{c_i}^{r_s} \geq 0$$

- *Router–core association constraints*
  - A core c_i can be associated with the router at location r_s (i.e., $cl_{c_i}^{r_s}$ can be set to 1) if the physical distance ($cdist_{c_i r_s}$) is less than L_{MAX}.

$$\forall c_i \in C, \forall r_s \in R, cdist_{c_i r_s} \times cl_{c_i}^{r_s} \leq L_{MAX}$$

  - Now, mapping of a core onto a router is possible if there is a link between them.

$$\forall c_i \in C, \forall r_s \in R, cl_{c_i}^{r_s} - m_{c_i}^{r_s} \geq 0$$

- *Inter-router link constraints*
  - Link can exist between routers, if the distance between them is less than the maximum link length L_{MAX}.

$$\forall r_i \in R, \forall r_j \in R, i \neq j, rdist_{r_i r_j} \times rl_{r_i r_j} \leq L_{MAX}$$

- *Constraints for core graph edges*
  - Each edge present in the core graph has to be mapped onto a path in the evolved topology.

$$\forall e_{ij} \in E, \forall r_s, r_t \in R, m_{c_i}^{r_s} + m_{c_j}^{r_t} - P_{c_i c_j}^{r_s r_t} \leq 1$$

and

$$2 \times P_{c_i c_j}^{r_s r_t} \leq m_{c_i}^{r_s} + m_{c_j}^{r_t}$$

- *Path constraints*
  - To find the path between two router locations r_s and r_t, it is required to identify the links and routers forming part of the path. The following conditions are imposed for this purpose.
    - Starting and ending routers must be part of the path.

$$\forall e_{ij} \in E, \forall r_s, r_t \in R, P_{c_i c_j}^{r_s r_t} - n_i^{r_s r_t} = 0$$

- To get starting link of the path, the following constraint is imposed. Starting router may have more than one outgoing link, out of which one has to be selected.

$$\forall e_{ij} \in E, \forall r_s, r_t \in R, P_{c_i c_j}^{r_s r_t} - \sum_{r_i, r_j \in R} l_{r_i r_j}^{r_s r_t} = 0$$

- If a link is part of the path, starting and ending routers of that link must also be part of the path.

$$\forall r_s, r_t \in R, \forall r_i, r_j \in R, 2 \times l_{r_i r_j}^{r_s r_t} - n_i^{r_s r_t} - n_j^{r_s r_t} \le 0$$

- Each router (except starting and ending ones) should have equal in and out degrees.

$$\forall r_s, r_t \in R, \forall r_i \in R, r_i \ne r_s, r_i \ne r_t, r_i \ne r_j, 2 \times n_i^{r_s r_t} - \sum_{\forall (r_i, r_j) \in R} l_{r_i r_j}^{r_s r_t} = 0$$

- A link can be part of the path if it would have satisfied the link length constraint.

$$\forall r_s, r_t \in R, \forall r_i, r_j \in R, rl_{r_i r_j} - l_{r_i r_j}^{r_s r_t} \ge 0$$

- The distance between routers can be calculated by using following equation:

$$\forall r_s, r_t \in R, D_{r_s r_t} - \sum_{\forall (r_i, r_j) \in R} l_{r_i r_j}^{r_s r_t} = 0$$

This completes the formulation. The objective function along with the constraint set can be fed to any ILP solver to get the router positions minimizing the communication cost for the synthesized NoC. However, excepting for very small NoCs, it takes huge amount of CPU time to arrive at the solution. In Section 9.6.2, a particle swarm optimization (PSO)-based technique has been discussed to explore the search space for synthesizing larger NoCs.

9.6.2 PSO for Flexible Router Placement

Particle Swarm Optimization (PSO) is an optimization technique developed around the idea of birds flocking. Originally proposed by Kennedy and Eberhart (1995), the technique has been applied in numerous optimization problem. In the following a PSO formulation has been presented as the flexible router placement problem.

9.6.2.1 Particle Structure and Fitness Function

Let *l* be the number of available router positions in the floorplan. For all these *l* available router positions, a particle is a permutation of numbers from 0 to $l - 1$. It is assumed that the router positions are numbered as 0 to $l - 1$. A particle identifies a set of router positions. The total communication cost forms the fitness function. While calculating the communication cost, we consider the locations from 0 to n ($n < l$) in the particle, where n represents the number of cores in the application. Thus, only those first n router positions are used for the mapping of cores.

Fitness of a particle P is the total communication cost due to the router positions specified by the particle. For every particle, its fitness is calculated as follows:

1. The distance between each pair of core and router is calculated.
2. A core is mapped to its nearest router. If the distance is more than L_{MAX}, the fitness of the particle is set to infinity.
3. Links are established between the routers, taking L_{MAX} constraint into account. No link can be of length larger than this.
4. For each edge in the core graph, the shortest path is found between the cores in the router graph.
5. Communication cost (fitness) is calculated using the formula:

$$\text{Communication cost} = \sum \left(\begin{array}{c} \text{Number of hops} \times \\ \text{Bandwidth between each pair} \\ \text{of cores in core graph} \end{array} \right)$$

It may be noted that while identifying the shortest paths between the cores, the capacities of constituent links and all the communications passing through the link are to be taken into consideration. The issues such as deadlock can be taken care of later either by adding virtual channels or via communication scheduling.

9.6.2.2 Local and Global Bests

Every particle has a *local best* (*pbest*), which is one set of router positions giving minimum communication cost, among all sets of router positions that the particle has seen so far in the evolution process. This local best partially guides the evolution of the particle. For a particular generation, the *global best* (*gbest*) is the particle resulting in the minimum communication cost for that generation. It also controls the evolution of particles. The local best of each particle and the global best are modified if the corresponding values in the current iteration are less than the values till the previous iteration.

9.6.2.3 Evolution of Generation

Evolution of the particles is done over generations to create new particles that are expected to give results closer to the optimum. To start with, the initial population is created randomly and the fitness of individual particles is evaluated. The local best (*pbest*) of each particle is initialized to be the same as that of the initial particle. The global best of the generation is initialized with the particle giving the least communication cost (smallest fitness function) in the generation. The second generation results through random exchange of router positions within the particles. The local best and global best values are updated if they give better fitness values. Further generations are created through a series of *swap* operations. The local best of each particle and the global best are modified if the corresponding values in the current generation are less than the values in the previous generation. The local best and the global best evolution thus center on the basic operator, swap, explained in Section 9.6.3.4.

9.6.2.4 Swap Operator

Each particle is a sequence of *l* probable router positions. Out of these, first *n* corresponds to the selected router positions. To effect a change in the particle, the swap operator is used. The operator takes two indices (say *i* and *j*) of particle *P* as inputs and creates a new particle P_1. The particle P_1 is the same as *P*, except that the positions *i* and *j* of *P* are interchanged in it.

Let the particle *P* be

r1	r3	r5	*r7*	r4	*r2*	r6	r8

where *rx* represents the router at position *x*. The indices of *r1*, *r3*, and *r5* are 0, 1, and 2, respectively. The swap operator SO(3, 5) swaps positions 3 and 5 in *P* to generate a new particle as shown in the following:

r1	r3	r5	*r2*	r4	*r7*	r6	r8

9.6.2.5 Swap Sequence

A swap sequence is a sequence of swap operators. For example, a swap sequence SS = {SO(1, 7), SO(3, 4)} creates particle P_{new} working on particle *P* in two steps as follows:

Let the particle *P* be

r3	*r6*	r8	r4	r1	r5	r2	*r7*

SO(1, 7) on particle *P* creates intermediate particle P_{int}.
Let the particle P_{int} be

r3	*r7*	r8	r4	r1	r5	r2	*r6*

SO(3, 4) on P_{int} results in new particle P_{new}.

Let the particle P_{new} be

r3	r7	r8	*r1*	*r4*	r5	r2	r6

In PSO, each particle tries to move toward the local best and the global best with some inertia of movement. After all particles have undergone the evolution, a new generation gets created. The best fitness of this generation gives the global best for the population. The PSO terminates if there is no improvement in the *gbest* value for a predefined number of generations, or the PSO has already iterated for a preset maximum number of generation. The best particle of this generation is taken as the solution to the flexible router placement problem.

9.7 Summary

In this chapter, we have seen a few techniques to synthesize ASNoC. The floorplan of the NoC along with router locations is evolved. The topologies generated are irregular and custom-made. Hence, they are expected to optimize system performance further. In Chapter 10, we will look into the reconfigurable NoC design that can run different applications at different time instants.

References

Ababei, C. 2010. Efficient congestion-oriented custom network-on-chip topology synthesis. *Proceedings of the International Conference on Reconfigurable Computing and FPGAs*, IEEE, pp. 352–357.

Abderazek, B. A., Akanda, M., Yoshinaga, T., and Sowa, M. 2007. Mathematical model for multiobjective synthesis of NOC architectures. *Proceedings of the Parallel Processing Workshops*, IEEE, p. 36.

Ahoen, T., David, A., Bin, H., and Nurmi, J. 2004. Topology optimization for application specific networks on chip. *Proceedings of the International Workshop on System level Interconnect Prediction*, IEEE, pp. 53–60.

Ar, Y., Tosun, S., and Kaplan, H. 2009. TopGen: A new algorithm for automatic topology generation for network on chip architectures to reduce power consumption. *Proceedings of the International Conference on Application of Information and Communication Technologies*, IEEE, pp. 1–5.

Ascia, G., Catania, V., and Palesi, M. 2004. Multi-objective mapping for mesh-based noc architectures. *Proceedings of the International Conference on Hardware/Software Codesign and System Synthesis*, IEEE, pp. 182–187.

Bei, Y., Dong, S., Chen, S., and Goto, S. 2010. Floorplanning and topology generation for application-specific network-on-chip. *Proceedings of the Asia and South Pacific Design Automation Conference*, IEEE, pp. 535–540.

Benini, L. 2006. Application specific NoC design. *Proceedings of Design, Automation and Test in Europe Conference and Exhibition*, IEEE, pp. 1–5.

Bertozzi, D., Jalabert, A., Murali, S., Tamhankar, R., Stergiou, S., Benini, L., and Micheli, G. D. 2005. NoC synthesis flow for customized domain specific multi-processor systems-on-chip, *IEEE Transactions on Parallel and Distributed Systems*, vol. 16, no. 2, pp. 113–129.

Binjie, S., Shan, Z., Ma, Y., Xu, N., and Wang, Y. 2011. Tree-based partitioning approach for network-on-chip synthesis. *Proceedings of the International Conference on Computer-Aided Design and Computer Graphics*, IEEE, pp. 465–470.

Chan, J. and Parameswaran, S. 2008. NoCOUT: NoC topology generation with mixed packet-switched and point-to-point networks. *Proceedings of the Asia and South Pacific Design Automation Conference Design Automation Conference*, IEEE, pp. 265–270.

Choudhary, N., Gaur, M. S., Laxmi, V., and Singh, V. 2010. Energy aware design methodologies for application specific NoC, *Proceedings of the NORCHIP*, IEEE, pp. 1–4.

Choudhary, N., Gaur, M. S., Laxmi, V., and Singh, V. 2011. GA based congestion aware topology generation for application specific NoC, *Proceedings of the IEEE International Symposium on Electronic Design, Test and Application*, IEEE, pp. 93–98.

Dumitriu, V. and Khan, G. N. 2009. Throughput-oriented NoC topology generation and analysis for high performance SoCs. *IEEE Transactions on Very Large Scale Integration (VLSI) Systems*, vol. 17, no. 10, pp. 1433–1446.

Filippopoulos, I., Anagnostopoulos, I., Bartzas, A., Soudris, D., and Economakos, G. 2010. Systematic exploration of energy-efficient application-specific network-on-chip architectures, *Proceedings of the IEEE Computer Society Annual Symposium on VLSI*, IEEE, pp. 133–138.

Garey, M. R. and Johnson, D. S. 1979. *Computers and Intractability: A Guide to the Theory of NP-Completeness*. Sanfrancisco, CA: Freeman.

Ho, R., Mai, K. W., and Horowitz, M. A. 2001. The future of wires. *Proceedings of the IEEE*, vol. 89, no. 4, pp. 490–504.

Hu, Y., Chen, H., Zhu, Y., Chien, A. A., and Cheng, C. K. 2005. Physical synthesis of energy-efficient networks-on-chip through topology exploration and wire style optimization. *Proceedings of the IEEE International Conference on Computer Design: VLSI in Computers and Processors*, IEEE, pp. 111–118.

Jalabert, A., Murali, S., Benini, L., and Micheli, G. D. 2004. xpipesCompiler: A tool for instantiating application specific networks on chip. *Proceedings of the Design, Automation and Test in Europe Conference and Exhibition*, IEEE, pp. 884–889.

Kennedy, J. and Eberhart, R. 1995. Particle swarm optimization. *Proceedings of the IEEE International Conference on Neural Networks*, vol. 4, IEEE, pp. 1942–1948.

Khan, G. N. and Tino, A. 2012. Synthesis of NoC interconnects for custom MPSoC architectures. *Proceedings of the IEEE/ACM International Symposium on Networks on Chip*, IEEE, pp. 75–82.

Kwon, S., Pasricha, S., and Cho, J. 2011. POSEIDON: A framework for application-specific network-on-chip synthesis for heterogeneous chip multiprocessors. *Proceedings of the International Symposium on Quality Electronic Design*, IEEE, pp. 1–7.

Korotkyi, I. and Lysenko, O. 2012. Application-specific network-on-chip with link aggregation, *Proceedings of the Mediterranean Conference on Embedded Computing*, IEEE, pp. 9–12.

Lai, G., Lin, X., and Lai, S. 2010. GA-based floorplan-aware topology synthesis of application-specific network-on-chip. *Proceedings of the IEEE International Conference on Intelligent Computing and Intelligent Systems*, IEEE, pp. 554–558.

Leary, G. and Chatha, K. S. 2010. A holistic approach to network-on-chip synthesis, *Proceedings of the IEEE/ACM/IFIP International Conference on Hardware/Software Codesign and System Synthesis*, IEEE, pp. 213–222.

Leary, G., Srinivasan, K., Mehta, K., and Chatha, K. S. 2009. Design of network-on-chip architectures with a genetic algorithm-based technique. *IEEE Transactions on Very Large Scale Integration (VLSI) Systems*, vol.17, no.5, pp. 674–687.

Li, X. and Hammami, O. 2010. Multi-objective network-on-chip synthesis with transaction level simulation. *Proceedings of the International Conference on Microelectronics*, IEEE, pp. 487–490.

Lin, S., Su, L., Haibo, S., Depeng J., and Zeng, L. 2008. Hierarchical cluster-based irregular topology customization for networks-on-chip. *Proceedings of the IEEE/IFIP International Conference on Embedded and Ubiquitous Computing*, IEEE, pp. 373–377.

M'zah, A. and Hammami, O. 2011. Area/delay driven NoC synthesis. *Proceedings of the International Conference on Microelectronics (ICM)*, IEEE, pp. 1–6.

Morgan, A. A., Elmiligi, H., El-Kharashi, M. W., and Gebali, F. 2009. Area-aware topology generation for application-specific networks-on-chip using network partitioning. *Proceedings of the IEEE Pacific Rim Conference on Communications, Computers and Signal Processing*, IEEE, pp. 979–984.

Morgan, A. A., Elmiligi, H., Watheq, E. M., and Gebali, F. 2008. Networks-on-Chip topology generation techniques: Area and delay evaluation. *Proceedings of the Design and Test Workshop*, IEEE, pp. 33–38.

Murali, S., Meloni, P., Angiolini, F., Atienza, D., Carta, S., Benini, L., Micheli, G. D., and Raffo, L. 2006. Designing application-specific networks on chips with floorplan information. *Proceedings of the IEEE/ACM International Conference on Computer-Aided Design Computer-Aided Design*, IEEE, pp. 355–362.

Murali, S. and Micheli, G. D. 2004. Bandwidth constrained mapping of cores onto NoC architectures. *Proceedings of Design, Automation and Test in Europe Conference and Exhibition*, pp. 896–901.

Ogras, U. Y. and Marculescu, R. 2005a. Application-specific network-on-chip architecture customization via long-range link insertion. *Proceedings of the IEEE/ACM International Conference on Computer-Aided Design*, IEEE, pp. 246–253.

Ogras, U. Y. and Marculescu, R. 2005b. Energy- and performance-driven NoC communication architecture synthesis using a decomposition approach. *Proceedings of the Design, Automation and Test in Europe*, IEEE, pp. 352–357.

Pinto, A., Carloni, L. P., and Sangiovanni-Vincentelli, A. L. 2003. Efficient synthesis of networks on chip, *Proceedings of the International Conference on Computer Design*, IEEE, pp. 146–150.

Ravi, R., Marathe, M. V., Ravi, S. S., Ronenkarntz, D. J., and Hunt, H. B. 2001. Approximation algorithms for degree-constrained minimum-cost network design problems. *Algorithmica*, vol. 31, no. 1, pp. 58–78.

Reza K. M., Angiolin, F., Murali, S., Pullini, A., Seiculescu, C., and Benini, L. 2009. A floorplan-aware interactive tool flow for NoC design and synthesis. *Proceedings of the IEEE International SOC Conference*, IEEE, pp. 379–382.

Roy, G. D. 1978. U-statistic hierarchical clustering, *Psychometrica*, vol. 4, no. 1, pp. 58–67.

Salminen, E., Kulmala, A., and Hmlinen, T. D. 2008. Survey of network-on-chip proposals, Technical Report, http://www.ocpip.org/socket/whitepapers.

Sait, S. M. and Youssef, H. 1994. *VLSI Physical Design Automation: Theory and Practice.* McGraw-Hill, New York.

Sami, J. H. and Mohammad, G. M. 2006. Q8WARE: Synthesis tool for network-on-chip applications. *Innovations in Information Technology*, pp. 1–5.

Shan, Y. and Lin, B. 2008. Application-specific network-on-chip architecture synthesis based on set partitions and steiner trees. *Proceedings of the Asia and South Pacific Design Automation Conference*, IEEE, pp. 277–282.

Soumya, J. and Chattopadhyay, S. 2013. Application-specific network-on-chip synthesis with flexible router placement, *Journal of Systems Architecture*, vol. 59, pp. 361–371.

Srinivasan, K. and Chatha, K. S. 2005. SAGA: Synthesis technique for guaranteed throughput NoC architectures. *Proceedings of the Asia and South Pacific Design Automation Conference*, IEEE, pp. 489–494.

Srinivasan, K., Chatha, K. S., and Konjevod, G. 2005. An automated technique for topology and route generation of application specific on-chip interconnection networks. *Proceedings of the IEEE/ACM International Conference on Computer-Aided Design*, IEEE, pp. 231–237.

Srinivasan, K., Chatha, K. S., and Konjevod, G. 2006. Linear-programming-based techniques for synthesis of network-on-chip architectures. *IEEE Transactions on Very Large Scale Integration (VLSI) Systems*, vol. 14, no. 4, pp. 407–420.

Stergiou, S., Angiolini, F., Carta, S., Raffo, L., Bertozzi, D., and Micheli, G. D. 2005. xpipes Lite: A synthesis oriented design library for networks on chips. *Proceedings of the Design, Automation and Test in Europe*, IEEE, pp. 1188–1193.

Tino, A. and Khan, G. N. 2011. Multi-objective tabu search based topology generation technique for application-specific network-on-chip architectures. *Proceedings of the Design, Automation & Test in Europe Conference & Exhibition*, IEEE, pp. 1–6.

Wang, K. P., Huang, L., Zhou, C. G., and Wei, P. 2003. Particle swarm optimization for traveling salesman problem. *Proceedings of the International Conference on Machine Learning and Cybernetics*, IEEE, pp. 1583–1585.

Zhong, W., Yu, B., Song, C., Yoshimura, T., Dong, S., and Goto, S. 2011. Application-specific network-on-chip synthesis: Cluster generation and network component insertion. *Proceedings of the International Symposium on Quality Electronic Design*, IEEE, pp. 1–6.

10

Reconfigurable Network-on-Chip Design

10.1 Introduction

Designing system containing multiple applications having almost common set of cores leads to reconfigurable computation. In such a system, the same hardware platform is utilized to implement the applications. In a network-on-chip (NoC) (Dally and Towles 2001; Atienza et al. 2008) environment, the intellectual property (IP) cores communicate with each other using an underlying router network. The routers are often connected in a predefined fabric. Commonly used network topologies include mesh, tree, and star. Out of these, mesh is the most widely used topology due to its regular structure and short interconnections. A reconfigurable computation in any such topology essentially means designing the NoC so that the performances of all applications become acceptable. The set of cores considering all applications is mapped to the routers in such a fashion that the stated objective is met.

A key point in optimizing the NoC power/performance is to place the cores that are communicating more frequently, close to each other. This is typically known as the application mapping problem, discussed in detail in Chapter 5. Application mapping has been a very well-researched domain. Many NoC mapping strategies are available in the literature. Most of the existing NoC mapping methods try to find an optimal mapping for the communication pattern of a single application. For a set of applications, the NoC architecture for the design should closely match the traffic characteristics and performance requirements of different applications. As different applications have different functionalities, the inter-IP communication characteristics can be very different across the applications. In general, a NoC that is designed to run exactly one application does not necessarily meet the design constraints of other applications.

NoC reconfiguration problem can be addressed at various levels. First of all, standard reprogrammable devices, such as *field-programmable gate array* (*FPGA*) can be used to change the entire NoC logic—cores, routers, and their interconnections can get modified per application. However, this is not feasible in Application Specific Integrated Circuit (ASIC) design environment. For ASIC-based NoCs, the suggested reconfigurations are as follows:

1. Local reconfiguration of core attachments to neighboring routers (Soumya et al. 2013). Multiplexers are introduced to change the attachments of cores to different routers for different applications.
2. Topology reconfiguration (Modarressi et al. 2011) via topology switches to evolve application-specific topologies for individual applications.
3. Router wrapper design (Stensgaard et al. 2008) to effect reconfiguration.
4. Link reconfiguration (Lan et al. 2011) to change directions of links connecting any two routers, based on traffic load in each direction.

In this chapter, these strategies will be discussed. However, to start with, Section 10.2 performs a brief review of the literature in the domain of reconfigurable NoC design.

10.2 Literature Review

A ReNoC architecture has been presented in Stensgaard et al. (2008) which enables the network topology to be configured by the application running on the SoC by using topology switches. Core mapping mechanisms for ReNoC architectures have been presented by Modarressi and Sarbazi-Azad (2007) and Modarressi et al. (2010, 2011), where reconfiguration is achieved via programmable switches in the network. The reconfigurability allows NoC to dynamically tailor its topology to the traffic pattern of different applications. Reconfigurable hybrid bus–network architecture has been proposed by Avakian et al. (2010), where the number of processor cores attached to each bus is reconfigurable and is dependent on the needs of the active processes and applications. Wu et al. (2011) proposed a dynamic bypass circuit and north-last-weave routing algorithm to realize dynamically reconfigurable ReNoCs. The network reconfigures itself for different applications at runtime. *OperaNP*, a ReNoC-based platform, has been proposed by Elmiligi et al. (2007) in which cores can be embedded on an array of programmable logic blocks while the data routing is done using another array of configurable routers. A topology allocation algorithm has been presented in Kunert et al. (2007) considering the demands of different applications. It selects topology depending on the application running on NoC. A hybrid-communication ReNoC (HCR-NoC) has been presented in Zheng et al. (2010) which dynamically reconfigures Multi-Processor System-on-Chip (MPSoC) architecture based on bus traffic. They have used a designable framework that enables topology reconfiguration upon some regular physical network topologies. It provides a customized domain-specific framework for on-chip interconnection, which applies both NoC-based and bus-based systems to fulfil specific application requirements. A runtime ReNoC framework has been presented in Rana et al. (2009) based on the partial dynamic reconfiguration capabilities of FPGAs. This framework dynamically creates or deletes express lines between SoC components (implementing dynamically circuit-switching channels) and performs runtime NoC topology and routing table reconfigurations to handle interconnection congestion. A flexible network design has been presented in Bartic et al. (2003, 2005) which is scalable and can be changed to accommodate various needs of applications. This design is realized as part of the platform for reconfigurable systems. It is suitable for building networks with irregular topologies. An architecture of dynamically reconfigurable NoC has been proposed by Ahmad et al. (2006) for MPSoC. It dynamically configures itself with respect to routing, switching, and data packet size with the change in communication requirements of the system at runtime. This work generates specific topology for the application running on NoC, which is a time-consuming and complex approach, as this consists of generating floorplan, network component placement, and deadlock free routing. Ding et al. (2012) has proposed a configuration algorithm based on a ReNoC by clustering the cores. Many cores are connected per router that may lead to increase in the complexity and power consumption of the routers. Only one application is taken at a time for mapping onto the reconfigured NoC. Dumitriu and Khan (2009) has presented an approach for throughput oriented NoC generation technique.

10.3 Local Reconfiguration Approach

In this section, a locally reconfigurable NoC architecture is presented (Soumya et al. 2013). The architecture is built around the one reported in the work of Ding et al. (2012) (shown in Figure 10.1).

Compared to many other reconfiguration topologies (Stensgaard et al. 2008; Modarressi et al. 2011) that attempt to reduce distances between communicating cores via the introduction of configurable switches, this architecture uses multiplexers. The cores have limited choice to get attached to the routers; however, the overall architecture remains a mesh with small regular interconnects, bounded delay, and the applicability of standard mesh routing algorithms for the resulting application's message communication.

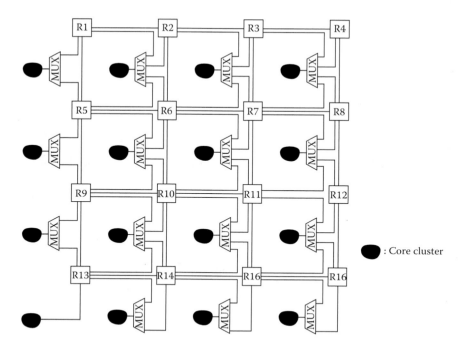

FIGURE 10.1
Locally reconfigurable architecture.

However, the work of Ding et al. (2012) allows a cluster of cores to be attached with individual routers, increasing the router complexities and their associated power consumptions. The work presented in the work of Soumya et al. (2013) and discussed in this section restricts the number of cores attached to any router to at most two. This brings more regularity in the resulting NoC. Connecting two cores to one router also enjoys the advantage that two highly communicating cores, if attached to the same router, will encounter very little delay in their communication. The modified architecture is shown in Figure 10.2 for a 4 × 4 network. The network architecture consists of routers, multiplexers, and selection logic. In the following sections, each of them will be elaborated.

10.3.1 Routers

Similar to the standard mesh topology, each router has at most four global ports connecting to the four neighbors. The boundary routers possess less number of neighbors, resulting in reduced number of global ports in them. Apart from that, each router has two local ports. A core can be attached to a

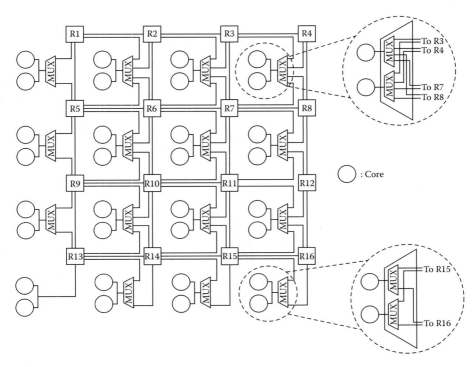

FIGURE 10.2
Another locally reconfigurable architecture.

local port. Thus, each router can have up to two cores connected to it. For an $M \times N$ mesh, the router requirements are as follows:

Six-port routers (four global, two local ports) = $(M - 2)(N - 2)$
Five-port routers (three global, two local ports) = $2(M + N - 4)$
Four-port routers (two global, two local ports) = 4

10.3.2 Multiplexers

To bring the flexibility of local reconfiguration of cores, multiplexers are introduced. As a result, a core can get attached to any of its neighboring routers. The cores in the leftmost column and the bottommost row can get attached to only two routers. The cores at the bottom-left corner can get attached to a single router; hence, no multiplexers are necessary for them. The remaining cores are connected to neighboring routers via multiplexers. Each 4:1 *MUX* block in Figure 10.2 consists of two 4:1 multiplexers, whereas a 2:1 *MUX* block contains two 2:1 multiplexers in it.

For an $M \times N$ network, the number of different types of multiplexers needed are as follows:

4:1 multiplexers $= (M - 1)(N - 1)$
2:1 multiplexers $= M + N - 2$

10.3.3 Selection Logic

Selection logic blocks are incorporated in each router to associate cores, connected to the router, to the individual local ports. We have kept full flexibility to connect a core to either of the local ports of a router. Figure 10.3 shows the position of a selection logic block for a router. Each nonboundary router has eight cores surrounding it, any two of which may be connected to its two local ports. Thus, the selection logic block has eight inputs (one for each core) and two outputs (one for each local port). For boundary routers, the structure is a bit simple, as each router has only four candidate cores to which it may be connected. Thus, four-input, two-output selection logic blocks will suffice for them.

For an $M \times N$ mesh network, the number of such selection logic blocks are as follows:

Four-input, two-output blocks: $M + N - 2$
Eight-input, two-output blocks: $(M - 1)(N - 1)$

10.3.4 Area Overhead

We next estimate the area overhead of the proposed architecture. For this, we have first computed the area of different types of routers used in the topology using the tool Orion 2.0. The corresponding parameters are noted

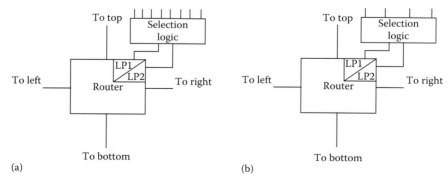

(a)   (b)

FIGURE 10.3
Selection logic in nonboundary routers (a) and boundary routers (b).

TABLE 10.1

Parameters for Orion

Parameter	Value
Technology	90 nm
Operating condition (transistor type)	NVT (normal)
Supply voltage	1.0 V
Frequency	1.5 GHz
Virtual channels	4
Input buffer	1 (present)
Input buffer depth	6
Buffer type	Register

NVT, nominal threshold voltage (V_t)

TABLE 10.2

Router Areas

Router Type	Area (μm^2)
Four-port	80,305.5
Five-port	190,930
Six-port	259,761

TABLE 10.3

Area for Other Module Types

Module Type	Area (μm^2)
4:1 Multiplexer	3237
2:1 Multiplexer	1618.5
Eight-input two-output selection logic	809.25
Four-input two-output selection logic	404.62

in Table 10.1. The router area values are noted in Table 10.2. Different types of multiplexers (4:1 and 2:1) and selection logic modules (eight-input two-output and four-input two-output) have been synthesized using Synopsys Design Vision tool with 90-nm library. The corresponding area values are noted in Table 10.3.

For an $M \times N$ NoC assuming two cores per router (i.e., a total of $2MN$ cores) and each core being of area x μm^2, the overhead of the proposed approach is given by the ratio of the total network area to the total core area. The total network area overhead is given by:

$$
\left[
\begin{array}{l}
(\text{Number of four-port routers}) \times (\text{Area of four-port routers}) \\
+(\text{Number of five-port routers}) \times (\text{Area of five-port routers}) \\
+(\text{Number of six-port routers}) \times (\text{Area of six-port routers}) \\
+(\text{Number of 4:1 multiplexers}) \times (\text{Area of 4:1 multiplexers}) \\
+(\text{Number of 2:1 multiplexers}) \times (\text{Area of 2:1 multiplexers}) \\
+(\text{Number of four-input two-output selection logic blocks}) \\
\times(\text{Area of four-input two-output selection logic blocks}) \\
+(\text{Number of eight-input two-output selection logic blocks}) \\
\times(\text{Area of eight-input two-output selection logic blocks})
\end{array}
\right]
$$
$$
\text{Total core area}
$$

$$
= \frac{\left[\begin{array}{c} 259761 \times (M-2)(N-2) + 2 \times 190930 \times (M+N-4) + 4 \times 80305.5 \\ + 3237 \times (M-1)(N-1) + 1618.5 \times (M+N-2) + 809.25 \\ \times (M-1)(N-1) + 404.62 \times (M+N-2) \end{array}\right]}{2MNx}
$$

$$
= \frac{\begin{array}{c} 259761 \times (M-2)(N-2) + 381860 \times (M+N-4) + 4046.25 \\ \times (M-1)(N-1) + 2023.12 \times (M+N-2) \end{array}}{2MNx}
$$

The area overheads of different architectures are shown in Table 10.4. It can be seen that the area overhead of the proposed reconfigurable scheme is 0.1% more than a simple mesh without reconfiguration. Thus, the proposed architecture is a feasible one and supports reconfiguration. The reconfiguration procedure has to identify the select lines of multiplexers (implemented by the configuration manager at the application layer) which will decide the connection between cores and routers for each application.

10.3.5 Design Flow

The strategy to design a reconfigurable NoC for a set of given applications, using the architecture proposed, has been discussed in this section. It is assumed that the applications have a number of tasks common between themselves; however, the communication pattern may change across the

TABLE 10.4

Area Overheads of Different Architectures

Number of Cores (M × N)	Simple Two-Core per Router Mesh (without Reconfiguration)	Reconfigurable Architecture
16 (2 × 4)	0.129	0.131
32 (4 × 4)	0.130	0.131
64 (4 × 8)	0.130	0.131
128 (8 × 8)	0.130	0.131
256 (8 × 16)	0.130	0.131
512 (16 × 16)	0.130	0.131

applications. This is justified as the applications constitute a full system. Further, the reconfiguration is assumed to be static in nature. Thus, for a task, the core accomplishing it is known at the system design time itself. Each task is bound to a fixed core. Cores may be multifunctional; thus, a number of tasks may be bound to a single core. It may be noted that dynamic task allocation to cores in NoC can also be performed for the proposed topology; however, this chapter does not address this issue. We rather restrict our attention to show the promise of the proposed architecture for a multiapplication environment with core sharing among applications.

Once the core for each application task has been discussed, the applications can be viewed as a set of core graphs defined next. Each core performs one or more tasks of one or more applications. The communication requirement of a pair of cores for each application is computed from the communication requirements of the corresponding communicating tasks in the application bound to these cores. All such task communications are added to get communication between the pair of cores. For application A_i, its core graph G_i consists of a number of nodes equal to the number of cores needed for A_i. There is an edge between the nodes c_{ij} and c_{ik} if there are communications between cores j and k in application A_i. The edge has a weight equal to the corresponding bandwidth requirement.

The design flow of the proposed reconfigurable NoC synthesis procedure (shown in Figure 10.4) can be divided into following three stages:

- Construction of combined core graph (CCG)
- Mapping of cores in CCG to the mesh network
- Configuration generation for each application

Each stage is described in Sections 10.3.5.1 through 10.3.5.3.

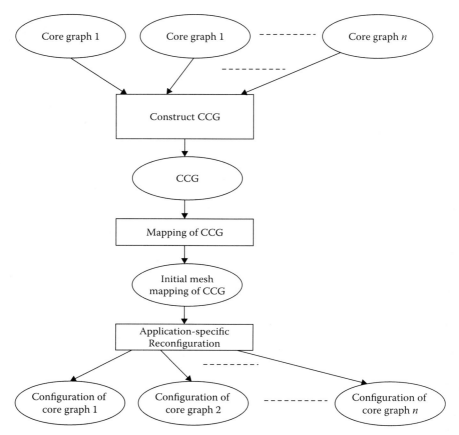

FIGURE 10.4
Design flow of ReNoC synthesis process.

10.3.5.1 Construction of CCG

Let $\{A_1, A_2, ..., A_n\}$ be the set of applications to be implemented in the NoC. Application A_i is represented by the core graph $G_i = (C_i, E_i)$ where C_i is the set of cores participating in the application and E_i is the set of edges representing the communication pattern of A_i. Each edge in E_i has a weight corresponding to the bandwidth requirement of the communication. The CCG $G = (C,E)$ is defined as follows:

The node set $C = C_1 \cup C_2 \cup ... C_n$ is the set of cores required for the entire set of applications. An edge e is included in E if and only if $e \in E_i$ for at least a single application A_i. The weight of the edge e is set to be the sum of weights of all such edges in the entire application set. That is,

$$\text{Weight}(e) = \sum_{i \in \{1..n\}} \left(\text{Weight of } e \text{ in } E_i\right)$$

10.3.5.2 Mapping of CCG

The CCG is now mapped onto the mesh topology. The cost of a mapping solution is computed by determining the total communication cost of the CCG. The communication cost corresponding to a pair of cores in CCG is computed as the product of the bandwidth requirement between the cores and the number of hops between the corresponding routers. Each router can have at most two cores connected to it. The communication cost between two such cores connected to the same router is taken as zero, as no router cycle is spent in the process.

For achieving this mapping, first an integer linear programming (ILP) formulation of the problem has been done. This produces optimal results, but could give solutions for small graphs only. Next, a particle swarm optimization (PSO) has been performed for obtaining mapping of larger CCGs.

10.3.5.3 Configuration Generation

The mapping stage attaches cores to routers taking a global view of the set of applications. The next task is to fine-tune the mapping for each application separately, and thus generate the corresponding configuration program for the application. As it can be noted in Figure 10.2, excepting the boundary routers, each core can be attached to any of the four routers surrounding it, by applying suitable controls to the multiplexers. Thus, it leads to a restricted version of the mapping problem that can be solved in the global mapping stage considering the CCG. The problem can be solved using the following three different strategies:

- ILP
- PSO
- An iterative improvement algorithm

The ILP and PSO formulations are similar to that in the previous phase, whereas the iterative approach is a new one. In Section 10.3.6, the ILP formulation to solve the mapping problem is presented. This can be used to get solution to both the CCG mapping and local configuration generation. Section 10.3.7 presents the PSO-based approach that may also be used to solve both global and local mapping. Section 10.3.8 presents the iterative approach for local mapping.

10.3.6 ILP-Based Approach

This section presents an ILP formulation for the problem of mapping and reconfiguration onto the proposed reconfigurable NoC architecture. First, formulation has been given for mapping problem, which has next been extended for reconfiguration.

10.3.6.1 Parameters and Variables

The parameters and variables used in the ILP formulation are noted in Table 10.5.

10.3.6.2 Objective Function

The objective is to minimize the communication cost by selecting suitable routers in the ReNoC for mapping and reconfiguration. The objective function can be formulated as follows: If cores c_i and c_j are mapped to routers r_s and r_t and a path exists between them in the network, $P_{c_i c_j}^{r_s r_t}$ is equal to 1. This multiplied by $D_{r_s r_t}$ gives the number of hops of the communication from c_i to c_j. The number of hops multiplied by bandwidth gives the communication cost, which has to be minimized over all the edges in the core graph.

$$\min \left[\sum_{e_{ij} \in E} BW_{ij} \left(\sum_{r_s, r_t \in R} D_{r_s r_t} \times P_{c_i c_j}^{r_s r_t} \right) \right]$$

10.3.6.3 Constraints

The following is the set of constraints framed to solve the mapping and reconfiguration problem:

1. Mapping constraints
   a. Each core has to be mapped onto only one router.

   $$\forall c_i \in C, \sum_{r_s \in R} m_{c_i}^{r_s} = 1$$

   b. Each router can have at most two cores mapped onto it.

   $$\forall r_s \in R, \sum_{c_i \in C} m_{c_i}^{r_s} \leq 2$$

TABLE 10.5

Parameters and Variables of ILP Formulation

Parameters and Variables	Definitions
$D_{r_s r_t}$	The precalculated Manhattan distance between the routers r_s and r_t
BW_{ij}	The bandwidth requirement from core c_i to core c_j
$m_{c_i}^{r_s}$	Binary variable. $m_{c_i}^{r_s} = 1$ if core C_i is mapped to the router r_s; otherwise, $m_{c_i}^{r_s} = 0$.
$P_{c_i c_j}^{r_s r_t}$	Binary variable. $P_{c_i c_j}^{r_s r_t} = 1$ if a path exists between routers r_s and r_t to which cores C_i and C_j have been mapped, respectively; otherwise, $P_{c_i c_j}^{r_s r_t} = 0$.

2. Constraints for core graph edges
   a. Each edge present in the core graph has to be mapped onto a path in the NoC considered.

$$\forall e_{ij} \in E, \forall r_s, r_t \in R, m_{c_i}^{r_s} + m_{c_j}^{r_t} - P_{c_i c_j}^{r_s r_t} \leq 1$$

$$2 \times P_{c_i c_j}^{r_s r_t} \leq m_{c_i}^{r_s} + m_{c_j}^{r_t}$$

This completes the formulation. The objective function along with the constraint set can be fed to any ILP solver to get mapping and reconfiguration for minimizing the communication cost of NoC.

For mapping of CCG onto the reconfigurable NoC, the equations noted so far used. The cores in the combined graph can be mapped onto any router in the network. Therefore, the router variables r_s and r_t in the equations can take any value from 1 to the number of routers present in the architecture. For the reconfiguration phase, each core cannot have the flexibility to move from its initial mapped position to any arbitrary router position in the network. As the output from the mapping approach is taken as input in the Mapping phase output is taken as input for the reconfiguration phase. Thus, the flexibility of the attachment of a core to a router in the reconfiguration phase depends heavily on the initial core-to-router attachment and the flexibility provided by the reconfiguration architecture (Figure 10.2). For example, if a core of combined graph is mapped onto router R1, through the multiplexer between R1 and R5, the mapped core can have the flexibility of moving to R5 or remain at R1 only. If a core is mapped onto router R1, through the multiplexer present between R1, R2, R5, and R6, the mapped core can move to any of these routers, and so on. Therefore, the flexibility of the core moving from its initial mapped position (output of mapping phase) depends upon onto which router it has been mapped in the initial phase of mapping. In ILP formulation, the routers in equations cannot take all the possible values in reconfiguration phase. The router position values that are allowed for each core can only be taken in ILP for the reconfiguration purpose. However, except for very small NoCs, it takes huge amount of CPU time to arrive at the solution. Hence, Soumya et al. (2013) has also proposed a PSO-based optimizer to find mapping for larger core graphs.

10.3.7 PSO Formulation

As noted in Chapter 5, PSO is a population-based stochastic technique developed by Eberhart and Kennedy in 1995, inspired by social behavior of bird flocking or fish schooling. In a PSO system, multiple candidate solutions coexist and collaborate simultaneously. Each solution, called a *particle*, flies in the problem space according to its own experience as well as the experience of neighboring particles. It has been successfully applied in many problem areas. In PSO, each single solution is a particle in the search space, having a fitness value. The quality of a particle is evaluated by its fitness. A discrete

PSO (DPSO) formulation has been developed in Soumya et al. (2013) for mapping the cores onto the NoC architecture and then reconfiguring them with the application considered.

10.3.7.1 Particle Formulation and Fitness Function

10.3.7.1.1 For Mapping Problem

A particle corresponds to a possible mapping of cores to the routers. An example of a particle structure is shown in Figure 10.5. The numbers shown in the boxes are the core numbers present in the CCG. The numbers outside the box are the router numbers of the NoC architecture. It is assumed that the routers are numbered in ascending order from the top-left to the bottom-right position as shown in Figure 10.2. If the number of the nodes (routers) present in the architecture is greater than half the number of cores present in the CCG, dummy nodes are added to the CCG to make the two numbers same. These nodes are connected to all core nodes and between themselves. The edges connecting a core node to dummy nodes and the edges between dummy nodes are assigned cost *zero*. Let N be the number of cores present in the CCG for mapping of cores onto the ReNoC architecture, after connecting the dummy nodes, if required. For these N cores, there are $N/2$ positions in the architecture. A particle is a permutation of numbers from 1 to N, which shows the placement of the cores to the node positions of the architecture. The overall communication cost is influenced by the position of the cores in a particle. In this formulation, the overall communication cost forms the fitness function. The fitness of a particle is equal to the overall communication cost after the placement of cores of the CCG to different routers as specified by the particle.

10.3.7.1.2 For Reconfiguration Problem

For the reconfiguration problem, a particle corresponds to a possible movement of cores from their initial mapped position to the available router positions in the reconfigurable architecture. The particle structure is similar to that of mapping problem, shown in Figure 10.5. The main difference between the particle structure in the mapping phase and this phase is in the possible router positions for mapping. In the mapping phase, each entry in the particle (core number) can choose any router position for mapping, whereas in the reconfiguration phase, it can choose from a limited number of router positions, which depends on the initial mapped position of the core (output of mapping phase).

1	5	6	8	9	3	7	10	4	2
1	1	2	2	3	3	4	4	5	5

FIGURE 10.5
A sample particle.

10.3.8 Iterative Reconfiguration

In this section, a heuristic algorithm is presented to perform reconfiguration. For each application, it attempts to finalize the positions of cores, so that the communication cost for the application can be reduced further. It may be noted that in the proposed architecture, a core can get attached to any of the four routers surrounding it. It is assumed that the global mapping phase has attached the core to the router at the upper right position. The local reconfiguration phase will evaluate other three positions and shift the core to the most suitable router.

The algorithm *Heuristic_Configure* performs this job. It picks up each application by turn and generates its reconfiguration information. For an application A_i represented by the core graph $G_i = (C_i, E_i)$, it first computes the communication cost of each of its edges. The edges are sorted in a descending order of the communication cost. The first edge $[e = (c_i^j, c_i^k)]$ is taken. Since it is having the highest communication cost, reducing distance between the corresponding cores is expected to have good impact on communication cost improvement of the application. It then attempts to find the best location for c_i^j. It can move to any of the four neighboring router positions. The core gets attached to the router resulting in the minimum cost of edge e. The core position gets locked, and in the corresponding router, only one more core position is available. The same operation is carried out with core c_i^k. The process continues till all cores get locked to some router positions.

Algorithm *Heuristic_Reconfigure*

Input: Mapping of core graphs $G_1, G_2, ... G_n$ corresponding to applications $A_1, A_2, ... A_n$
Output: Reconfigured mappings
Begin

    For each application core graph $G_i = (C_i, E_i)$ of A_i do
    Begin

        Mark all cores of C_i as unlocked
For each edge $e = (c_i^j, c_i^k) \in E_i$ do
        Begin

            Communication cost of e = bandwidth (e) * hop-distance between routers to which cores c_i^j and c_i^k are mapped

        End For
        Sort all edges in E_i on decreasing communication cost
        For each edge $e = (c_i^j, c_i^k) \in E_i$ picked up in order do
        Begin

            If c_i^j and c_i^k are both locked **then** continue with next edge
            If c_i^j and c_i^k are both mapped to same router **then** Mark both cores as locked

                              Else
                              Begin
                                        If c_i^j is unlocked then
                                        Begin
                                                  Find candidate positions with
                                                  minimum cost of c_i^j
                                                  Min_pos = Candidate position
                                                  with minimum cost of e
                                                  Map c_i^j to Min_pos
                                                  Lock core c_i^j
                                        End
                                        If c_i^k is unlocked **then**
                                        Begin
                                                  Find candidate positions with
                                                  minimum cost of c_i^k
                                                  Min_pos = Candidate position
                                                  with minimum cost of e
                                                  Map c_i^k to **Min_pos**
                                                  Lock core c_i^k
                                        End
                              End
                    End
          End
End

Section 10.4 looks into another reconfiguration strategy, commonly known as topology reconfiguration. Unlike the strategy discussed in this section, the reconfiguration can result in long interconnects. Thus, if communication can take place over these longer links within the router clock period, the strategies can be adopted without introducing further delays into the network.

10.4 Topology Reconfiguration

Standard topologies for NoC, such as mesh and tree, and their variants provide the flexibilities to the designers, as physical design issues can be resolved once and reused for several designs. However, ASNoCs are designed for some particular application(s). Topology reconfiguration-based strategies are intermediary between these two. These techniques essentially put some switches in the network, so that the topology can be dynamically changed to suit the requirements of a particular application. For the next application, the switch configurations can be changed, thus leading to a different topology altogether. In the following, two such techniques will be discussed that provide the facility of topology reconfiguration. The first approach (Stensgaard et al. 2008) adds some topology switches around the individual routers. The second one (Modarressi et al. 2011) introduces similar switches into the topology, though not exclusively earmarked for a particular router.

10.4.1 Modification around Routers

The strategy ReNoC (Stensgaard et al. 2008) combines the flexibilities of packet switching and circuit switching into NoC. Figure 10.6 shows a physical architecture of the process, on top of which logical architectures can be developed. The *network nodes* are connected by *links* in a two-dimensional mesh fashion. Each network node consists of a conventional router wrapped with a *topology switch*. The physical architecture (shown in Figure 10.7a) can produce, for example, two different configurations shown in Figure 10.7b and c. Long logical links can be formed connecting directly two IP blocks, two routers, or between a router and a core.

A multiplexer-based implementation of such a topology switch is shown in Figure 10.8. It connects four links, an IP block, and a five-port router. It is also capable of connecting links directly to each other or to a port of the router. The following alternating connections are possible.

- Incoming links can be connected directly to an outgoing link, bypassing the router.
- Incoming link may be connected to a router port.
- Ports of the router can be connected to outgoing links.

Next, a reconfiguration architecture that puts explicit configuration switches between routers will be explored.

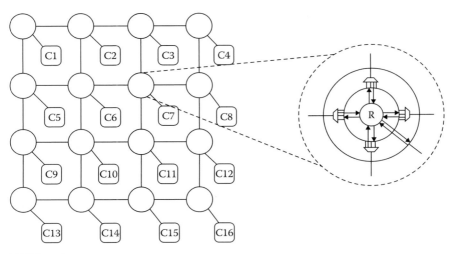

FIGURE 10.6
A physical architecture with routers wrapped by topology switches.

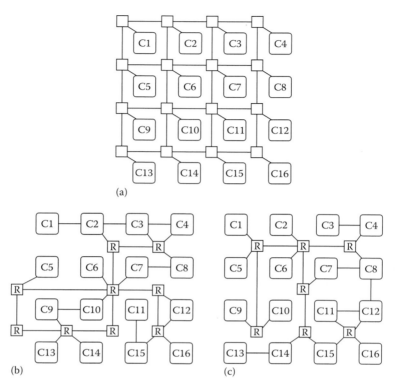

(a)

(b)   (c)

FIGURE 10.7
(a) Physical architecture; (b) logical topology 1; (c) logical topology 2.

10.4.2 Reconfiguration Architecture

Figure 10.9 shows an $m \times n$ (4 × 4, in the example) network with nodes arranged in a two-dimensional mesh fashion (Modarressi et al. 2010). Similar to normal mesh, the square boxes correspond to routers to which individual cores can be attached. The horizontal and vertical lines are the links between routers. However, unlike a general mesh, routers are not connected directly; rather the links have *configuration switches* in between. These switches can be configured to create different types of topologies in which the routers get connected. For example, Figure 10.9 shows the architecture configured as a mesh. Figure 10.10 configures it as a binary tree. The configuration consists of simple pass transistor switches to establish connection between incoming and outgoing links. Though only a single connection is shown between two ports of a switch, each link is bidirectional and can be controlled separately to connect to two different ports. For example, the incoming north link may be connected to the outgoing south link of a switch, whereas the outgoing north link may get connected by the incoming east link. It should be noted that long links may get created via the reconfiguration process. It degrades

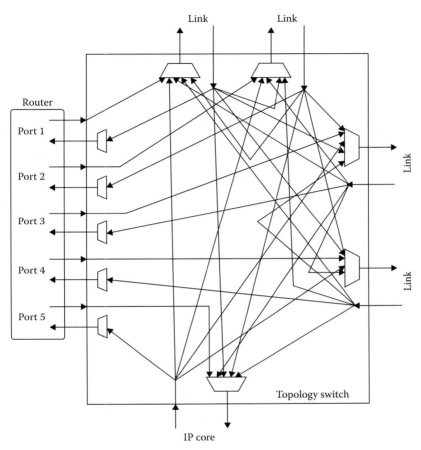

FIGURE 10.8
A multiplexer-based implementation.

the NoC clock frequency, as flits may not be able to cross a long link within a single router clock. For solving these problems, long links are segmented into fixed-length links connected by a register (one-flit buffer). Data may be sent over such a link in a pipelined fashion.

10.4.2.1 Application Mapping

The overall mapping problem for this architecture can be stated as follows:
  Given a set of input applications using a specific set of IP cores,

1. Attach the cores to different routers in the architecture.
2. Determine the customized topology for each application, based on mapping and application traffic characteristic.

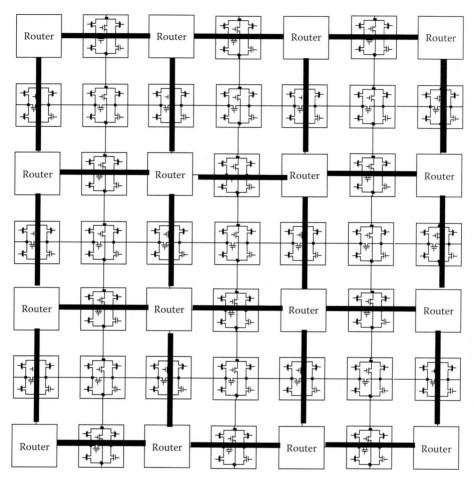

FIGURE 10.9
A reconfigurable topology.

3. Find a route for the traffic flows of each application. The routers are assumed to follow table-driven routing; hence, specific routing algorithms are not necessary.

The problem can be solved in a two-stage fashion. In the first stage, core-to-network mapping is done, whereas topology and route generation are performed in the second stage. Each application is described as a *communication task graph* (CTG), a directed graph $G(V,E)$. Each node $v_i \in V$ represents a task, whereas an edge $e_{i,j} \in E$ represents communication between nodes v_i and v_j. The edge is labeled by $t_{i,j}$, the communication volume (bits per second) between the tasks. Tasks are assumed to be nonmigratory and already mapped to IP cores.

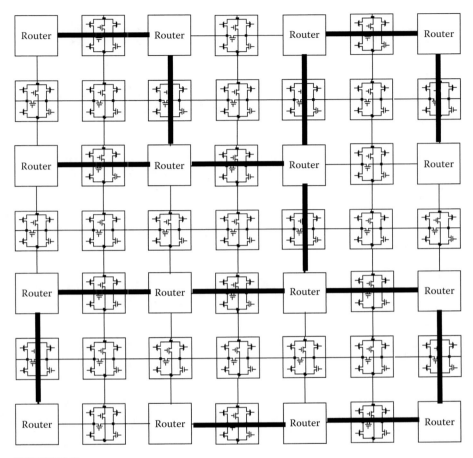

FIGURE 10.10
Another configuration of the topology in Figure 10.9.

10.4.2.2 Core-to-Network Mapping

A weight is assigned to each task graph based on its criticality. Criticality of an application is defined as the percentage of time the application runs on the NoC. It is assumed to be assigned by the designer of the NoC and is taken as an input for the mapping problem. Mapping is performed by first constructing a synthetic/average task graph, considering task graphs of all applications and their criticality values. This average graph has nodes of all applications. Every pair of nodes has an edge between them. The weight of edge $e_{x,y}$, $t_{x,y}^{\mathrm{avg}}$, is calculated as

$$t_{x,y}^{\mathrm{avg}} = \frac{\sum_{\forall i \in \mathrm{applications}} t_{x,y}^{i} \times W_{i}}{n}$$

where:

W_i is the weight of ith task graph

$t_{x,y}^i$ and $t_{x,y}^{avg}$ denote the communication volume of edge $e_{x,y}$ in the ith application and the average graph, respectively

The number of input task graphs is n. Once the graph has been constructed, mapping techniques noted in Chapter 5 can be utilized to get a mapping solution.

10.4.2.3 Topology and Route Generation

In this step, suitable topologies and routes are generated for each individual application. Due to their varying communication requirements, the applications may have inclinations to different topologies. For a particular application, the configuration switches are set such that the number of hops between source and destination routers for high-volume communications is as small as possible. The basic idea is to select the heaviest communication flow yet to be assigned a route and find a minimum hop count path for it.

All edges of an application are sorted in decreasing order of communication volume. All configuration switches are initially unconfigured. For each edge of the application, a branch-and-bound algorithm is used to choose the path with least cost. The communication cost component due to flow through a router can be taken as 5, whereas that through a switch is 1. The branch-and-bound steps for reconfiguration are carried out as follows:

1. *Branch*: Every path starts at the source node, which happens to be a router. A new branch to the path is created by adding a router or a configuration switch adjacent to the current node in the partial path. The added node must belong to the shortest path area— routers and configuration switches located along one of the shortest paths between source and destination nodes, as well as the neighboring configuration switches. That is, for a router node, the path is extended by including neighboring configuration switches along the shortest path. If the node is a configuration switch, the path is extended by adding neighboring routers and configuration switches along the shortest path. This, of course, is restricted by the situation in which the switch has already been configured. In this case, the path can be extended and also constrained by the direction determined by the current configuration.

2. *Bound*: A path may be bounded (i.e., discarded) if an addition of a new node violates the bandwidth constraint of the newly added link. In general, the bandwidth constraint of each link must be satisfied. Also, if the cost of partial path reaching a particular node is larger than the already known partial paths to that node, this path is

discarded. The completed paths between a pair of nodes can also be used to reject more costly partial paths. Moreover, if the current path configures switches in such a way that all possible paths between the source and destination nodes for at least one unmapped edge are blocked, the partial path gets bounded.

10.5 Link Reconfiguration

In traditional NoC architectures, neighboring routers are connected via a pair of unidirectional links. Each link is hardwired to carry traffic in one direction only. However, depending upon the network traffic, it may be possible that one channel is overflowed with heavy traffic, whereas the traffic in the other direction is almost zero. This results in inappropriate resource utilization, leading to performance loss. A strategy has been presented in Lan et al. (2011) that supports dynamic self-reconfiguration of links, resulting in a bidirectional NoC architecture, aptly named as *BiNoC*. Adjacent routers negotiate the flow directions of connecting links using a *channel direction control (CDC)* protocol.

10.5.1 Estimating Channel Bandwidth Utilization

Bandwidth utilization is defined as the percentage of time data channels that are kept busy during the execution of an application.

$$U = \frac{\sum_{t=1}^{T} N_{\text{busy}}(t)}{T \times N_{\text{total}}}$$

where:
T is the total execution time
N_{total} is the number of channels available to transmit data
$N_{\text{busy}}(t)$ is the number of channels busy at clock cycle t

It is obvious that the ideal value for U is 1. However, practical simulation with different traffic patterns and routing algorithms leads to an interesting observation. The channel bandwidth utilization peaks at 45% and 40% for uniform traffic with XY and odd–even routing, respectively, at a heavy traffic load. For transpose traffic under XYY routing, U falls below 20%. Thus, it can be inferred that even with two channels between a pair of routers, at most one channel is kept busy on an average. The other channel remains idle, and thus can be utilized to transport traffic in the other direction, improving performance.

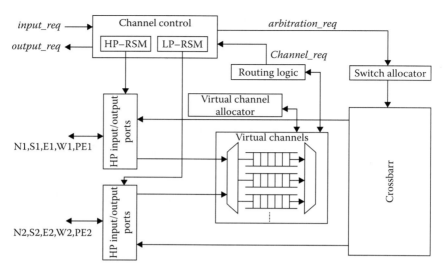

FIGURE 10.11
Modified router architecture.

A modified router architecture for this purpose is shown in Figure 10.11 (Lan et al. 2011). The distinct components of the architecture are as follows:

1. *Reconfigurable input/output ports*: One of the input/output ports is marked as a high-priority (HP) port and the other one as a low-priority (LP) port. Transmission directions for each channel between a pair of routers are determined individually via a channel control protocol.

2. *Channel control module*: This module determines the direction of each channel at runtime, and also sends an arbitration request signal to the switch allocator module. Two finite-state machines (FSMs), shown in Figure 10.12, control the HP and LP ports of adjacent routers. Each FSM consists of the following three states:

    a. *Free state*: The channel is available for data output to adjacent router.

    b. *Idle state*: The channel is ready to input data from the adjacent router.

    c. *Wait state*: An intermediate state prepares the transition from *idle* state to *free* state.

10.6 Summary

In this chapter, various schemes have been discussed for ReNoC design. The reconfiguration can be obtained through topology changes, core attachment patterns for routers, and the link reconfiguration. Local reconfiguration

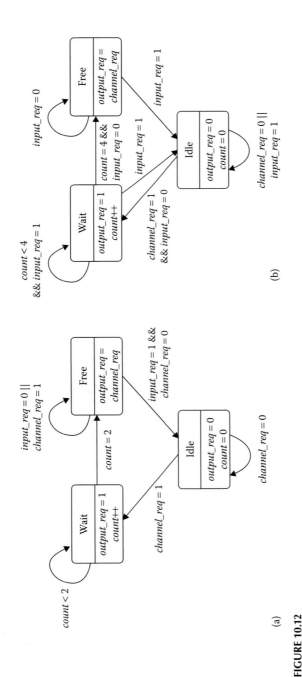

FIGURE 10.12
FSM for HP port (a) and LP port (b). (Adapted from Lan, Y. C., et al., A bidirectional NoC (BiNoC) architecture with dynamic self-reconfigurable channel, *IEEE Transactions on Computer Aided Design of Integrated Circuits and Systems*, pp. 427–440, 2011.)

strategies attempt to relocate cores to the neighboring routers of the one to which it was originally mapped. These strategies provide the advantages of shorter link lengths. However, topology reconfiguration-based approaches may create long links. Suitable arrangement needs to be done to take care of the link delays. The link reconfiguration-based approaches can change the link directions between routers to cater to heavy traffic flow in one direction, whereas in the other direction traffic may be negligible. However, the strategy needs good amount of modification to the router architecture.

References

Ahmad, B., Erdogan, A. T., and Khawam, S. 2006. Architecture of a dynamically reconfigurable NoC for adaptive reconfigurable MPSoC. *NASA/ESA Conference on Adaptive Hardware and Systems*, IEEE, pp. 405–411.

Atienza, D., Angiolini, F., Murali, S., Pullini, A., Benini, L., and De Micheli G. 2008. Network-on-chip design and synthesis outlook, *Integration—The VLSI Journal*, vol. 41, no. 3, pp. 340–359.

Avakian, A., Nafziger, J., Panda, A., and Vemuri, R. 2010. A reconfigurable architecture for multicore systems. *IEEE International Symposium on Parallel & Distributed Processing*, IEEE, pp. 1–8.

Bartic, T. A., Mignolet, J. Y., Nollet, V., Marescaux, T., Verkest, D., Vernalde, S., and Lauwereins, R. 2003. Highly scalable network-on-chip for reconfigurable systems. *International Symposium on System-on-Chip*, IEEE, pp. 79–82.

Bartic, T. A., Mignolet, J. Y., Nollet, V., Marescaux, T., Verkest, D., Vernalde, S., and Lauwereins, R. 2005. Topology adaptive network-on-chip design and implementation. *IEE Proceedings Computers and Digital Techniques*, vol. 152, no. 4, pp. 467–472.

Dally, W. J. and Towles, B. 2001. Route packets, not wires: On-chip interconnection networks, *Design Automation Conference*, IEEE, pp. 684–689.

Ding, H., Gu, H., Li, B., and Du, K. 2012. Configuring algorithm for reconfigurable network-on-chip architecture. *International Conference on Consumer Electronics, Communications and Networks*, IEEE, pp. 222–225.

Dumitriu, V. and Khan, G. N. 2009. Throughput-oriented NoC topology generation and analysis for high performance SoCs, *IEEE Transactions on Very Large Scale Integration (VLSI) Systems*, vol. 17, no. 10, pp. 1433–1446.

Elmiligi, H., El-Kharashi, M. W., and Gebali, F. 2007. Introducing OperaNP: A reconfigurable NoC-based platform. *Canadian Conference on Electrical and Computer Engineering*, IEEE, pp. 940–943.

Kunert, K., Wecksten, M., and Jonsson, M. 2007. Algorithm for the choice of topology in reconfigurable on-chip networks with real time support. *International Conference on Nano-Networks*, ACM.

Lan, Y. C., Lin, H. A., Lo, S. H., Hu, Y. H., and Chen, S. J. 2011. A bidirectional NoC (BiNoC) architecture with dynamic self-reconfigurable channel. *IEEE Transactions on Computer Aided Design of Integrated Circuits and Systems*, IEEE, pp. 427–440.

Modarressi, M. and Sarbazi-Azad, H. 2007. Power-aware mapping for reconfigurable NoC architectures. *International Conference on Computer Design*, IEEE, pp. 417–422.

Modarressi, M., Sarbazi-Azad, H., and Tavakkol, A. 2010. An efficient dynamically reconfigurable on-chip network architecture. *ACM/IEEE Design Automation Conference*, IEEE, pp. 166–169.

Modarressi, M., Tavakkol, A., and Sarbazi-Azad, H. 2011. Application-aware topology reconfiguration for on-chip networks. *IEEE Transactions on Very Large Scale Integration (VLSI) Systems*, vol. 19, pp. 2010–2022.

Rana, V., Atienza, D., Santambrogio, M.D., Sciuto, D., and Micheli, G. De. 2009. A reconfigurable network-on-chip architecture for optimal multi-processor SoC communication, *VLSI-SOC*, IEEE, pp. 1–20.

Soumya, J., Sharma, A., and Chattopadhyay, S. 2013. Multi-application network-on-chip design using global mapping and local reconfiguration. *ACM Transactions on Reconfigurable Technologies and Systems,* vol. 7, no. 2, p. 2014.

Stensgaard, M. B. and Sparso, J. 2008. ReNoC: A network-on-chip architecture with reconfigurable topology. *The 2nd ACM/IEEE International Symposium on Networks-on-Chip*, IEEE, pp. 55–64.

Wu, L. W., Tang, W. X., and Hsu, Y. 2011. A novel architecture and routing algorithm for dynamic reconfigurable network-on-chip. *International Symposium on Parallel and Distributed Processing with Applications*, IEEE, pp. 177–182.

Zheng, L., Jueping, C., Ming, D., Lei, Y., and Zan, L. 2010. Hybrid communication reconfigurable network on chip for MPSoC. *International Conference on Advanced Information Networking and Applications*, IEEE, pp. 356–361.

11

Three-Dimensional Integration of Network-on-Chip

11.1 Introduction

On-chip networks can enhance the communication bandwidth among the individual functional blocks of an integrated system, but at the same time, speed and power consumed by the networks are eventually limited by the delay of the wires connecting the network links. In the upcoming technologies, it is hardly sufficient to support the ever-increasing performance demand, without raising the energy consumption. Since long wire is the major bottleneck in designing a network-on-chip (NoC)-based system in deep submicron (DSM) technology, in terms of performance, speed, and energy consumption, researchers are trying to find out the alternatives to it. Current-mode signaling (Bashirullah et al. 2003; Nigussie et al. 2007), wave pipelined interconnects (Deodhar and Davis 2005), and low-swing signaling (Zhang et al. 2000) are the incremental techniques to address this problem to some extent. Three emerging technologies have been envisioned to address the above issues: three-dimensional (3D) NoC, photonic NoC, and wireless NoC (Carloni et al. 2009).

1. *3D NoC*: After the advent of 3D integrated circuit (IC), as it exhibits higher performance and lesser energy consumption, researchers have amalgamated NoC with 3D IC that gives birth to a new technology, 3D NoC. In 3D IC, multiple silicon layers are integrated vertically in a stack where interlayer distance is taken as tens of microns. Copper through-silicon via (TSV) is the mostly used interconnect between two adjacent silicon layers (Savidis et al. 2010). Copper wire is also used as an intralayer interconnect.

2. *Photonic NoC*: The major challenges in adopting photonic communication in NoC design are flit buffering and header processing (Shacham et al. 2007; Petracca et al. 2008). Moreover, integrating a modulator and detector onto the silicon within a standard complementary metal oxide semiconductor (CMOS) process is also a difficult task (Haurylau et al. 2006). In addition, the detector and

modulator should exhibit performance characteristics that ensure that the optical links outperform the electrical interconnect (Chen et al. 2007). The integration of silicon photonic devices with CMOS ICs for chip-to-chip communication became commercially available (Gunn 2006). This remarkable achievement has paved the path to design a 3D IC-based photonic NoC where the top layer is used for high-bandwidth circuit-switched optical network and the bottom layer is used for low-bandwidth packet-switched electronic network (Ye et al. 2009). The cores are placed at the electronic layer and are connected to the optical layer through TSVs and via electro-optical/optoelectronic interfaces. The header information (source and destination addresses) is routed through the electronic layer to set up the optical path between source and destination. The payload information is transmitted along the reserved optical path at very high speed without buffering, whereas the tailer is used to release the path. It leverages two important advantages of optical communication: (1) the energy dissipation is essentially independent of the bit rate and (2) the energy dissipation is independent of transmission distance. Hence, photonic NoCs can deliver very high bandwidth and offer a low-power communication medium (Carloni et al. 2009). Photonic NoC research is now growing extensively and a number of implementations have already been reported in literature (Vantrease et al. 2008; Cianchetti et al. 2009; Pan et al. 2009). Gu et al. (2009) proposed the design of optical router for photonic NoC.

3. *Wireless NoC*: Another promising alternative to two-dimensional (2D) NoC is the use of radio frequency (RF)/wireless interconnects for signal transmission. Unlike photonic and 3D NoCs, NoC with RF interconnects can be built using existing 2D CMOS technology. But it requires long on-chip transmission lines that serve as wave guides. It achieves an on-chip effective speed of light signal propagation and also saves power consumption (Chang et al. 2008).

This chapter describes the 3D integration of NoC in detail. The rest of the chapter is organized as follows: Section 11.2 describes the pros and cons of 3D integration. Section 11.3 describes the design and evaluation of 3D NoC architecture. Performance and cost evaluation of 3D NoC architecture is performed with self-similar and application-specific traffic and compared with that of 2D NoC counterpart. Finally, Section 11.4 summarizes this chapter.

11.2 3D Integration: Pros and Cons

In the many-core era, integrating large number of cores on a 2D IC has limited floorplanning choices. Although the size of an individual core is reduced up to a certain level due to technology shrinking, chip sizes may increase due to the incorporation of huge number of cores on single silicon die. The 3D IC technology that stacks multiple layers of active silicon using special vertical interconnects, known as *through-silicon vias*, is an attractive solution to this problem. A survey of existing 3D fabrication technologies were carried out by Beyne (2006). The pros and cons of going in vertical direction were investigated by Davis et al. (2005). Three-dimensional IC has higher integration density and smaller form factor. However, due to low thermal conductivity of dielectric materials inserted between two adjacent layers, high-temperature zones will get created, particularly in layers away from heat sink. This will necessitate better cooling arrangement. Dissipated heat at any layer is conducted to the ambient through vertical and horizontal flows. However, larger length of a wafer than its thickness makes vertical heat flow to be dominant. Figure 11.1a shows a cross section of vertically stacked tiles for an *n*-layer 3D IC. Each layer consists of a silicon layer (bottom), an interlayer dielectric (ILD) and Cu layer, and a glue layer. Figure 11.1b depicts the dissipated heat of a tile at layer *i* that flows to layer $(i - 1)$ through silicon at layer *i*, glue at layer $(i - 1)$, and insulator at layer $(i - 1)$. This unequal heat dissipation has negative impacts on system reliability and performance. Over the past few years, 3D IC has evolved into a design paradigm. The salient features and important challenges of 3D integration are briefly reviewed in Sections 11.2.1 and 11.2.2.

11.2.1 Opportunities of 3D Integration

- *Decrease in interconnect length*: A common metric to characterize the longest interconnect is to assume that its length is equal to the summation of length and breadth of the die. Hence, for a 2D square-shaped die of area *A*, the length of the longest interconnect is $L_{2D\ max} = 2\sqrt{A}$. Implementing the same design in *n*-layered 3D IC requires an area of (A/n) in each plane keeping the total area of the system remains same with 2D case. Hence, for an *n*-layered 3D IC, the length of the longest interconnect is $L_{3D\ max} = 2\sqrt{(A/n)}$, considering that each plane is square in shape. The actual benefit of 3D IC relies on the fact that the relatively long wires (approximately in millimeters) of 2D IC can be replaced by the interlayer TSVs whose lengths are about tens of microns. This considerable decrease in interconnect length minimizes the link delay and link energy consumption significantly and at the same time more immunity to noise (Topol et al. 2006; Flic and Bertozzi 2010). Due to increased connectivity, 3D ICs have the potential for enhancing system performance, achieving better functionality, and for producing higher packaging density compared to its traditional 2D counterpart (Davis et al. 2005).

- *Heterogeneous and multifunctional SoC design*: Unlike 2D planer ICs, 3D ICs offer increased system integration either by increasing

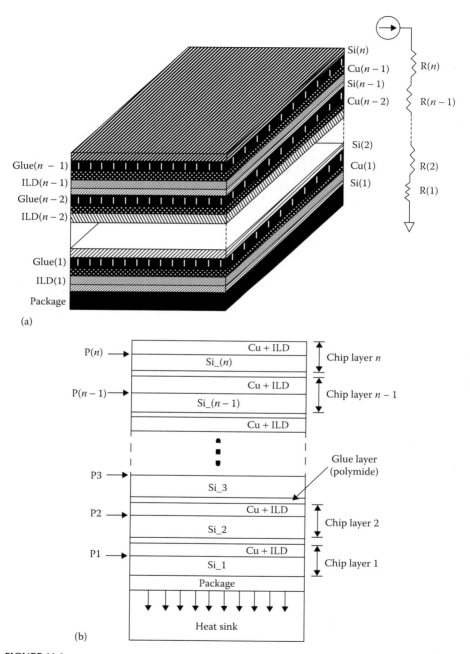

FIGURE 11.1
(a) A cross section of vertically stacked layers for an *n*-layer 3D IC; (b) heat dissipation in an *n*-layer 3D IC.

functionality or by combining different technologies. Currently, SoC solutions limit designers to one fabrication technology for both analog and digital circuits. Usage of 3D ICs allows integrating the best technology for a particular portion of an application into the chip cube (Davis et al. 2005). In a typical 3D SoC, optical devices, analog circuitry, and digital circuitry can be implemented in separate layers. This defining feature of 3D ICs offers unique opportunities for highly heterogeneous and multifunctional systems.

11.2.2 Challenges of 3D Integration

- *Thermal effects*: One of the major concerns in 3D IC design is thermal effects. Although shorter interconnect length causes decrease in power consumption, the power density is more in 3D IC compared to that in 2D IC due to lesser footprint area. As the power density increases, the temperature of those planes not adjacent to the heat sink of the package will rise. Each 10°C increase in operating temperature increases delay by almost 5%. Doubling the heat density without any improvement in cooling capacity will lead to more than 30% degradation in performance (Davis et al. 2005). While performance benefit is the major aspect in 3D IC, performance degradation due to temperature increment is the main bottleneck.

- *Interconnect design*: In 3D IC, due to integration of different fabrication process or disparate technologies in different layers, interconnect design is the major design challenge in 3D IC. In these diverse systems, global interconnect such as clock distribution grows in interest. Figure 11.2 shows different clock distribution structures for 3D IC-based systems. In Figure 11.2a,

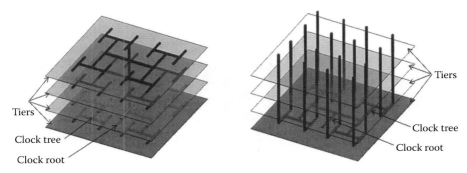

FIGURE 11.2
Clock distribution structures for 3D IC-based systems. (a) H-Tree is at each layer of 3D IC; (b) H-Tree is at ground layer of 3D IC.

H-tree is replicated in each layer of 3D IC, whereas clock root can be at any layer (ground layer in the figure). The clock signal is propagated in each layer through TSV. The impedance of the TSV will cause clock skew between the layers. Furthermore, due to temperature difference, the clock skew between the layers will be more prominent. Figure 11.2b depicts another scenario where H-tree is at the same layer with the clock root (ground layer in the figure). From each leaf of this H-tree, interlayer TSVs are propagated across the layers. In this structure, clock skew due to temperature difference between the layers can be mitigated at the cost of more number of TSVs.

- *Reliability*: The primary failure mechanisms for TSVs are misalignments and random (complete or partial) open defects (Patti 2007). Misalignments are due to imprecise wafer alignment prior to and during wafer bonding (Figure 11.3), which results in shifts of the bonding pads from their nominal positions. Random defects comprise a variety of physical phenomena during, for example, the thermal compression process used in wafer stacking, eventually leading to opens along TSVs (Loi et al. 2011).

- *CAD tools*: CAD algorithm and tool development for 3D IC is another challenge to design 3D NoC-based system. Until now, CAD tools have been mostly the outcome of academic research. Industry contribution to this problem is still in its infancy.

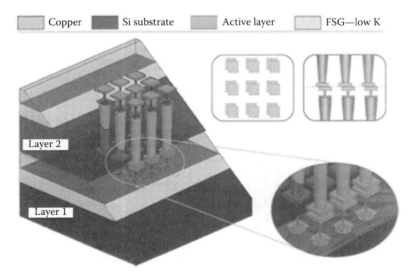

FIGURE 11.3
Cross section of a vertical link (TSV) across two layers and worst-case misalignment scenario. FSG, fluorinated silicate glass.

11.3 Design and Evaluation of 3D NoC Architecture

Similar to 2D NoC, mesh topology is a popular network structure in 3D NoC. Pavlidis and Friedman (2007) showed a different NoC architecture on 3D IC (e.g., 3D IC–2D NoC and 3D IC–3D NoC) as shown in Figure 11.4. In 3D IC–2D NoC (Figure 11.4a), the interconnection network (router) is contained within one physical plane, while each processing element (PE) is integrated in multiple planes. In 3D IC–3D NoC (Figure 11.4b), both the interconnection network and the PEs can span more than one physical plane of the stack. In 3D IC–2D NoC, due to the reduction in horizontal link length, zero-load latency and energy consumption can be lesser than the conventional 2D IC–2D NoC. With 3D IC–3D NoC, greatest saving in both energy and zero-load latency can be achieved, due to reduced average distance and horizontal link length.

Although the nodes in each layer of 3D IC–3D NoC communicate through packet-switched network, due to the small interlayer spacing, Li et al. (2006) suggested the usage of buses for vertical communication having single-hop delay between the layers. Jacob et al. (2005) also showed that employing wide buses for vertical communication in 3D IC is beneficial to improve the overall system performance. Feero and Pande (2009) compared the performance and energy consumption of different types of mesh-based 3D NoC structures, namely, fully connected 3D mesh, stacked mesh, and ciliated 3D mesh (Figure 11.5), by applying self-similar traffic in a cycle-accurate NoC simulator. A fully connected 3D mesh structure (Figure 11.5b) employs a seven-port router. Each node is connected to a single core, four cardinal directions (north, south, east, and west), and vertically adjacent layers (up and down) by using point-to-point unidirectional opposite links. In a stacked mesh (Figure 11.5c), instead of using point-to-point vertical links, bus (32- or 128-bit) is used from each node for interlayer communication. A switch in a stacked mesh network has, at most, six ports: one to the IP, one to the bus, and four for the cardinal directions (Figure 11.5c). In a ciliated 3D mesh (Figure 11.5d), two cores are connected to a single router. Cores residing in different layers are connected to the router via point-to-point links.

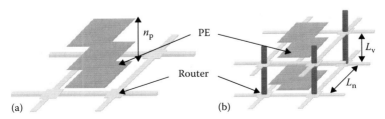

FIGURE 11.4
Various NoC topologies on 3D IC: (a) 3D IC–2D NoC; (b) 3D IC–3D NoC.

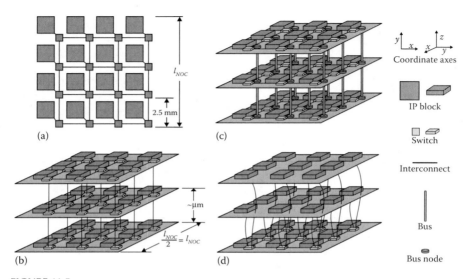

FIGURE 11.5
Mesh-based NoC architectures: (a) 2D mesh; (b) fully connected 3D mesh; (c) stacked mesh; (d) ciliated 3D mesh.

In the ciliated 3D mesh network, each switch contains seven ports (one for each cardinal direction, one either up or down, and one to each of the two IP blocks), as shown in Figure 11.5d.

Feero and Pande (2009) reported that a ciliated 3D mesh structure has slightly higher throughput than a 2D mesh-based NoC, but considerably lesser throughput than that of fully connected 3D mesh and stacked mesh structures. The stacked mesh structure, while employing a 32-bit bus for vertical communication, shows worse performance than the fully connected 3D mesh-based NoC. This performance gap can be diminished by using a 128-bit bus for vertical communication. On the energy front, the ciliated 3D mesh structure consumes the least average energy per packet due to lesser number of links and switches. The stacked mesh structure with 128-bit bus has higher average energy consumption per packet than the fully connected 3D mesh-based NoC. The average energy consumption profile per cycle is the highest in stacked mesh and the least in ciliated 3D mesh. The energy profile of a fully connected 3D mesh structure is almost similar to the 2D NoC implementation. On the area front, due to higher connectivity of the routers, the fully connected 3D mesh-based NoC occupies the largest area among all the 3D mesh-based NoCs. Moreover, on the floorplanning aspect, minimizing the number of TSVs between two adjacent layers reduces the fabrication cost and silicon area (Pavlidis and Friedman 2009). In a four-layered ciliated 3D mesh structure, TSVs connect between *layer 1* and *layer 2* and *layer 1* and *layer 3* through *layer 2*. Thus, between layer 1 and layer 2, the number of TSVs is large, which may cause floorplanning problems. To reduce the number of interlayer vias, usage of a combination of 2D and 3D mesh-based network routers was proposed by Pavlidis and Friedman (2009). Depending on the position of vertical interconnection links, a number of mesh-based heterogeneous 3D NoC structures were developed. The performance, area requirement, and energy consumption of these heterogeneous networks were evaluated and compared with those of the fully connected 3D mesh-based NoC.

Most of the research works in 3D NoC are based on mesh topology. There are very few works reported on tree topologies. In any tree-based network, the length of the interconnection link increases toward the root of the tree, whereas the mesh structure has a uniform wire length. In a torus network, the length of the end-around connection increases with increasing network size. Therefore, torus and tree-based topologies may not be a good choice for NoC designers while attempting large number of cores in a 2D IC. In 3D IC implementation, due to the shorter TSVs, the intrinsic problem of having long interconnection wires in tree-based topologies gets significantly resolved. Matsutani et al. (2008) instantiated a number of existing tree-based topologies in 3D platform and showed the benefit of energy reduction over their 2D implementation. Similar to Matsutani et al. (2008), instead of proposing any new 3D tree-based topology, Feero and Pande (2009) also instantiated the already existing butterfly fat tree (BFT) and the fat tree structures in 3D platform to show the energy reduction over their 2D implementations. However, mesh-of-tree (MoT) topology in 3D context is not included in any of the existing studies. This chapter proposes an extension of MoT topology for the 3D environment and carries out performance and cost benefits of the proposed 3D structure over its 2D counterpart. Detailed performance evaluation, energy consumption, and area estimation have been carried out for the proposed structure and compared with BFT and two variants of mesh networks for equal number of cores in 3D NoC context. The salient contributions of this chapter are as follows:

1. A new 3D MoT topology has been proposed. Expressions for the number of directed edges and the average distance in an $M \times N \times Z$ MoT have been formulated.

2. Performance and cost of proposed MoT-based 3D NoC have been evaluated under self-similar traffic. The results have been compared with BFT and two variants of mesh networks, having the same number of cores, in 3D NoC context. Simulation results show MoT's applicability as communication infrastructure design of 3D NoC.

3. Performance and cost benefits of all 3D NoC structures have been shown over their 2D counterparts.

4. The MoT-based 3D NoC also works fine under real benchmark applications such as dual video object plane decoder (DVOPD) having 32 cores.

11.3.1 3D Mesh-of-Tree Topology

A 2D $M \times N$ MoT topology can be extended to a 3D structure by connecting multiple $M \times N$ 2D MoTs via $M \times N$ number of vertical trees. The number of *leaf* nodes in each vertical tree of an $M \times N \times Z$ MoT is Z, where Z is the number of $M \times N$ 2D MoTs. Figure 11.6 shows a $2 \times 4 \times 4$ MoT structure having four layers of 2×4 2D MoTs, each of which has two row trees of depth 2 and four column trees of depth 1. For each row tree, RS and RR denote the *row stem* and *row root* nodes, respectively, whereas for each column tree, CR denotes a *column root* node. The ZS and ZR are the *stem* and *root* nodes in the vertical trees, respectively. The *leaf* nodes (L) are common to all the three types of trees. Two cores (not shown in Figure 11.6) are attached with each *leaf* node. The *stem* (RS and ZS) and *root* (RR, CR, and ZR) nodes are not having any core attached to them. Thus, in an $M \times N \times Z$ MoT, $2 \times (M \times N \times Z)$ number of cores can be attached. In general, an $M \times N \times Z$ MoT has the following properties:

1. Number of nodes = $4 \times (M \times N \times Z)-(M \times Z + N \times Z + M \times N)$
2. Diameter = $2 \lfloor (\log_2 M) \rfloor + 2 \lfloor (\log_2 N) \rfloor + 2 \lfloor (\log_2 Z) \rfloor$
3. Bisection width = $\min(M \times Z, N \times Z, M \times N)$
4. Symmetric and recursive structure

In Sections 11.3.1.1 and 11.3.1.2, a general formulation of the number of directed edges and the average distance for an $M \times N \times Z$ MoT is presented where all the trees are complete binary trees.

11.3.1.1 Number of Directed Edges

The number of undirected edges in any complete binary tree having k leaf nodes is $(2k - 2)$. In an $M \times N \times Z$ MoT structure, the number of *leaf* nodes in each row tree, column tree, and vertical tree are N, M, and Z, respectively. Thus, the number of edges ($\acute{E}$) of an undirected $M \times N \times Z$ MoT graph can be formulated as

$$\acute{E} = \left[M \times Z \times (2N - 2) + N \times Z \times (2M - 2) + M \times N \times (2Z - 2) \right]$$

In the proposed 3D MoT structure, adjacent nodes are connected by two unidirectional opposite edges. Thus, the number of directed edges can be written as

$$E = 2\acute{E} = 12 \times M \times N \times Z - 4(M \times Z + N \times Z + M \times N) \qquad (11.1)$$

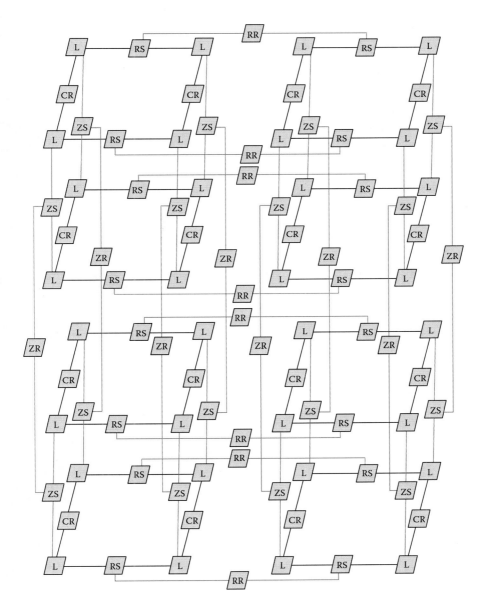

FIGURE 11.6
A 2 × 4 × 4 MoT topology.

11.3.1.2 Average Distance

In a complete binary tree having N leaf nodes, where two cores are connected to each leaf, distribution of destination cores from a specific source core is as follows: single core at distance 0, two cores at distance 2, four cores at distance 4, eight cores at distance 6, and so on as depicted in Chapter 2. The summation of minimum distances to all the destination cores from a specific source core in a complete binary tree, noted in Equation 2.4, is reproduced in Equation 11.2 for the sake of continuity.

$$S_b = [4N\log_2 N - 4(N-1)] \tag{11.2}$$

A $(2^1 \times N \times 2^0)$ MoT consists of two row-wise binary trees of depth $\log_2 N$ and N column-wise binary trees of depth 1 (which can be written as $\log_2 2^1$). Each row tree consists of $(2^1 \times N)$ cores. Thus, the summation of minimum distances of $(2^1 \times N)$ cores lying in the second row tree from any specific core of the first row tree can be written as $\left[2^0 \times S_b + (2^1 \times N) \times (2\log_2 2^1)\right]$. Hence, the summation of minimum distances to all the destination cores from a specific source core in a $(2^1 \times N \times 2^0)$ MoT is

$$S_M(2^1 \times N \times 2^0) = S_b + 2^0 \times S_b + (2^1 \times N) \times (2\log_2 2^1) \tag{11.3}$$

In the same way, a $(2^2 \times N \times 2^0)$ MoT consists of two $(2^1 \times N \times 2^0)$ MoTs where each $(2^1 \times N \times 2^0)$ MoT contains 2^1 row-wise binary trees. The depth of each column tree of $(2^2 \times N \times 2^0)$ MoT is 2 (which can be written as $\log_2 2^2$). Thus, the summation of minimum distances of $(2^1 \times N)$ cores lying in the second $(2^1 \times N \times 2^0)$ MoT from any specific core of the first $(2^1 \times N \times 2^0)$ MoT is equal to $\left[2^1 \times S_b + (2^2 \times N) \times (2\log_2 2^2)\right]$. Thus, the summation of minimum distances to all the destination cores from a specific source core in $(2^2 \times N \times 2^0)$ MoT is as follows:

$$S_M(2^2 \times N \times 2^0) = S_M(2^1 \times N \times 2^0) + 2^1 \times S_b + (2^2 \times N) \times (2\log_2 2^2) \tag{11.4}$$

In general, a $(2^{\log_2 M} \times N \times 2^0)$ MoT can be split into two $(2^{(\log_2 M)-1} \times N \times 2^0)$ MoTs, where each $(2^{(\log_2 M)-1} \times N \times 2^0)$ MoT consists of $2^{(\log_2 M)-1}$ row-wise binary trees. The depth of each column tree of $(2^{\log_2 M} \times N \times 2^0)$ MoT is $\log_2(2^{\log_2 M})$. Thus, the summation of minimum distances of $(2^{(\log_2 M)} \times N)$ cores lying in the second $(2^{(\log_2 M)-1} \times N \times 2^0)$ MoT from any specific core of the first $(2^{(\log_2 M)-1} \times N \times 2^0)$ MoT is equal to $\{2^{(\log_2 M)-1} \times S_b + (2^{\log_2 M} \times N) \times [2\log_2(2^{\log_2 M})]\}$. Thus for a $(2^{\log_2 M} \times N \times 2^0)$ MoT, the summation of minimum distances to all the destination cores from a specific source core can be written as

$$S_M(2^{\log_2 M} \times N \times 2^0) = S_M(2^{(\log_2 M)-1} \times N \times 2^0)$$
$$+ \left\{2^{(\log_2 M)-1} \times S_b + (2^{\log_2 M} \times N) \times \left[2\log_2(2^{\log_2 M})\right]\right\} \tag{11.5}$$

After simplification, the above equation becomes

$$S_M(M \times N \times 2^0) = \left[4 \times M \times N \times \log_2(M \times N) - 8 \times M \times N + 4(M + N)\right] \tag{11.6}$$

While considering the third dimension, a $(M \times N \times 2^1)$ MoT can be split into two $(M \times N \times 2^0)$ MoTs, where each $(M \times N \times 2^0)$ MoT is connected by a vertical tree of depth 1. Thus, for an $(M \times N \times 2^1)$ MoT, the summation of minimum distances to all the destination cores from a specific source core can be written as

$$S_M(M \times N \times 2^1) = S_M(M \times N \times 2^0) + 2^0 \times S_M(M \times N \times 2^0)$$
$$+ 2^1 \times M \times N \times (2\log_2 2^1) \tag{11.7}$$

In the similar fashion, an $(M \times N \times 2^{\log_2 Z})$ MoT can be split into two $(M \times N \times 2^{(\log_2 Z)-1})$ MoTs. The summation of minimum distances to all the destination cores of an $(M \times N \times 2^{\log_2 Z})$ MoT from a specific source core can be written as

$$S_M(M \times N \times 2^{\log_2 Z}) = S_M(M \times N \times 2^{(\log_2 Z)-1})$$
$$+ \left[\begin{array}{c} 2^{(\log_2 Z)-1} \times S_M(M \times N \times 2^0) + \\ 2^{\log_2 Z} \times M \times N \times (2\log_2 2^{\log_2 Z}) \end{array} \right] \tag{11.8}$$

After simplification, the above equation becomes

$$S_M(M \times N \times Z) = 4M \times N \times Z \times \left[\log_2(M \times N \times Z) - 3 \right]$$
$$+ 4(M \times Z + N \times Z + M \times N) \tag{11.9}$$

Due to the symmetric structure of 3D MoT topology, the summation of minimum distances to all the destination cores from any source core is always same. Hence, the average distance of an $M \times N \times Z$ MoT network connecting C number of cores can be written as

$$D_M = \frac{\left[\begin{array}{c} 4M \times N \times Z \times \left[\log_2(M \times N \times Z) - 3 \right] \\ + 4(M \times Z + N \times Z + M \times N) \end{array} \right]}{C - 1} \tag{11.10}$$

In general, for an $M \times N \times \lceil C/(2 \times M \times N) \rceil$ MoT network (C being the total number of cores attached) having two cores connected to each *leaf* node, the number of directed edges (Equation 11.1) and the average distance (Equation 11.10) can be written, respectively, as

$$E_M = 12M \times N \times \lceil C/(2MN) \rceil - 4 \left(MN + M \times \lceil C/2MN \rceil + N \times \lceil C/2MN \rceil \right)$$

$$D_M = \frac{\left[\begin{array}{c} 4MN \times \lceil C/2MN \rceil \times \left(\log_2 \left(MN \times \lceil C/2MN \rceil \right) - 3 \right) \\ + 4 \left(MN + M \times \lceil C/2MN \rceil + N \times \lceil C/2MN \rceil \right) \end{array} \right]}{(C - 1)}$$

As E/D is a good indicator for the throughput of a network without considering contentions between packets, it can be shown by taking partial derivatives that the value of (E_M/D_M) reaches its maximum and D_M reaches its minimum when the condition $M = N = \lceil C/2\,MN \rceil$ holds. This implies that 3D MoT network will show maximum throughput and minimum latency in a congestion-free environment when the number of row trees, that of column trees, and that vertical trees are same. To work with the proposed 3D MoT topology, an addressing scheme and a deterministic routing algorithm is presented in Section 11.3.1.2.1.

11.3.1.2.1 *Addressing Scheme and Routing Algorithm*

The addressing scheme for each individual node of a 2D $M \times N$ MoT has been described in Chapter 2, where the address of each node consists of four fields: *row number* (RN), *column level* (CL), *column number* (CN), and *row level* (RL). The same scheme has been extended to address every individual node of a 3D $M \times N \times Z$ MoT with an additional field, *layer number* (LN). In the $M \times N \times Z$ MoT with Z number of layers in stack, each layer consists of a 2D $M \times N$ MoT. The number of bits required for addressing each core of an $M \times N \times Z$ MoT is shown in Table 11.1. For example, in a $4 \times 4 \times 4$ MoT, each core needs a 15-bit address.

In 3D NoC, similar to 2D NoC, every two adjacent routers are connected to each other via two unidirectional opposite links, each one with its own data, framing, and flow control signals. Message passing communication is followed by wormhole switching approach, where messages are sent by means of packets, which are further decomposed into *flits* (flow control units). A flit can be classified as header, payload, tailer, and invalid flit. Header flit carries information about the source and destination addresses, whereas payload and tailer flits contain the actual data.

A deadlock- and livelock-free dimension order routing algorithm for 2D $M \times N$ MoT has been proposed in Chapter 2. Routing decision is taken by *leaf* and *stem* routers, whereas each *root* router is replaced by a first-in first-out (FIFO). In 3D MoT, from any source, a packet will first traverse through the vertical tree to reach a *leaf* node whose layer number is same as that of destination core. After reaching that layer, the packet will traverse through the column tree to reach a *leaf* node whose RN is the same as that of the destination node. After matching the RN, the packet will traverse through the row tree to reach a *leaf* node whose CN is the same as that of the destination node. The packet will next go to the destination core depending on the *Core-ID* bit. To implement the routing scheme of the proposed 3D MoT structure in hardware, wormhole router has been designed as described in Chapter 3.

TABLE 11.1

Bits Required for Addressing a Core in $M \times N \times Z$ MoT

Core-ID	LN	RN	CL	CN	RL
1	$\lceil \log_2 Z \rceil$	$\lceil 2(\log_2 M) \rceil$	$\lceil \log_2(\log_2 2M) \rceil$	$\lceil 2(\log_2 N) \rceil$	$\lceil \log_2(\log_2 2N) \rceil$

11.3.2 Performance and Cost Evaluation

For evaluating the performance of 3D NoC, a SystemC-based cycle-accurate simulator has been developed as described in Chapter 4. In this work, similar to Chapter 4, each core is inserted in a tile of dimension 2.5 mm × 2.5 mm. Due to increased burden of placing multilayer IPs in a limited number of layers, as depicted in the work of Feero and Pande (2009), this work assumes each core to be placed in a single silicon layer. A thorough evaluation of performance of 3D MoT structure along with its energy consumption and area overhead is performed. The results are compared with 3D BFT and 3D mesh-based networks having the same number of cores. Here, performance and cost of all the networks are compared for a 32-core-based system. For a fair comparison with 3D MoT having two cores at each *leaf* level router, another variation of 3D mesh topology, connecting two cores to each router, is also included in this comparative study. To reduce the number of interlayer vias and to simplify the floorplanning problem, unlike ciliated 3D mesh structure, two cores are placed in a single layer. In this work, it has been considered that single link traversal of length approximately 2.5 mm can be completed in a single clock cycle. The length of the core-to-router link is taken to be 1.25 mm. Similar to Chapter 4, Figures 11.7 through 11.10 show the possible distributions of cores, routers, and links for chip area estimation. These diagrams enable us to compare the area overheads of alternate NoC topologies under consideration.

In general, in 3D mesh structure having a single core attached with each router, the middle layers have three types of routers: (1) *center* having node degree 7, (2) *edge* having node degree 6, and (3) *corner* having node degree 5. For a mesh structure with 32 cores, a probable distribution of cores, routers, and links of a 2 × 4 × 4 network is shown in Figure 11.7 with bisection width 8. Depending on the connectivity, the middle layer of this network has two types of routers: (1) *edge* having node degree 6 and (2) *corner* having node degree 5. The length of the links between the two rows and the last two columns in each layer is taken as tens of micron, whereas the rest of the inter-router links in a single layer is 2.5 mm long, as shown in Figure 11.7. The length of the vertical links between two adjacent layers is taken to be 20 µm, as in the work of Feero and Pande (2009).

In the second variant of mesh network, having two cores connected with each router, the middle layer consists of three types of routers: (1) *center* having node degree 8, (2) *edge* having node degree 7, and (3) *corner* having node degree 6. For a 32-core-based system, a similar distribution of cores, routers, and links of such 2 × 2 × 4 mesh architecture is shown in Figure 11.8. The middle layer of this network has only one type of router with node degree 6. The number of routers required and the bisection width of this network are half of those for the first variant of 3D mesh network. As wire delay increases exponentially with its length, the links having more than 2.5 mm length are pipelined. The registers used for pipelining are shown as small white nodes in Figure 11.8. In this structure, the links between two adjacent rows in a

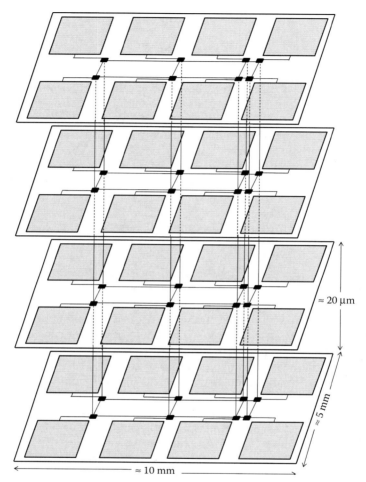

$\approx 20\ \mu m$

$\approx 5\ mm$

$\approx 10\ mm$

FIGURE 11.7
A 3D $2 \times 4 \times 4$ Mesh-1 network.

layer are few micrometers long, whereas between the two adjacent layers, the links are 20 μm long. Here, mesh topology having a single core attached to each router is termed as Mesh-1 network, whereas that having two cores attached to each router is termed as Mesh-2 network.

A BFT-based NoC with four cores attached in each *leaf* level router is shown in Figure 11.9. The BFT-based network connecting 32 cores also has three types of routers: (1) *leaf* having node degree 6, (2) *stem* having node degree 6, and (3) *root* having node degree 2. The routing decisions are taken by *leaf* and *stem* routers, whereas the *root* router is replaced by a FIFO and is used to pipeline the links. In 2D NoC implementation, as shown in Chapter 4, the length of the longest wire of the BFT-based network is $\min(l_1, l_2)/2$, where l_1 and l_2 are

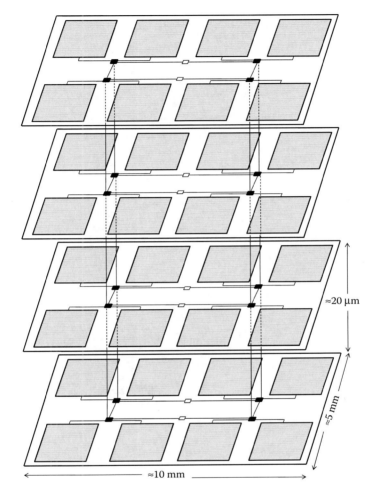

FIGURE 11.8
A 3D 2 × 2 × 4 Mesh-2 network.

the length and breadth of the chip, respectively. Hence, the links need to be pipelined to restrict its delay within the clock cycle budget. In a 3D environment, when the same BFT network is mapped onto a four-layer 3D IC, the longest interswitch wire length is reduced by a factor of 4, approximately. This reduced wire length leads to the reduction in the number of pipelining stages and also reduces the link energy consumption. As the interlayer distance is very less, there is no need to pipeline the interlayer links. Hence, the *root* router (shown as dotted node in Figure 11.9) is bypassed.

A probable distribution of cores, routers, and links of the proposed MoT-based 3D NoC with two cores attached to each *leaf* level router is shown in Figure 11.10. This network also has three types of routers: (1) *leaf* having

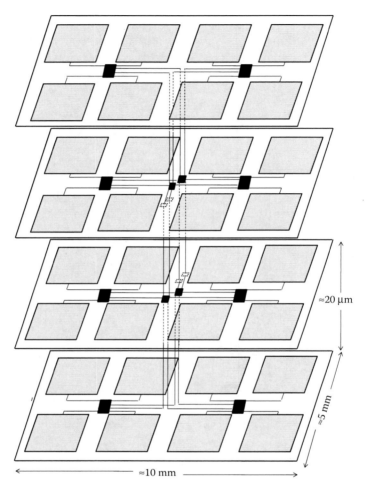

FIGURE 11.9
A 3D BFT network.

node degree 5, (2) *stem* having node degree 3, and (3) *root* having node degree 2. The routing decisions are taken at *leaf* and *stem* routers, whereas the *root* routers are replaced by FIFOs and are used to pipeline the links. In 2D NoC implementation, as shown in Chapter 4, the longest edge of MoT topology is the connection between a *stem* and its corresponding *root* router. The length can be estimated as $\max(l_1, l_2)/4$, where l_1 and l_2 are the length and breadth of the chip, respectively.

As the wire delay increases with its length, it is essential to pipeline the links after a certain length, such that its delay does not fall into the critical path of the design. When the proposed MoT network is mapped onto a four-layer 3D IC, the longest interswitch wire is reduced by about a factor

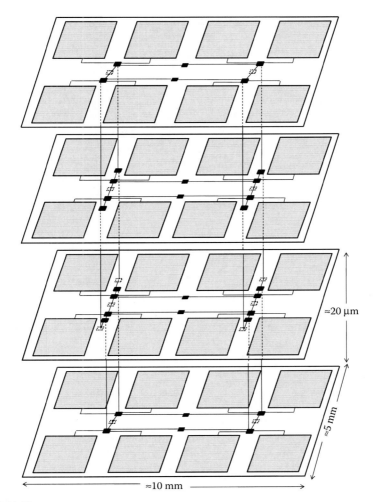

FIGURE 11.10
A 3D 2 × 2 × 4 MoT network.

of 2 compared to its 2D structure. It can be observed from Figure 11.10 that the links between *leaf* and *root* routers of the column tree in each layer are tens of micrometers long. The length of the interlayer links is 20 μm. Hence, like BFT, the *root* nodes of the column trees and the vertical trees (shown as dotted node in Figure 11.10) of the proposed MoT-based 3D NoC is bypassed. Here, it is customary to say that for larger number of cores in each layer, mesh structures are planer, whereas BFT and MoT networks require two metal layers to route the links. Moreover, the wire density of mesh networks is more uniform across any given cross section than that of BFT and MoT.

However, the TSV process does not scale with the CMOS technology. TSV diameters and pitches are 2–3 times bigger than transistor gate lengths. This implies that, even moving to newer technologies, the intrinsic cost for vertical interconnect does not change. For this geometry and sizing, the TSV inductance and inductive coupling becomes negligible and the intrinsic TSV delay can be assumed as a function of resistance and capacitance only.

Routers for all the above-mentioned networks have been designed in Verilog HDL and synthesized using *Synopsys Design Vision* supporting 90 nm technology. For a specific network, critical path delay of a router increases as the routing logic and arbitration complexity increase with increasing connectivity. Hence, in a network, the router with the highest connectivity has the minimum frequency. To support mesochronous clocking, the clock with the minimum frequency is applied to all the routers of a network. Table 11.2 depicts the clock frequencies of the routers used in the middle layers of the 3D NoC structures under consideration. Due to their least connectivity requirements, MoT routers can run at a higher frequency compared to others. However, in this work, to provide a consistent comparison with other networks, all the routers are driven at 1.5 GHz clock. The worst-case link delay (as shown in Chapter 4) is much lesser than the router clock period of 666 ps. The delay of interlayer vias and the links having a length of tens of microns is also very less. Hence, those links do not come into the critical path of the overall NoC.

11.3.2.1 Network Area Estimation

Table 11.2 contains the silicon area required by each type of router from its gate-level netlist to implement the middle layer of different four-layered 3D NoCs by taking 32-bit flit size. For a fair comparison of the networks, this section revisits all the four-layered networks taken here into consideration. Although the dimension of each tile is taken as a square of side 2.5 mm, inter-tile spacing in each layer varies significantly with underlying topology due to varying sizes of routers, which in turn causes variations in layer dimension. For larger network dimension in each layer, the number of links running through inter-tile spaces of a specific layer varies in different networks. For larger number of cores in a single layer, while mesh structure (both Mesh-1 and Mesh-2) has uniform wiring density, MoT and BFT have nonuniform wire densities in each layer and use flyover links over the top of another router as shown in Chapter 4. Thus, to compare the dimension of the middle layer in different topologies, this work has taken uniform channel width of 32 bits for all the networks. The width of each wire and inter-wire spacing are taken to be the same and equal to 0.25 μm, as mentioned earlier. The dimension of each router is assumed to be a perfect square.

The dimensions of the middle layer of a mesh-based network can be estimated as follows: The routers in each row are placed between two

TABLE 11.2

Router Connectivity, Silicon Area, and Frequency in the Middle Layer of Four-Layered 3D NoC with Different Types of Networks under Consideration

	Type 1 Router			Type 2 Router				Type 3 Router			
Networks	Position	Connectivity	Area (mm^2)	Frequency (GHz)	Position	Connectivity	Area (mm^2)	Frequency (GHz)	Position	Connectivity	Area (mm^2)	Frequency (GHz)
Mesh-1	Center	7	0.132	1.52	Edge	6	0.115	1.55	Corner	5	0.079	1.6
Mesh-2	Center	8	0.15	1.50	Edge	7	0.132	1.52	Corner	6	0.115	1.55
BFT	Leaf	6	0.115	1.52	Stem	6	0.115	1.52	Root[a]	4	0.06	1.66
MoT	Leaf	5	0.072	1.60	Stem	3	0.041	1.70	Root	2	0.022	1.90

[a] In BFT network with 2^n cores, the connectivity of root routers is 4 when n is even and is 2 when n is odd.

row-wise adjacent tiles such that the breadth of the layer increases just because of channel width. The length of the layer will increase by the length of the routers that occupy maximum area in each column. Thus, in a 3D NoC having eight cores in each layer, the dimension of the middle layer of $2 \times 4 \times 4$ Mesh-1 network considering unidirectional opposite links gets incremented from 10 mm $\times$ 5 mm to 11.5 mm $\times$ 5.064 mm, and that of $2 \times 2 \times 4$ Mesh-2 network becomes 10.678 mm $\times$ 5.064 mm. For a BFT-based network, the length of the middle layer increases due to *leaf*, *stem*, and *root* routers, whereas its breadth increases just because of channel width. The dimension of the middle layer of a four-layered BFT network having eight cores in each layer becomes 11.017 mm $\times$ 5.064 mm. For a 3D MoT-based network, all routers and repeaters of a row tree in each layer are also placed between row-wise adjacent tiles such that they do not increase the breadth of the layer. For a $2 \times 2 \times 4$ MoT network having eight cores in each layer, the dimension of the middle layer becomes 10.684 mm $\times$ 5.064 mm. For larger number of cores in each layer, it can be noted that only the *stem* routers of the column trees of each 2×2 MoT subnetwork will increase the breadth of the layer. The length of the layer will be increased due to the routers and repeaters of the row tree. Taking all these factors into account, the dimension of the middle layer of a $4 \times 4 \times 4$ MoT network having 32 cores in each layer increases from 20 mm $\times$ 10 mm to 21.624 mm $\times$ 10.404 mm. Assuming each router to be a perfect square, the length of each side of the *stem* router is found to be 200 µm which is almost 6 times wider than the cross section of two opposite unidirectional 32-bit links. Thus, unlike mesh and BFT networks, MoT network connects up to 16 cores in a single row tree in each layer, and its channel width will not increase the breadth of the layer. In the same way, it can be shown that the dimension of the middle layer of a $4 \times 8 \times 4$ Mesh-1 and a $4 \times 4 \times 4$ Mesh-2 (both having 32 cores in each layer) network becomes 22.856 mm $\times$ 10.128 mm and 21.5 mm $\times$ 10.128 mm, respectively. For a four-layered BFT based network, the dimension of the middle layer having 32 cores becomes 22.33 mm $\times$ 10.128 mm. For a 1024-core system where 256 cores are residing in each layer, the area of $16 \times 16 \times 4$ Mesh-1, $16 \times 8 \times 4$ Mesh-2, $16 \times 8 \times 4$ MoT, and BFT networks are incremented from 40 mm $\times$ 40 mm to 45.76 mm $\times$ 40.512 mm, 43.048 mm $\times$ 40.512 mm, 43.504 mm $\times$ 41.616 mm, and 46.38 mm $\times$ 40.768 mm, respectively.

Table 11.3 depicts the area overhead of underlying networks for 32-, 128-, and 1024-core-based systems where the number of cores residing in each layer is 8, 32, and 256, respectively. It may be noted that in all cases the area occupied by the MoT network is lesser than that of Mesh-1 and BFT-based networks, but higher than that of Mesh-2 network. Although there exists a possibility of trading-off this additional area for energy/performance benefits, in this work, this avenue has not been explored as it goes deep into the physical design issues of systems involving these NoC topologies.

TABLE 11.3

Area Overhead of the Middle Layer of Different Four-Layered 3D NoCs Having 8-Core, 32-Core, and 256-Core in Each Layer

	8 Cores/Layer		32 Cores/Layer		256 Cores/Layer	
Networks	Area Required (mm²)	Overhead (%)	Area Required (mm²)	Overhead (%)	Area Required (mm²)	Overhead (%)
Mesh-1	58.24	16.47	231.49	15.75	1853.83	15.86
Mesh-2	54.07	8.14	217.75	8.88	1743.96	8.99
BFT	55.79	11.58	226.16	13.08	1890.82	18.17
MoT	54.10	8.20	224.98	12.49	1810.46	13.15

11.3.2.2 Network Aspect Ratio

Besides channel width and flit size, the network aspect ratio has also an important role in determining the overall performance and cost of NoC, as described in Chapter 4. In general, for an $M \times N \times C/(M \times N)$ Mesh-1 network (where M and N being the number of nodes in each row and column, respectively, and C being the total number of cores attached), the average distance (D) and the number of directed edges (E) can be written respectively, as (Pavlidis and Friedman 2007)

$$D = \frac{M \times N \times \left(\lceil C/MN \rceil\right) \times \left[M + N + \left(\lceil C/MN \rceil\right)\right] - \left(\lceil C/MN \rceil\right) \times (M+N) - MN}{3\left[M \times N \times \left(\lceil C/MN \rceil\right) - 1\right]}$$

$$E = 2 \times \left\{ \begin{array}{c} M \times N \times \left(\lceil C/MN \rceil - 1\right) + N \times \lceil C/MN \rceil \times (M-1) \\ + M \times \lceil C/MN \rceil \times (N-1) \end{array} \right\}$$

It can be shown that the value of E/D reaches its maximum and the value of D reaches its minimum when the condition $M = N = (C/M \times N)$ is held. This signifies that a cubic 3D mesh network with equal number of rows, columns, and vertical layers will show better throughput and lesser latency than a cuboidal structure having the same area. This statement is also true for a Mesh-2 network.

Table 11.4 shows the different topological parameters such as diameter, average distance in hops (D), and number of directed edges (E) of all the four-layered 3D NoCs under consideration, having eight cores in each layer. It also compares with their 2D networks for connecting 32 cores.

In the simulation, the packet length is fixed to 64 flits, as in the work of Pande et al. (2005). The packet injection is continued for the entire simulation time of 200,000 cycles of the routers' clock including 10,000 cycles to make the network stable from the initial transient effects. The following section

TABLE 11.4

Topological Parameters of Different Networks with 32 Cores

Networks	Number of Edges (E)		Average Distance (D)		E/D		Diameter	
	2D	3D	2D	3D	2D	3D	2D	3D
Mesh-1	104	128	4.00	3.10	26.00	41.33	10	7
Mesh-2	48	56	2.65	2.40	18.15	23.33	6	5
BFT	40	40	2.84	2.84	14.08	14.08	4	4
MoT	96	112	5.16	4.65	18.60	24.11	8	8

compares the performance and cost of the proposed 3D MoT-based network with other network topologies under consideration. For deterministic routing in 3D mesh networks, ZXY routing is adopted, whereas a least common ancestor (LCA) routing (Pande et al. 2003) is used for BFT-based networks.

11.3.3 Simulation Results with Self-Similar Traffic

11.3.3.1 Accepted Traffic versus Offered Load

The accepted traffic depends on the rate at which the cores inject traffic into the network as discussed in Chapter 4. Figure 11.11 compares the throughput of all the 3D networks under consideration, each with 32 cores, by applying a uniformly distributed self-similar traffic. For determining network throughput, besides E and D, the network bisection width has also an important role to play. A network with higher bisection width is expected to perform better. The bisection width of a 3D $2 \times 4 \times 4$ Mesh-1 network is 8, whereas for other 3D networks under consideration, the value is 4. Table 11.5 shows that the value of E/D is the highest in 3D Mesh-1 network and the least in BFT network. In the proposed 3D MoT network, after bypassing the *root* of the column trees and vertical trees of $2 \times 2 \times 4$ network as shown in Figure 11.10, the values of E, D, and E/D become 88, 3.61 and 24.37, respectively. However, the value of E/D for a $2 \times 2 \times 4$ Mesh-2 network is 23.33. Therefore, in a contention-free environment, the throughput of 3D Mesh-1 is expected to be the highest and that of 3D BFT be the least, whereas the throughput of 3D MoT network is higher than that of 3D Mesh-2 network. In the simulation, similar responses have been observed for all 3D networks under consideration by applying a uniformly distributed traffic, as shown in Figure 11.11.

Next, we will show the throughput gains of the 3D networks over their 2D counterparts. For 2D structures, the dimensions of the networks are taken to be 4×8 for Mesh-1, 4×4 for Mesh-2, and 4×4 for MoT. 2D BFT and 2D MoT have been shown in Chapter 4. The bisection width of all 2D networks under consideration is 4. Due to higher E/D value, 3D Mesh-1, 3D Mesh-2, and 3D MoT networks are expected to show better throughput than their

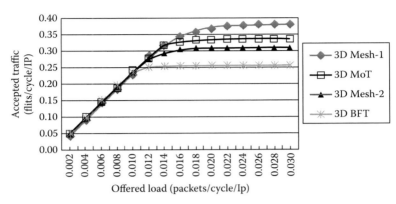

FIGURE 11.11
Accepted traffic with uniformly distributed offered load in different 3D networks under consideration.

TABLE 11.5

Throughput Variation with Locality Factor in 2D and 3D Networks

	BFT		Mesh-1		Mesh-2		MoT	
Locality Factor	2D	3D	2D	3D	2D	3D	2D	3D
0.0	0.25	0.25	0.26	0.38	0.27	0.31	0.29	0.34
0.3	0.27	0.27	0.30	0.42	0.31	0.37	0.36	0.4
0.5	0.34	0.34	0.35	0.45	0.35	0.42	0.40	0.45
0.8	0.43	0.43	0.44	0.54	0.61	0.64	0.65	0.66

2D counterparts. Table 11.5 shows a comparison of throughputs of different 2D and 3D networks with varying locality factor. For BFT network, although the same topology is mapped onto four-layered 3D IC, the number of pipe-lined registers gets reduced as discussed earlier. Moreover, the *root* routers are bypassed as shown in Figure 11.9. Therefore, the values of E, D, and E/D of the optimized 3D BFT network become 32, 2.32, and 13.79, respectively. As the values of E/D in 2D and 3D networks are very close to each other, in actual traffic condition, their throughput values are almost identical.

11.3.3.2 Throughput versus Locality Factor

The effect of traffic spatial localization on network throughput is shown in Table 11.5. It can be observed that localization of traffic has a significant impact in all the 3D networks as it enhances the network throughput. As the locality factor increases, more traffic are directed toward their local clusters, thus traversing lesser hops, which in turn increases throughput.

In BFT, localized traffic is constrained within a cluster consisting of a single subtree having four cores. It can be observed that the throughput of BFT-based networks is the least among all the networks under uniformly distributed traffic. For localized traffic, although its throughput increases with increasing locality factor, it has the minimum value compared to other networks. This is due to the fact that BFT-based networks are more congested as there are three destination cores in the local cluster and lesser number of edges.

In Mesh-1 network, localized traffic is constrained within the destinations placed at the shortest Manhattan distance, whereas Mesh-2 network enjoys the advantage of having only a single core in its local clusters. In case of highly localized traffic, the benefit of connecting two cores in each router becomes clearly visible, as depicted in Table 11.5. The proposed MoT-based NoC also has a single destination core in its local cluster. Thus, at highly localized traffic, more packets will reach their destinations, resulting in higher throughput.

Table 11.5 also compares the throughput of 3D networks with their 2D counterparts. Intuitively, it can be stated that as the number of packets traversing toward the local cluster increases with increasing locality factor, the difference of average hop count between 2D and 3D structures converges. Hence, at highly localized traffic, increment in throughput of 3D networks is lesser than that of 2D networks.

11.3.3.3 Average Overall Latency under Localized Traffic

In a contention-free environment, zero-load latency (in cycles) is another widely used performance metric. Zero-load latency of a network is the latency where only one packet traverses through the network (Pavlidis and Friedman 2007). Table 11.6 shows the zero-load latency of all the networks including the cycle delay of source router. According to wormhole router architecture, each router has a two-cycle latency (one cycle in each input buffer [IB] and switch arbiter [SA] unit), whereas routers having node degree 2 have single-cycle latency. Cycle latency of inter-router link traversal of all the networks is taken from Figures 11.7 through 11.10.

TABLE 11.6

Number of Inter-Router Links, FIFOs, and Zero-Load Latency (in Cycle) of the Networks under Consideration with 32 Cores

| | | | Number of Inter-Router Links | | | | | |
| | Zero-Load Latency (Cycle) | | Few Micrometers | | ≈2.5 mm | | Number of FIFOs | |
Networks	2D	3D	2D	3D	2D	3D	2D	3D
Mesh-1	10.00	8.20	24	96	80	32	136	160
Mesh-2	8.45	7.74	8	40	64	32	80	88
BFT	10.00	6.65	0	0	96	32	80	72
MoT	12.32	8.71	32	56	80	32	128	120

Under actual traffic scenario, where contention of packets is a major challenge, latency of any network depends on both offered load and locality factor. Here, simulation has been carried out to estimate the average overall latency for all the networks with uniformly distributed and localized load as shown in Figures 11.12 through 11.15. It shows that at lower load, the latency variation is not significant. This is because at lower traffic, contention in the network is less. The contention increases as the offered load increases, which in turn increases the latency. Simulation results show that as the offered load increases toward the network saturation point, latency increases exponentially, which signifies that packets will take much longer time to reach their destinations. Therefore, it is always desirable to operate any network below its saturation point.

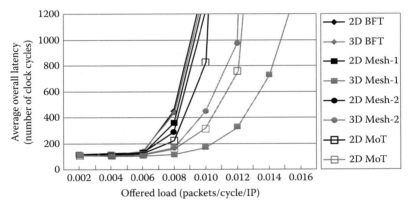

FIGURE 11.12
Latency variation in different 2D and 3D networks under consideration with uniformly distributed offered load.

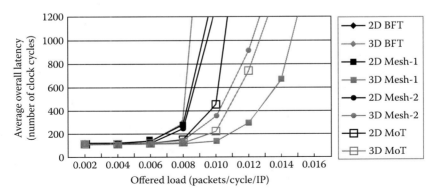

FIGURE 11.13
Latency variation in different 2D and 3D networks under consideration with offered load at locality factor of 0.3.

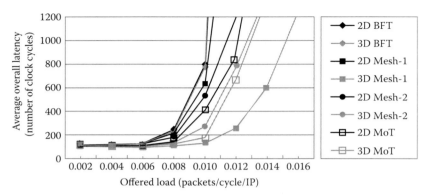

FIGURE 11.14
Latency variation in different 2D and 3D networks under consideration with offered load at locality factor of 0.5.

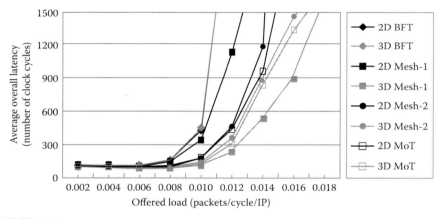

FIGURE 11.15
Latency variation in different 2D and 3D networks under consideration with offered load at locality factor of 0.8.

Although the zero-load latency of 3D MoT-based network is the maximum as shown in Table 11.6, in actual traffic condition, 3D MoT network experiences lesser contention than 3D Mesh-2 and 3D BFT networks. This happens as 3D MoT network has more inter-router links than 3D BFT and 3D Mesh-2 networks. Thus, it encounters lesser contention and has a better latency profile under uniformly distributed traffic, as shown in Figure 11.12. However, due to more interconnection links in 3D Mesh-1 network, it experiences lesser contention than 3D MoT network. Thus, 3D Mesh-1 shows the best latency profile among all the topologies. Figure 11.12 also shows the improvement of latency profile in 3D networks over their 2D counterparts. Table 11.6 depicts the difference of zero-load latencies between 2D and 3D networks. In BFT networks, due to the elimination of pipelined registers and bypassing of the root routers in a 3D environment, the value of zero-load latency is lesser than in 2D networks. But due to lesser number of inter-router links in BFT networks, in actual traffic condition, simulation results show that both the networks have almost similar latency profile under uniformly distributed and localized traffic conditions.

The effect of traffic spatial localization on the average overall latency has also been studied for all the networks under consideration (shown in Figures 11.13 through 11.15). The average overall latency of all the networks decreases with increasing locality factor. As the locality factor increases, more traffic will go to their local clusters. Hence, packets traverse lesser number of hops and will create lesser contention in the network. It can be observed that the latency profile of both Mesh-2 and MoT networks in a 3D environment becomes closer to their 2D counterparts with increasing locality factor. This is due to the fact that the contention in all these networks becomes almost identical at highly localized traffic as they have single destination cores in their local clusters.

From the graphs shown in Figures 11.14 and 11.15, it can be observed that the latency profile of the 3D $2 \times 4 \times 4$ Mesh-1 network improves significantly over the 2D 4×8 Mesh-1 network. Due to the rectangular structure of 2D Mesh-1 network, packets traverse more hops in row-wise direction under uniform distribution. Thus, the network suffers from more contention. In 3D topology, due to the square structure in vertical surface, the contention is less. It has been observed in simulation that 3D Mesh-1 network has the best latency profile under uniformly distributed and localized traffic condition.

11.3.3.4 Energy Consumption

Energy consumption in NoC is the summation of the energy consumed by the routers and the communication links. Both these factors are network topology dependent. Energy consumption of the proposed 3D MoT-based network after applying clock gating in the FIFO under uniformly distributed self-similar traffic is shown in Figure 11.16 for 200,000 cycles.

It can be observed that the network energy consumption increases linearly with the offered load but saturates as the offered load increases to the throughput limit, similar to 2D NoC. Beyond saturation, no additional packets can be injected successfully into the network and, consequently, no additional energy is consumed. Simulation result shows that after gating the write clock, the total energy consumption by all the FIFOs is about 35% of the overall network energy consumption, whereas all the links consume almost 50% of it. The combined energy consumption by the routing logics arbiters and control logic is about 15% of the total energy consumption.

Table 11.7 presents a comparison of average energy consumption at saturation for all the networks connecting 32 cores in 2D and 3D platforms. In 3D MoT network with dimension $2 \times 2 \times 4$, the number of *leaf* routers having connectivity 5 is 16 and that of *stem* routers (in vertical direction) having connectivity 3 is 8. In 4×4 2D MoT network, the number of *leaf* routers having

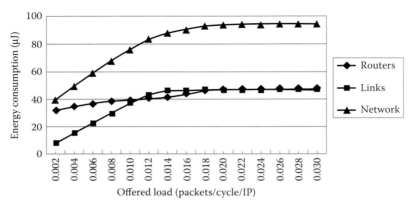

FIGURE 11.16
Network energy consumption of 3D MoT network under uniformly distributed offered load.

connectivity 4 is 16 and that of *stem* routers having connectivity 3 is also 16. In both 2D and 3D networks, the number of *root* routers having connectivity 2 is 8. Due to higher connectivity, 3D *leaf* routers consume more energy than 2D *leaf* routers. However, as the number of *stem* routers is less in 2 × 2 × 4 3D MoT network, the total energy consumption by all the *stem* routers is lesser than its 2D counterpart. Moreover, from Figure 11.10 it can be intuitively said that in 3D MoT network, lesser traffic will pass through the *root* routers than in 2D under uniformly distributed traffic. Hence, the energy consumption by the *root* routers is also less. It has been found that the total energy consumption by all the routers of 2 × 2 × 4 MoT network is lesser than that of 4 × 4 MoT network.

For a system with 128 or more number of cores, Table 11.8 shows that the number of FIFOs required to implement 3D MoT network is significantly higher than its 2D implementation. As FIFOs are the most energy-hungry components of a router, the total router energy consumption of 3D structures will be more than its 2D counterpart for networks with large number of cores. In BFT network, although the same topology is mapped onto

TABLE 11.7

Average Energy Consumption per Cycle at Saturation by Different Network Structures Connecting 32 Cores at 2D and 3D Platforms

	Average Energy (pJ/cycle) at Saturation under Uniform Distribution					
	Routers		Links		Networks	
Networks	2D	3D	2D	3D	2D	3D
Mesh-1	292.70	323.36	355.71	252.35	648.41	575.71
Mesh-2	205.13	207.67	325.37	228.28	530.50	435.95
BFT	202.21	174.54	368.16	210.89	570.37	385.43
MoT	305.79	237.07	370.29	232.29	676.08	469.36

TABLE 11.8

Number of Inter-Router Links and FIFOs Required to Implement 2D NoC and Four-Layered 3D NoC with 64 and 128 Cores

| | Number of Inter-Router Links | | | | | Number of FIFOs | |
| | Few Micrometers | | ≈2.5 mm | | Number | | |
Networks	2D	3D	2D	3D	of TSVs	2D	3D
			64 Cores				
Mesh-1	32	64	192	128	96	288	352
Mesh-2	8	16	136	96	48	168	192
BFT	0	0	256	64	48	160	160
MoT	64	96	192	96	48	272	272
			128 Cores				
Mesh-1	48	96	416	320	192	592	736
Mesh-2	16	32	320	256	96	352	416
BFT	0	0	640	256	96	352	336
MoT	128	192	512	320	96	576	672

3D structure, due to the elimination of *root* routers (Figure 11.9), the total energy consumption by the routers is reduced in 3D network. However, the total energy consumption by the routers in both 3D Mesh-1 and 3D Mesh-2 networks is more than that of its 2D structures due to more FIFOs that exist in the design (Tables 11.6 and 11.8) and more packet reception. While comparing with the proposed 3D MoT structure, the number of FIFOs is less in 3D BFT and 3D Mesh-2 networks for any number of cores, as shown in Table 11.8, which in turn causes lesser energy consumption by the routers. The energy consumption by the routers of 3D Mesh-1 is higher than that of 3D MoT as it requires more number of FIFOs for any number of cores.

Interconnection links, as shown in Figure 11.16, are the most energy-hungry components at higher offered load. The core-to-router links of length 1.25 mm and the inter-router links of length 2.5 mm are the major energy-consuming links. Link energy consumption in all the 3D networks, due to lesser number of 2.5 mm links compared to 2D structures (Table 11.6), is drastically reduced which causes energy reduction in the overall network. Table 11.9 shows the average energy consumption per cycle by the routers, links, and network at saturation for all the NoC structures under uniformly distributed self-similar traffic. With increasing locality factor, as more packets are sent to their local clusters, energy consumption of the local links increases, whereas that of the inter-router links decreases. The toggling of data in those ports of router that connect to cores increases, whereas the toggling of data in other ports decreases with increasing locality factor.

A network with more links will definitely have higher network throughput but at the cost of more energy consumption. Therefore, the average energy

TABLE 11.9

Average Energy Consumption per Packet at Saturation by Different Network
Structures Connecting 32 Cores at 2D and 3D Platforms

	Average Energy per Packet (nJ) at Saturation							
	Locality Factor = 0.0		Locality Factor = 0.3		Locality Factor = 0.5		Locality Factor = 0.8	
Networks	2D	3D	2D	3D	2D	3D	2D	3D
Mesh-1	5.15	3.01	4.48	2.76	3.85	2.63	2.92	2.21
Mesh-2	4.02	2.79	3.32	2.41	3.05	2.07	2.02	1.72
BFT	4.47	3.03	3.97	2.91	3.42	2.54	2.54	2.09
MoT	4.52	2.77	3.69	2.38	3.30	2.18	2.27	1.74

consumption per packet reception is a meaningful metric while comparing
the energy consumption of various network structures. Table 11.9 shows
the reduction of average energy per packet at saturation in 3D networks
over their 2D structures for a 32-core-based system. It can be observed
that 3D Mesh-1 network, due to higher connectivity of the routers, consumes the highest average energy per packet at saturation at all locality
factors. As the throughput of the BFT structure is the least, its average
packet energy is also higher than the proposed 3D MoT network under
all localized conditions. 3D Mesh-2 network, due to its lesser energy consumption and lesser throughput at uniformly distributed and low localized traffic, shows similar average packet energy consumption as 3D MoT.
At highly localized traffic, due to the single core in the local clusters, both
3D MoT and 3D Mesh-2 networks show similar energy consumption per
packet.

In this work, as links are pipelined after every 2.5 mm, Table 11.8 shows
the number of 2.5 mm links for all the 2D and 3D networks for systems with
larger number of cores. In 2D NoC platform, it can be observed that MoT
requires more 2.5 mm links than Mesh-1 network with increasing number of
cores. BFT requires the maximum number of 2.5 mm links, whereas Mesh-2
network requires the least. It can be seen that Mesh-1 requires more number
of FIFOs than MoT network in all the cases, whereas Mesh-2 and BFT require
comparatively lesser number of FIFOs. In a 3D environment, the requirement of 2.5 mm links in Mesh-1 and MoT networks is higher than in BFT and
Mesh-2. As interconnection links and FIFOs are the major energy-consuming
components in any network, due to more FIFOs and 2.5 mm links, it can be
apparently estimated that the total energy consumption of 3D Mesh-1 network will be more than that of 3D MoT network at any localized condition.
However, due to lesser number of FIFOs and 2.5 mm links in 3D BFT and
3D Mesh-2 networks, they are expected to consume lesser energy than that
of the proposed 3D MoT structure. Moreover, the number of TSVs required
for 3D Mesh-1 network is double that for other 3D networks taken here into

consideration. From performance side, due to lesser interconnection links, the throughput of 3D BFT and 3D Mesh-2 networks will be lesser under uniformly distributed and low localized traffic.

11.3.4 Simulation Results with Application-Specific Traffic

For evaluating the MoT network under real benchmark application, this chapter considers a DVOPD application consisting of 32 cores where two VOPDs are running in parallel. The core graph of DVOPD application has been shown in Chapter 4. Due to unavailability of mapping algorithm in the literature for 3D NoC structures, this chapter uses hand mapping of cores for all the 3D NoCs taken here into consideration. Table 11.10 shows the hand mapping of cores with their names and coordinates.

The performance and cost of MoT network is evaluated and compared with other networks taken here into consideration. Here, traffic generation is done in a self-similar manner. However, the communication requirements of the tasks in the application have been taken into consideration. The total traffic generated per unit time confirms with the bandwidth requirement specified for the edges of the task graph.

In the simulation, the parameters such as packet length, flit size, link width, core size, and operating frequency of the networks are taken as same as before. Table 11.11 presents the simulation results of different 3D NoC structures. As the average overall latencies of all the topologies are well below the network saturation point, it can be stated that the injection loads to the network are less. Figure 11.16 shows that at low offered load, router energy dominates over the link energy. Thus, due to low offered load and high-connectivity routers in Mesh-1 network, its energy consumption is the highest. From the simulation results, it can be stated that the hand mapping solution for MoT is comparable with the other 3D NoC structures.

TABLE 11.10

Hand Mapping of Cores in Four-Layered 3D SoC

Coordinate	Cores Mapped in Different Active Silicon Layers			
	Layer 0	**Layer 1**	**Layer 2**	**Layer 3**
0,0	*vop mem1*	*arm2*	*vld2*	*ac/dc pred2*
0,1	*vop rec1*	*sh mem2*	*sh mem1*	*stripe mem2*
0,2	*down samp2*	*arith dec1*	*inv scan1*	*idct2*
0,3	*arith dec2*	*pad2*	*rld1*	*iquant2*
1,0	*pad1*	*arm1*	*vld1*	*rld2*
1,1	*mem2*	*down samp1*	*stripe mem1*	*inv scan2*
1,2	*up samp1*	*vop rec2*	*ac/dc pred1*	*mem1*
1,3	*idct1*	*vop mem2*	*iquant1*	*up samp2*

TABLE 11.11

Simulation Results of All 3D NoCs under Consideration for DVOPD Application

Networks	Average Overall Latency (Cycle)	Total Energy Consumption (µJ)
Mesh-1	93.98	59.36
Mesh-2	94.53	46.71
BFT	96.61	42.49
MoT	95.47	47.25

Evolution of better mapping algorithm for all the networks will definitely improve the mapping solution. However, we do not address the issue in this work.

11.4 Summary

In this chapter, we have proposed a 3D MoT topology and applied it in 3D NoC design. The performance and cost of the proposed MoT-based 3D NoC have been evaluated by applying a self-similar traffic and compared with a well-known tree-based topology, BFT, and two variants of mesh topology connecting single or two cores to each router. For uniformly distributed and less localized traffic condition, the throughput and latency values obtained for MoT are better than all other topologies excepting Mesh-1. However, at highly localized traffic condition, both Mesh-2 and MoT perform equally well, as both of them are having a single destination core in their local clusters. The area overhead of 3D MoT network is lesser than those of 3D BFT and 3D Mesh-2 structures. Moreover, for a 32-core-based system, MoT shows the least average packet energy consumption, almost similar to 3D Mesh-2 network. Thus, taking performance and cost into consideration, MoT appears to be a very competitive topology among the alternatives proposed in the literature. The MoT network has also been evaluated and compared with other topologies under a real benchmark application, DVOPD. The comparative study shows that MoT-based 3D NoC also works fine under an application-specific traffic. On the architecture front of the wormhole router, due to lesser connectivity of MoT routers, synthesis result (Table 11.2) shows that they can be operated at a higher frequency than other networks, thus increasing the speed of the overall network. However, for a system with large number of cores, like other tree-based topologies, MoT will also suffer from the large number of pipelining stages required for the longer edges in each silicon layer. Adopting current mode signaling in NoC link and usage of photonic interconnects in 3D NoC are expected to alleviate this bottleneck, making MoT a more acceptable topology for larger core-based 3D NoC design.

References

Bashirullah, R., Liu, W., and Cavin, R. K. 2003. Current-mode signaling in deep submicrometer global interconnects. *IEEE Transactions on Very Large Scale Integration (VLSI) Systems*, vol. 11, no. 3, pp. 406–417.

Beyne, E. 2006. The rise of the 3rd dimension for system integration. *Proceedings of IEEE International Interconnect Technology Conference*, Burlingame, CA, IEEE, pp. 1–5, June 5–7.

Carloni, L. P., Pande, P. P., and Yuan, X. 2009. Networks-on-Chip in emerging interconnect paradigms: Advantages and challenges. *Proceedings of ACM/IEEE International Symposium on Networks-on-Chips*, San Diego, CA, IEEE, pp. 93–102, May 10–13.

Chang, M. F. et al. 2008. CMP network-on-chip overlaid with multi-band RF interconnect. *Proceedings of International Symposium on High-Performance Computer Architecture*, Salt Lake City, UT, IEEE, pp. 191–202, February 16–20.

Chen, G., Chen, H., Haurylau, M., Nelson, N. A., Albonesi, D. H., Fauchet, P. M., and Friedman, E. G. 2007. Predictions of CMOS compatible on-chip optical interconnect. *Integration, the VLSI Journal*, vol. 40, no. 4, pp. 434–446.

Cianchetti, M. J., Kerekes, J. C., and Albonesi, D. H. 2009. Phastlane: A rapid transit optical routing network. *International Symposium on Computer Architecture* Austin, Texas, USA, ACM, pp. 441–450, 20–24 June.

Davis, W. R., Wilson, J., Mick, S., Xu, J., Hua, H., Mineo, C., Sule, A. M., Steer, M., and Franzon, P. D. 2005. Demystifying 3D ICs: The pros and cons of going vertical. *IEEE Design and Test of Computers*, IEEE, vol. 22, no. 6, pp. 498–510, November–December.

Deodhar, V. V. and Davis, J. A. 2005. Optimization of throughput performance for low-power VLSI interconnects. *IEEE Transactions on Very Large Scale Integration (VLSI) Systems*, vol. 13, no. 3, pp. 308–318.

Feero, B. S. and Pande, P. P. 2009. Networks-on-chip in a three dimensional environment: A performance evaluation. *IEEE Transactions on Computers*, vol. 58, no. 1, pp. 32–45.

Flic, J. and Bertozzi, D. 2010. *Designing Network On-Chip Architectures in the Nanoscale Era*. Chapman & Hall/CRC Computational Science, Boca Raton, FL.

Gu, H., Xu, J., and Wang, Z. 2008. A low-power low-cost optical router for optical network-on-chip in multiprocessor system-on-chips. *Proceedings of IEEE Computer Society Annual Symposium on VLSI (ISVLSI)*, Tampa, Florida, IEEE, pp. 19–24, May 13–15.

Gunn, C. 2006. CMOS photonics for high-speed interconnects. *IEEE Micro*, IEEE, vol. 26, no. 2, pp. 58–66, March–April.

Haurylau, M., Chen, G., Chen, H., Zhang, J., Nelson, N. A., Albonesi, D. H., Friedman, E. G., and Fauchet, P. M. 2006. *IEEE Journal of Selected Topics in Quantum Electronics*, IEEE, vol. 12, no. 6, pp. 1699–1705.

Jacob, P., Erdogan, O., Zia, A., Belemjian, P. M., Kraft, R. P., and McDonald, J. F. 2005. Predicting the performance of a 3D processor-memory stack. *IEEE Design and Test of Computers*, IEEE, vol. 22, no. 6, pp. 540–547, November–December.

Li, F., Nicopoulos, C., Richardson, T., Xie, Y., Narayanan, V., and Kandemir, M. 2006. Design and management of 3D chip multiprocessors using network-in-memory. *Proceedings of IEEE International Symposium on Computer Architecture*, Boston, MA, IEEE, pp. 130–142, June 17–21.

Loi, I., Angiolini, F., Fujita, S., Mitra, S., and Benini, L. 2011. Characterization and implementation of fault-tolerant vertical links for 3-D networks-on-chip. *IEEE Transactions on Computer-Aided Design of Integrated Circuits and Systems*, vol. 30, no. 1, pp. 124–134.

Matsutani, H., Koibuchi, M., Hsu, F., and Amano, H. 2008. Three-dimensional layout of on-chip tree-based networks. *Proceedings of International Symposium on Parallel Architectures, Algorithms, and Networks*, Sydney, NSW, IEEE, pp. 281–288, May 7–9.

Nigussie, E., Lehtonen, T., Tuuna, S., Plosila, J., and Isoaho, J. 2007. High-performance long NoC link using delay-insensitive current-mode signaling, *Journal of VLSI Design*, Hidwai Publishing Corporation, Article ID 46514, pp. 1–13.

Pan, Y., Kumar, P., Kim, J., Memik, G., Zhang, Y., and Choudhary, A. 2009. Firefly: Illuminating future network-on-chip with nanophotonics. *International Symposium on Computer Architecture*, Austin, Texas, ACM, pp. 429–440, June 20–24.

Pande, P. P., Grecu, C., Jones, M., Ivanov, A., and Saleh, R. 2005. Performance evaluation and design trade-offs for MP-SOC interconnect architectures. *IEEE Transactions on Computers*, vol. 54, no. 8, pp. 1025–1040.

Pande, P. P., Grecu, C., Ivanov, A., and Saleh, R. 2003. High-throughput switch-based interconnect for future SoCs. *Proceedings of IEEE International Workshop on System-on-Chip for Real Time Applications*, pp. 304–310.

Patti, R. 2007. Impact of wafer-level 3-D stacking on the yield of ICs. *Future Fab Int.* http://www.futurefab.com/documents.asp?d_id = 4415.

Pavlidis, V. F. and Friedman, E. G. 2007. 3-D Topologies for networks-on-chip. *IEEE Transactions on VLSI Systems*, vol. 15, no. 10, pp. 1081–1090.

Pavlidis, V. F. and Friedman, E. G. 2009. *Three-Dimensional Integrated Circuit Design*. Morgan Kaufmann Publishers, Burlington, MA.

Petracca, M., Bergman, K., and Carloni, L. P. 2008. Photonic networks-on-chip: Opportunities and challenges. *Proceedings of IEEE International Symposium on Circuits and Systems*, Seattle, WA, IEEE, pp. 2789–2792, May 18–21.

Savidis, I., Alam, S. M., Jain, A., Pozder, S., Jones, R. E., and Chatterjee, R. 2010. Electrical modeling and characterization of through-silicon vias (TSVs) for 3-D integrated circuits. *Microelectronics Journal*, Elsevier, vol. 41, pp. 9–16.

Shacham, A., Bergman, K., and Carloni, L. P. 2007. The case for low-power photonic networks-on-chip. *Proceedings of Design and Automation Conference (DAC)*, San Diego, California, ACM, pp. 132–135, June 4–8.

Topol, A. W., Tulipe, D. C. L., Shi, L., Frank, D. J., Bernstein, K., Steen, S. E., Kumar, A., Singco, G. U., Young, A. M., Guarini, K. W., and Ieong, M. 2006. Three-dimensional integrated circuits. *IBM Journal of Research and Development*, vol. 50, nos. 4/5, p. 491.

Vantrease, D., Schreiber, R., Monchiero, M., Mclaren, M., Jouppi, N. P., Fiorentino, M., Davis, A., Binkert, N., Beausoleil, R. G., and Ahn, J. H. 2008. Corona: System implications of emerging nanophotonic technology. *International Symposium on Computer Architecture*, Beijing, Peoples Republic of Chins, IEEE, pp. 153–164, June 21–25.

Ye, Y., Duan, L., Xu, J., Ouyang, J., Hung, M. K., and Xie, Y. 2009. 3D optical networks-on-chip (NoC) for multiprocessor systems-on-chip (MPSoC). *Proceedings of IEEE International Conference on 3D System Integration*, San Francisco, CA, IEEE, pp. 1–6, September 28–30.

Zhang, H., George, V., and Rabaey, J. M. 2000. Low swing on-chip signaling techniques: Effectiveness and robustness. *IEEE Transactions on Very Large Scale Integration (VLSI) Systems*, vol. 8, no. 3, pp. 264–272.

12

Conclusions and Future Trends

12.1 Conclusions

Network-on-chip (NoC) has evolved as a viable solution to the communication problem between cores in a system-on-chip (SoC). System cost, in terms of area, delay, power, and so on, has contributions from both computation and communication requirements. Individual cores can be designed efficiently to make computation faster, but communication may become a bottleneck. As discussed in all chapters, the solution is influenced by several factors. The first and foremost issue is the topology in which routers are to be connected. While regular topologies, such as mesh and tree, make the design process simpler with predictable link delay and power consumption values, irregular application-specific topologies are expected to produce better performance. Individual routers should be simple with modules, such as ports, routing logic, arbiter, and channel allocator. Design of routing algorithm plays an important role, as it has to be free from deadlock and livelock problems, which may route in a shortest path through the network. Performance of such a network is often evaluated with the help of simulators and traffic generators. Apart from application-specific traffic, networks are often evaluated using uniform, self-similar, hot spot, and other types of traffic. For regular topologies, cores of an application are mapped onto individual routers using some mapping techniques. The mapped core gets attached to that router and all communication to and from the core is made through that router. The mapping problem is NP-hard; however, many heuristic strategies have been developed for the same. However, to achieve better performance, application-specific NoC architectures are evolved. Multiapplication NoC design calls for a reconfigurable architecture, in which the same network resources are reused for different applications. The power consumed by the NoC in either case can be reduced using various strategies, such as encoding, serialization, and clock gating. A related issue is that of reliability of the system. Electromagnetic interference, synchronization failure, and soft errors come up as challenges to a reliable system operation. Testing of such system requires testing individual cores, routers, and links.

12.2 Future Trends

One of the upcoming NoC architectures discussed in the book is three-dimensional (3D) NoC. The NoC layers grow vertically with reduced communication overhead. The other directions, which are in progress, include the following:

- Photonic NoC
- Wireless NoC

12.2.1 Photonic NoC

Photonic communication can provide large data transfers with minimal power consumption. Photonic NoC provides the following two major advantages:

- Multiple terabits per second (Tbps) communication on a single waveguide (link) with limited power dissipation
- Power consumption that is independent of the link length and scales only with link transmission interface circuitry, such as modulators, drivers, and receivers

A major problem with the implementation of photonic NoC is the lack of optical memory and impracticality of optical processing. Shacham et al. (2007) proposed a hybrid approach for this situation. An optical plane is used for high-bandwidth multiwavelength transmission links, whereas an electronic plane performs network management and control functions (Figure 12.1). The communication takes place as follows:

1. A photonic circuit is reserved by a source core by sending a path setup packet over the electronic network to the destination core. The destination replies with a short acknowledgment pulse over the photonic network.
2. The source sends data over the photonic circuit, combining the time-division multiplexing (TDW) and wavelength-division multiplexing (WDM).
3. The communication is terminated by the source transmitting a tear-down packet, commonly known as *path teardown process*.

12.2.2 Wireless NoC

Deb et al. (2012) has shown that silicon-integrated antennas can operate in a millimeter-wave range of few tens to 100 GHz. Carbon nanotubes (CNTs) show excellent emission and absorption characteristics leading to an antenna-like

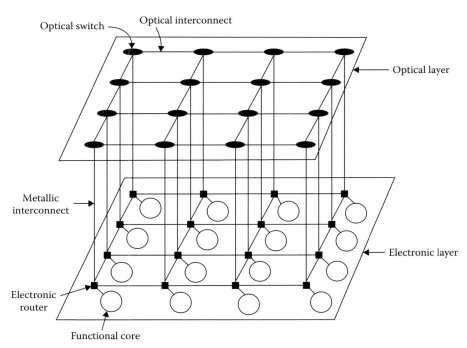

FIGURE 12.1

A 4 × 4 optical NoC structure.

behavior that can operate at optical frequencies. These two factors couple up and open the direction of designing wireless NoCs (WiNoCs) with on-chip antennas and suitable transceivers. The wireless communication alleviates the latency and energy dissipation issues of conventional technologies, and also solves the complex interconnect routing and placement problems. Multihop communication of traditional technologies can be converted into a single hop, resulting in significant saving in delay for communication. An ultrawideband (UWB) 4 × 4 two-dimensional (2D) mesh architecture-based WiNoC is shown in Figure 12.2. The processor tiles access the network via radio frequency (RF) nodes. Packets are delivered to destinations through single or multiple hops.

12.3 Comparison between Alternatives

The three emerging NoC architectures can be compared at different angles. As far as the design requirements are concerned, 3D NoCs contain multiple layers of active devices, while optical NoCs require silicon photonic components and WiNoCs have on-chip metal or CNT-based antennas. 3D NoCs

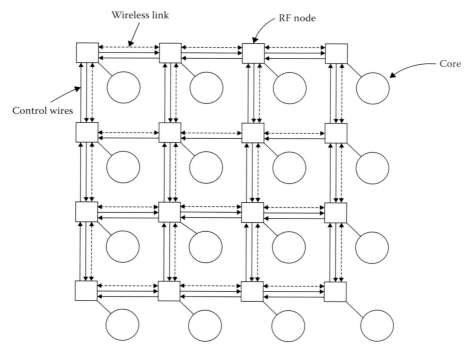

FIGURE 12.2
A 4 × 4 WiNoC structure.

enjoy higher connectivity and reduced hop count compared to 2D NoCs. Bandwidth advantages of optical NoCs come from the usage of high-speed optical devices and links. WiNoCs use single-hop high-bandwidth wireless links. Power dissipation is low in 3D NoC due to shorter average path length. Optical NoCs dissipate negligible power in optical data transport, whereas in WiNoCs, multihop paths can be replaced by single-hop links. Reliability may suffer in 3D NoCs due to the failure of vertical via. However, temperature sensitivity of photonic components and noisy wireless channels can be the sources of failure in optical NoCs and WiNoCs, respectively. The major challenge with 3D NoC is to handle the thermal problems due to higher power densities, particularly for the layers away from the heat sink. Integration of on-chip photonic components is the major challenge in photonic NoCs. Design of low-power millimeter-wave transceivers and control over CNT growth is the challenge for making WiNoC a success.

References

Deb, S., Ganguly, A., Pande, P. P., Belzer, B., and Heo, D. 2012. Wireless NoC as interconnection backbone for multicore chips: Promises and challenges. *IEEE Journal on Emerging and Selected Topics in Circuits and Systems*, pp. 228–239.

Petracca, M., Lee, B. G., Bergman, K., and Carloni, L. P. 2009. Photonic NoCs: System-level design exploration, *IEEE Micro*, IEEE Computer Society, pp. 74–84, August.

Shacham, A., Bergman, K., and Carloni, L. P. 2007. On the design of a photonic network-on-chip. *Proceedings of the NOCS*. Washington DC, pp. 53–64.

Ye, Y., Duan, L., Xu, J., Ouyang, J., Hung, M. K., and Xie, Y. 2009. 3D optical network-on-chip (NoC) for multiprocessor systems-on-chip (MPSoC). *Proceedings of the 3D System Integration*.

Index

Note: Locators followed by "*f*" and "*t*" denote figures and tables in the text

3D integrated circuit (3D IC), 317
  clock distribution structures for,
    321, 321*f*
  n-layer
    cross section of stacked layers,
      319, 320*f*
    heat dissipation in, 319, 320*f*
    NoC topologies on, 323, 323*f*
3D integration, 318–322
  challenges of, 321–322
    CAD tools, 322
    interconnect design, 321–322
    reliability, 322
    thermal effects, 321
  opportunities of, 319–321
    decrease in interconnect
      length, 319
    heterogeneous and
      multifunctional SoC design,
      319, 321
3D NoC architecture, design and
    evaluation of, 323–350
  application-specific traffic,
    simulation results with,
    349–350, 350*t*
  hand mapping of cores, 349, 349*t*
  MoT topology
    2 × 4 × 4, 327*f*
    average distance, 327–330
    number of directed edges, 326
    properties, 326
  performance and cost evaluation,
    331–340
    network area estimation, 336, 337*t*,
      338, 339*t*
    network aspect ratio, 339–340, 340*t*
  self-similar traffic, simulation results
    with
    accepted traffic *vs.* offered load,
      340–341, 341*f*
    energy consumption, 345–349,
      346*f*, 346*t*, 347*t*, 348*t*

L_{avg} under localized traffic,
    342–345, 342*t*, 343*f*, 344*f*
throughput *vs.* locality factor,
    341–342, 341*t*
1500 wrapper, 246, 247*f*

A

ACK/NACK flow control protocol,
    44, 45*f*
Adaptive router architecture design,
    70–73
Adaptive voltage scaling (AVS), 160
Aggressor lines, 81, 237
Ant colony optimization (ACO), 123
Application Specific Integrated Circuit
    (ASIC), 290
Application-specific NoC (ASNoC), 263
  literature survey, 265–267
  synthesis problem, 264–265
  synthesis with flexible router
    placement, *see* ASNoC
    synthesis with flexible router
    placement
Architecture-aware analytic mapping
    algorithm (A3MAP), 149
ASNoC synthesis with flexible router
    placement, 277–284
  communication trace graph, 277*f*
  ILP for
    constraints, 279–281
    objective function, 279
    variables, 278–279
  PSO for
    evolution of generation, 283
    local and global bests, 282
    particle structure and fitness
      function, 282
    swap operator, 283
    swap sequence, 283–284
  router locations, 277*f*
  routers at corners, 277*f*

Asynchronous FIFO design, 54–56
    dual-clock, 56*f*
    FIFO memory, 56*f*
    scalable gray code concept, 55, 55*f*
Automated test equipment (ATE), 245
Automated test pattern generation
        (ATPG), 239
Average overall latency (L_{avg}), 80
    under localized traffic
        in 3D NoC with self-similar
            traffic, 342–345, 342*t*, 343*f*, 344*f*
        in MoT with VC router, 110–111,
            110*f*, 111*f*, 112*f*
    MoT with WH router
        at different locality factors,
            86–87, 87*f*
        under localized traffic, 99–101, 99*t*,
            100*f*, 101*f*

B

Bandwidth (BW), 80
    utilization, 311
Best effort (BE) service, 45–46
Best fit decreasing (BFD) heuristics, 247
BI Hamming (BIH) code, 223
Binary and gray coding scheme,
        170, 170*t*
Binomial mapping (BMAP), 129
    flowchart, 130*f*
    OCN design flow in, 129*f*
    stages of OCN synthesis process
        in, 129
Bisection width, 75
Bit error (ε), probability, 228
Bit error rate (BER), 174, 228, 228*f*
Bose–Chaudhuri–Hocquenghem (BCH)
        code, 220
Boundary shift code (BSC), 224
Built-in self-repair (BISR)
        mechanism, 204
Built-in self-test (BIST), 254, 257
Built-in soft error resilience (BISER)
        technique, 206
Burst word error, probability, 229
Bus-invert (BI) code, 168, 222
    hardware of, 168, 169*f*
    scheme with 8-bit data bus, 169*t*

Butterfly fat tree (BFT) network, 75, 325
    2D, 21, 21*f*
    3D, 333, 334*f*
    distribution of cores, routers, and
        links in, 77*f*
    router classification, 76

C

Capacitive crosstalk avoidance
        techniques
    CAC, 212–216
    driver strength, 211
    increase inter-wire spacing, 211
    OCS, 211–212
    usage of shielding and duplicating
        wire, 210, 211*t*
Channel direction control (CDC)
        protocol, 311
Channel power consumption, 255
Channel width and flit size, selection, 84
Chip-Level Integration of
        Communicating
        Heterogeneous Elements
        (CLICHÉ), 14–15
Chip multiprocessing (CMP)
        system, 1, 9
Clustering-based approach, 276*f*
Communication cost (fitness), 282
Communication dependence
        and computation model
        (CDCM), 138
Communication fabric, testing, 236–244
    NoC links, 237–238
    NoC switches, 238–239
    test data transport, 239–241
    transport time minimization,
        241–244
        multicast test scheduling, 244
        unicast test scheduling, 242–244
Communication task graph (CTG), 308
Communication weighted model
        (CWM), 137
Complementary metal oxide
        semiconductor (CMOS)
        technology, 63, 77, 156, 317
Concentrated mesh (CMESH)
        topology, 16

Constructive heuristics
  for application mapping, 128–134
    binomial merging iteration,
      130–131, 132*f*
    hardware cost optimization,
      132–134
    topology mapping and traffic
      surface creation, 131–132
  with iterative improvement, 134–141
    initialization phase, 134–135
    iterative improvement phase,
      136–137
    other constructive strategies,
      137–141
    shortest path computation,
      135–136
Core graph, 120
Cores, testing, 245–260
  core wrapper design, 246–250
    1500 wrapper, 246, 247*f*
    algorithm, 249*t*
    two wrappers, 248*f*
  heuristic algorithms, 253–258
  ILP formulation, 250–253
  PSO-based strategy
    evolution of generations, 259–260
    particle structure and fitness,
      258–259, 259*f*
Crosstalk avoidance and double error
      correction (CADEC) codes,
      226–227, 226*f*
Crosstalk avoidance code (CAC), 210,
      212
  FOC, 212–213, 213*t*
  FPC, 214–215
  FTC, 213–214
  OLC, 216
Crosstalk delay, 197
  to MAF model, 195, 196*f*
  types of, 199*t*
Custom interconnection topology and
      route generation, 271–277
  constraints, 274–277
    latency, 276
    node-to-port mapping, 274
    port capacity, 274
    port-to-port mapping, 274
    traffic routing, 274–276
  objective function, 273–274
  router allocation for, 272*f*
  variables
    derived, 273
    independent, 272–273
Cyclic redundancy check (CRC), 217
Cyclic redundancy codes (CRCs), 218

D

DAP bus-invert (DAPBI) code, 227
Data link layer, 4
Deep submicron (DSM) technology, 75,
      191, 237, 317
  of *n* interconnects, 167, 167*f*
Design-for-testability (DfT) logic, 246
Discrete PSO (DPSO) technique,
      mapping using, 141–149
  augmentations to, 144–149
  convergence of DPSO, 143
  evolution of generations, 142–143
  overall PSO algorithm, 144
  particle structure, 141–142, 142*f*
Double error detection (DED) codes, 220
Double-switching errors, 198, 198*f*
  false/double clocking due to, 199*f*
Drain-to-source current (I_{DS}), 157
Duato's protocol, 41
Duplicate-add-parity (DAP) code, 224
Dynamic adaptive–deterministic
      (*DyAD*) routing, 70–73, 71*f*
Dynamic frequency scaling (DFS), 155
  architecture of, 180*f*
  in system-level power reduction,
      179–185
    characteristics, 184*t*
    DFS algorithm, 183
    history-based DFS, 181–182,
      182*f*, 184*t*
    link controller, 183–184
Dynamic voltage and frequency scaling
      (DVFS), 160
Dynamic voltage scaling (DVS), 155,
      172–179
  characteristics, 173
  components of links, 173*f*
  hardware implementation, 178, 178*f*
  history-based, 174–178

E

Edge (*e*), weight of, 298, 309–310
Electromagnetic interference
    (EMI), 2, 194
Energy and reliability trade-off in
    coding technique, 227–230
Energy overhead, 173
Energy reduction, 230
Error-correcting code (ECC), 220
Extended-BFT interconnection (EFTI)
    network, 21, 22*f*
Extended generalized fat-tree (XGFT), 9

F

Failures in time (FIT), 193
Faults in NoC fabric, sources of,
    193–204
  due to aging effects
    HCI, 195
    NBTI, 194
  permanent faults, 194
  transient faults
    capacitive crosstalk, 195, 196*f*,
      197–199
    other sources of, 203–204
    soft errors, 199–203
Fiduccia–Mattheyses (FM) partitioning
    algorithm, 265
Field-programmable gate array
    (FPGA), 289
Finite-state machines (FSMs), 312, 313*f*
First-in first-out (FIFO), 9, 34, 236, 330
  asynchronous design, 54–55
  design of memory, 56*f*
  dual-clock asynchronous, 54, 56*f*
  gating write clock of, 158*f*
  impact of size and placement in
    energy and performance of
    network, 90–93, 90*f*
  component-wise energy
    consumption, 92*f*, 93*f*
  network energy consumption,
    91*f*, 93*f*
Flow control
  protocol, 43–45
  signals, 58

Forbidden overlap codes (FOCs),
    212–213
  combining adjacent subchannels
    in, 214*f*
  truth table, 213*t*
Forbidden pattern code (FPC), 214–215
  combining adjacent subchannels
    in, 216*f*
  truth table, 215*t*
Forbidden pattern condition, 214
Forbidden transition codes (FTCs),
    213–214
  combining adjacent subchannels
    in, 215*f*
  truth table, 214*f*
Forbidden transition condition, 213
Forward error correction (FEC)
    technique, 220

G

GALS style of communication, 57, 57*f*
Gaussian pulse *Q*(*x*), 228
Generic core interface (GCI), 46
Genetic algorithm (GA), 123, 266
Globally asynchronous locally
    synchronous (GALS) style, 3
go-back-N retransmission, 219
Greedy incremental (GI) heuristics, 138
Ground bounce, 203
Guaranteed throughput (GT) service,
    45–46

H

Hamming code, 220, 221*f*
Head-of-line (HoL), 30
Hop-by-hop (HBH) error control, 219
Hot carrier injection (HCI), 195
Hurst parameter (HP), 83
Hybrid-communication ReNoC
    (HCR-NoC), 290–291

I

Idle periods (*P_i*), 83
Input buffer age, 175
Input buffer utilization, 175

Input FIFO buffer (IB), 58
Input flow controller (IFC), 58
Input read switch (IRS), 58
Integer linear programming (ILP), 123
  -based approach in local
      reconfiguration, 299–301
    constraints, 300–301
    objective function, 300
    parameters and variables,
      300, 300t
  for flexible router placement
    constraints, 279–281
    objective function, 279
    variables, 278–279
  formulation, 123–127
    other, 127–128
  in testing cores, 250–253
Intellectual property (IP), 119, 235
  calculate ranking, 131
  sets, 130
    merging of, 131, 133f
    refreshing, 131
Interconnection network, 13f
Internal power consumption, 156
Iteration (T_i), idle periods in, 83

K

Kernighan–Lin (K–L) partitioning
    scheme, 140–141

L

Largest communication first (LCF)
    heuristics, 138
Latency, 80; *see also* Average overall
    latency (L_{avg})
Least common ancestor (LCA)
    algorithm, 33, 93, 340
Linear feedback shift register (LFSR),
    218, 219f
Link-based mapping (LBMAP), 139
Link reconfiguration
  estimating channel bandwidth
    utilization, 311–312
  modified router architecture,
    312, 312f
Link utilization, 174, 181–182, 182f

Locality factor, 85
  MoT with WH router
    energy consumption at different,
      88–90, 88f, 89f
    L_{avg} at different, 86–87, 87f
    throughput *vs.*, 85–86, 87f
  throughput *vs.*
    in 3D NoC with self-similar
      traffic, 341–342, 341t
    MoT with VC router,
      109–110, 110f
    MoT with WH router, 98–99, 98f
Local link, 76
Local reconfiguration approach,
    291–304, 292f, 293f
  area overhead, 294–296
    of different architectures, 297t
    for module types, 295, 295t
    parameters for Orion, 294, 295t
    router areas, 295, 295t
  design flow, 296–299, 298f
    configuration generation, 299
    construction of CCG, 298
    mapping of CCG, 299
  ILP-based approach
    constraints, 300–301
    objective function, 300
    parameters and variables,
      300, 300t
  iterative reconfiguration, 303–304
  multiplexers, 293–294
  PSO formulation, 301–302
  routers, 292–293
  selection logic, 294, 294f
Look-up table (LUT), 61, 62f
Low-power code (LPC), 168–170, 221
Low-power methods for NoC links,
    166–172
  bus energy model, 167–168
  low-power coding, 168–170
  low-swing signaling, 171–172, 172f
  on-chip serialization, 170–171, 171f
Low-power methods for NoC routers
  clock gating, 158–159
    router-level, 159f
    write clock of FIFO, 158, 158f
  gate-level power optimization,
    159–160, 160f

Low-power methods for NoC routers
(*Continued*)
  multivoltage design, 160–164
    placement of level shifter, 164, 164*f*
    short-circuit current flow, 161–163,
      161*f*, 162*f*, 163*f*
  multi-V_T design, 164–165
  power gating
    architectural trade-offs, 165
    challenges, 166
    leakage power-saving profile
      using, 165*f*

M

Mapping problem, 120–123
  constraints for, 124–127
  onto mesh topology, 121*f*
Mapping using DPSO, 141–149
  augmentations to
    initial population generation,
      145–147
    multiple PSO, 144
    other evolutionary approaches,
      148–149
  convergence of DPSO, 143
  evolution of generations, 142–143
  overall PSO algorithm, 144
  particle structure, 141–142, 142*f*
Maximum aggressor fault (MAF)
    model, 195, 237
  effect of crosstalk, 195, 196*f*
  state machine for, 238*f*
Mean time between failures (MTBF),
    155, 193
Mesh-1 network, 76, 332, 332*f*
Mesh-2 network, 76, 332, 333*f*
Mesh-of-tree (MoT) network, 22, 23*f*, 26*f*
  deterministic routing in $M \times N$,
    33–41
    addressing scheme, 34, 35*f*
    avoidance of routing-dependent
      deadlock, 37–38, 40–41
    proof for shortest path, 37
    routing algorithm, 34, 36–37
  distribution of cores, routers, and
    links
    4 × 4 mesh structure, 77*f*
    4 × 4 MoT structure, 78*f*

4 × 8 mesh structure, 76*f*
    BFT networks, 77*f*
    labeling of channels in, 39*f*, 40*f*
  performance and cost
    comparison of
    with NoC structures having VC
      router, 109–114
    with NoC structures having WH
      router, 93–103
  simulation results and analysis of
    with VC router, 103–109
    with WH router, 84–90
  topology in 3D NoC, 326–330, 335*f*
    2 × 4 × 4, 327*f*
    properties, 326
Mixed integer linear programming
    (MILP)-based approach, 127,
    268, 276–277
Modified dual-rail (MDR) code, 224
MoT network, simulation results and
    analysis of
  with VC router, 103–109
    accepted traffic
      comparison, 104*f*
    area required, 108–109, 108*t*
    energy consumption, 105–108,
      107*f*, 108*f*
    latency *vs.* offered load, 104–105,
      105*f*, 106*f*, 107*f*
    throughput *vs.* offered load,
      104, 105*f*
  with WH router, 84–90
    accepted traffic *vs.* offered load,
      85, 85*f*
    energy consumption at
      different locality factors, 88–90,
      88*f*, 89*f*
    L_{avg} at different locality factors,
      86–87, 87*f*
    throughput *vs.* locality factor,
      85–86, 87*f*
MoT with NoC, performance and cost
    comparison of
  having VC router, 109–114
    accepted traffic *vs.* offered load,
      109, 109*f*
    area overhead, 113–114, 114*t*
    energy consumption,
      111–113, 113*f*

L_{avg} under localized traffic,
110–111, 110*f*, 111*f*, 112*f*
throughput *vs.* locality factor,
109–110, 110*f*
having WH router, 93–103
accepted traffic *vs.* offered load,
97, 97*f*
energy consumption, 102–103,
102*f*, 103*f*
L_{avg} under localized traffic, 99–101,
99*t*, 100*f*, 101*f*
network area estimation, 94–95,
94*t*, 95*t*
network aspect ratio, 96–97, 96*t*
throughput *vs.* locality factor,
98–99, 98*f*
Multicast test cost, 244
Multicast wrapper unit (MWU), 240
Multicast mode, 239
data transfer, 240*f*
transport, 241*f*
Multilevel voltage scaling (MVS), 160
Multiobjective adaptive immune
algorithm (MAIA), 149
Multiple error-correcting (MEC)
codes, 220
Multiple supply voltage (MSV), 160
Multiprocessor system-on-chip
(MPSoC) architecture, 1,
13, 291

N

Negative-bias temperature instability
(NBTI), 194
Network assignment (NA), 148
Network diameter, 14
Network interface (NI) module, 3, 31,
46–48, 47*f*, 127
Network layer, 5
Network-on-chip (NoC), 3
3D, 317
4 × 4 optical, 355*f*
abstraction layers, 4–5
development research issues, 5–8, 7*f*
application mapping, 7
communication infrastructure, 5
communication paradigm, 6
evaluation framework, 6

evaluation methodologies of, 75–81
cost metrics, 80–81
performance metrics, 78, 80
examples, 8–10
interconnect
bus encoding scheme for, 212*f*
low-swing signaling in, 172*f*
mesh-based, 323–324, 324*f*
paradigm, 4*f*
photonic, 317–318, 354
reconfigurations for ASIC-based, 290
SoC to, 3–5
testing
with input/output cores for,
245, 245*f*
issues, 236
wireless, 318, 354–355, 356*f*
Network processors, 127
Network topologies, 14–29
average distance, 25–29
BFT network, 21, 21*f*
binary tree network, 19, 20*f*
CMESH network, 17–18, 18*f*
EFTI network, 21, 22*f*
Flattened BFT, 23, 24*f*
folded torus network, 16, 17*f*
mesh network, 15*f*
MoT network, 22, 23*f*, 26*f*
number of edges, 25
octagon network, 17, 18*f*
spidergon network, 19, 19*f*
SPIN network, 20, 20*f*
torus network, 15, 16*f*
NoC links
four node with unidirectional,
243, 244*f*
low-power methods for, 166–172
bus energy model, 167–168
low-power coding, 168–170
low-swing signaling, 171–172
on-chip serialization, 170–171
testing, 237–238
NoC routers, low-power methods for
clock gating, 158–159, 158*f*
gate-level power optimization,
159–160
multivoltage design, 160–164
multi-V_T design, 164–165
power gating, 165–166

O

Odd-even turn model rules, 40, 41*f*
oe-fixed router, 71
On-chip network (OCN), 129
On-chip serialization (OCS) technique, 211–212
One lambda coding (OLC), 216
Open Core Protocol (OCP), 46
Output FIFO buffer (OB), 58
Output flow control (OFC), 58
Output read switch (ORS), 58

P

Packet disassembler (PD), 46–47
Packet maker (PM), 46–47
Particle swarm optimization (PSO), 123, 281
    for flexible router placement
        evolution of generation, 283
        local and global bests, 282
        particle structure and fitness function, 282
        swap operators, 283
        swap sequence, 283–284
    formulation in local reconfiguration, 301–302
Path teardown process, 354
Permanent fault controlling techniques, 204
Photonic NoC, 317–318, 354
Physical layer, 4
Pilot signal, 171
Power bounce, 203

Q

Q8WARE, 265
Quality of service (QoS), 6, 45–46, 127

R

Reconfigurable NoC (ReNoC)
    design flow of, 296–297, 298*f*
    literature review, 290–291
Reed–Solomon (RS) code, 220
Round-robin arbiter, 61, 62*f*
Router power consumption, 255

Routers classification
    4 × 4 network, 76
    4 × 8 network, 76
    BFT-based network, 76
    MoT-based network, 77
Routing computation (RC) unit, 58
Routing logic block (RLB), 240
Routing strategies, 30–43
    avoidance of message-dependent deadlock, 41–43
        request–response, 42*f*
        solutions to, 43*f*
    classification based on
        adaptability, 31
        single packet, 30
    routing-dependent deadlock, 31–41, 32*f*, 33*f*

S

Scalable, programmable, integrated network (SPIN), 9, 20, 20*f*, 75
Self-similar traffic
    in 3D NoC architecture simulation results with
        accepted traffic *vs.* offered load, 340–341, 341*f*
        energy consumption, 345–349, 346*f*, 346*t*, 347*t*, 348*t*
        L_{avg} under localized traffic, 342–345, 342*t*, 343*f*, 344*f*
        throughput *vs.* locality factor, 341–342, 341*t*
    algorithm, 83*f*
    MoT with NoC structures having WH router, comparison of, 93–103
        accepted traffic *vs.* offered load, 97, 97*f*
        energy consumption, 102–103, 102*f*, 103*f*
        L_{avg} under localized traffic, 99–101, 99*t*, 100*f*, 101*f*
        network area estimation, 94–95, 94*t*, 95*t*
        network aspect ratio, 96–97, 96*t*
        throughput *vs.* locality factor, 98–99, 98*f*

Single-chip cloud computer (SCC), 9
Single error correction (SEC), 220
Single-event transient (SET) pulse, 201
Single-event upset (SEU), 200
Soft error rate (SER), 201
Soft errors, 199–203
  in back-to-back inverter, 200, 200*f*
  classification of effect, 201
  correction, 206–210
    combinational logic, 207–210,
      208*f*, 209*f*
    in latches, 207*f*
    using time-shifted output, 209*f*
  in D-type latch, 200, 201*f*
Sort-based mapping (SBMAP), 139
STALL/GO flow control protocol, 44, 44*f*
Static voltage scaling (SVS), 160
Store-and-forward (SAF) packet
    switching techniques, 29
Subthreshold leakage current (I_{sub}), 156
Swap operation, 259–260
Switch allocation, 64
Switch arbiter (SA), 58, 205
Switching power consumption, 155–156
Switching techniques, 29–30
  and packet format, 53–54, 54*f*
Switch-to-switch flow control
    schemes, 218
System-level floorplanning, 268–271
  constraints, 269–270
  mesh topology constraints,
    270–271, 270*f*
  objective function, 269
  variables
    dependent, 268–269
    independent, 268
System-level power reduction, 172–188
  DFS, 179–185
    characteristics, 184*t*
    DFS algorithm, 183
    history-based DFS, 181–182, 182*f*
    link controller, 183–184
  DVS, 172–179
    characteristics, 173
    components of links, 173*f*
    hardware implementation, 178, 178*f*
    history-based DVS, 174–178
  runtime power gating, 186–188, 188*f*
  VFI partitioning, 185–186

System-on-chip (SoC), 1, 75, 129, 191, 235,
    319, 321
  categories, 1
  hand mapping of cores in four-
    layered 3D, 349, 349*t*
  integration and its challenges, 1–2
  to NoC, 3–5

T

TD factor, 202
Template-based efficient mapping
    (TEM) algorithm, 139
T-Error flow control protocol, 44, 45*f*
Test access mechanism (TAM), 235
Three-dimensional (3D) NoC, 317
Three-wire model
  energy obtained from HSPICE, 82*t*
  parasitic capacitance, inductance,
    and resistance of, 81*t*
Throughput, 78
  vs. locality factor, 85–86, 87*f*, 98–99,
    98*f*, 109–110, 110*f*, 341–342, 341*t*
  vs. offered load, 104, 105*f*
Through-silicon vias (TSVs), 8, 317, 319
  misalignments, 322, 322*f*
Time division multiplexed access
    (TDMA), 138
Time division multiplexing (TDW), 354
Topology graph, 120, 122
Topology reconfiguration, 304–311,
    308*f*, 309*f*
  architecture, 306–311
    application mapping, 307–308
    core-to-network mapping,
      309–310
    and route generation, 310–311
    routers wrapped by
      switches, 305*f*
  logical, 306*f*
  modification around routers, 305
  multiplexer-based implementation,
    307*f*
Traffic modeling, 81–84
Transient fault controlling techniques
  inter-router link error control,
    210–221
    capacitive crosstalk avoidance
      techniques, 210–216

Transient fault controlling techniques
    (*Continued*)
    error correction, 220–221
    error detection and
        retransmission, 216–220, 217*f*
    intra-router error control, 205–210
Transport layer, 5
Travelling salesman problem (TSP), 141
Tree-based topologies, limitations of,
    114–115
Triple modular redundancy (TMR)
    technique, 204, 218
Triplication error correction coding,
    225, 225*f*
Two-dimensional integrated circuit
    (2D IC), 8

U

Ultra-deep submicron (UDSM), 155, 220
Unicast cost function, 242
Unicast mode, 239
    data transfer, 240*f*
    transport, 241*f*
Unified coding framework,
    221–227, 222*f*
    joint CAC and ECC scheme
        (CAC + ECC), 224–227, 225*t*
    joint CAC and LPC scheme
        (CAC + LPC), 222–223, 223*f*
    joint CAC, LPC, and ECC scheme
        (CAC + LPC + ECC), 227
    joint LPC and ECC scheme
        (LPC + ECC), 223–224, 224*f*

V

VC allocation, 64
VC allocator (VCA), 205
    error, 206
VC router architecture design, 63–70
    input channel module, 65, 66*f*
    modified, 65*f*
    nonspeculative, 64*f*
    output links, 66–70
        (P − 1)*V, 66, 67*f*
        switch allocator, 69–70, 69*f*
        VC allocator, 66, 68*f*, 69

Very large scale integration (VLSI),
    158, 191
Victim line, 237
Video object plane decoder
    (VOPD), 119
    application graph for, 121*f*
    block diagram of, 120*f*
Virtual channel (VC), 187
Virtual cut-through (VCT) packet
    switching techniques, 29
Virtual ground (VGND), 187
Voltage–frequency island (VFI)
    2D mesh network with, 185*f*
    interface between voltages, 187*f*
    partitioning, 155, 185–186

W

Wavelength-division multiplexing
    (WDM), 354
Wire
    parasitic components of,
        192–193, 193*f*
    self-capacitance, 191–192
    test sequence for, 238*t*
Wireless NoC (WiNoC), 318, 354–355,
    356*f*
Wormhole (WH) router, 78
    architecture design, 57–63
        connections for, 59*f*
        data path, 63*f*
        input channel module, 58
        leaf level nodes, 60*f*
        output channel module, 58, 61–63
        priority logic, 61, 61*f*
    connectivity, number and frequency
        of, 79*t*
    MoT with NoC having, comparison
        of, 93–103
        accepted traffic *vs.* offered load,
            97, 97*f*
        energy consumption, 102–103,
            102*f*, 103*f*
        L_{avg} under localized traffic, 99–101,
            99*t*, 100*f*, 101*f*
        network area estimation, 94–95,
            94*t*, 95*t*
        network aspect ratio, 96–97, 96*t*

throughput *vs.* locality factor,
   98–99, 98*f*
MoT with, simulation results and
   analysis of, 84–90
   accepted traffic *vs.* offered load,
     85, 85*f*
   energy consumption at different
     locality factors, 88–90, 88*f*, 89*f*

L_{avg} at different locality factors,
   86–87, 87*f*
throughput *vs.* locality factor,
   85–86, 87*f*

X

XY routing in 2D mesh topology, 32, 33*f*